STABILITY AND SWITCHING IN CELLULAR DIFFERENTIATION

ADVANCES IN EXPERIMENTAL MEDICINE AND BIOLOGY

Editorial Board:

NATHAN BACK, *State University of New York at Buffalo*

NICHOLAS R. DI LUZIO, *Tulane University School of Medicine*

EPHRAIM KATCHALSKI-KATZIR, *The Weizmann Institute of Science*

DAVID KRITCHEVSKY, *Wistar Institute*

ABEL LAJTHA, *Rockland Research Institute*

RODOLFO PAOLETTI, *University of Milan*

Recent Volumes in this Series

Volume 150
IMMUNOBIOLOGY OF PROTEINS AND PEPTIDES – II
Edited by M. Z. Atassi

Volume 151
REGULATION OF PHOSPHATE AND MINERAL METABOLISM
Edited by Shaul G. Massry, Joseph M. Letteri, and Eberhard Ritz

Volume 152
NEW VISTAS IN GLYCOLIPID RESEARCH
Edited by Akira Makita, Shizuo Handa, Tamotsu Taketomi, and Yoshitaka Nagai

Volume 153
UREA CYCLE DISEASES
Edited by A. Lowenthal, A. Mori, and B. Marescau

Volume 154
GENETIC ANALYSIS OF THE X CHROMOSOME: Studies of Duchenne Muscular
Dystrophy and Related Disorders
Edited by Henry F. Epstein and Stewart Wolf

Volume 155
MACROPHAGES AND NATURAL KILLER CELLS: Regulation and Function
Edited by Sigurd J. Normann and Ernst Sorkin

Volume 156
KININS – III
Edited by Hans Fritz, Nathan Back, Günther Dietze,
and Gert L. Haberland

Volume 157
HYPERTHERMIA
Edited by Haim I. Bicher and Duane F. Bruley

Volume 158
STABILITY AND SWITCHING IN CELLULAR DIFFERENTIATION
Edited by R. M. Clayton and D. E. S. Truman

STABILITY AND SWITCHING IN CELLULAR DIFFERENTIATION

Edited by
R. M. Clayton
and
D. E. S. Truman
University of Edinburgh
Edinburgh, United Kingdom

PLENUM PRESS • NEW YORK AND LONDON

Library of Congress Cataloging in Publication Data

International Workshop on the Regulability of the Differentiated State (1981: Edinburgh, Lothian)
Stability and switching in cellular differentiation.

(Advances in experimental medicine and biology; v. 158)
"Proceedings of the International Workshop on the Regulability of the Differentiated State, held September 1-5, 1981, in Edinburgh, United Kingdom"—P.
Includes bibliographical references and index.
1. Cell differentiation—Congresses. 2. Cellular control mechanisms—Congresses. I. Clayton, R. M. II. Truman, D. E. S. (Donald Ernest Samuel), 1936-   . III. Title. IV. Series. [DNLM: 1. Cell differentiation—Congresses. 2. Gene expression regulation—Congresses. W1 AD559 v. 158/QH 607 S775 1981]
QH607.I57 1981   599′.087612   82-18053
ISBN 0-306-41181-4

Proceedings of the international workshop on the Regulability of the
Differentiated State, held September 1-5, 1981, in Edinburgh, United Kingdom

©1982 Plenum Press, New York
A Division of Plenum Publishing Corporation
233 Spring Street, New York, N.Y. 10013

All rights reserved

No part of this book may be reproduced, stored in a retrieval system, or transmitted in any form or by any means, electronic, mechanical, photocopying, microfilming, recording, or otherwise, without written permission from the Publisher

Printed in the United States of America

PREFACE

The international workshop on the Regulability of the Differentiated State was planned as a satellite meeting associated with the IXth International Congress of the International Society of Developmental Biologists held in Basel, Switzerland from August 28th to September 1st 1981. The workshop held in Edinburgh from September 1st to 5th 1981 was able to benefit from the presence in Europe of a number of developmental biologists from Japan and the United States. The workshop was intended to be an opportunity for a limited number of workers from a variety of areas in developmental biology to spend a short time exchanging data and a more prolonged time developing the ideas that arose from the data. Free-ranging discussion was intended right from the initial stages of planning the meeting and the preparation of the proceedings of the workshop gives an opportunity to others to see the directions taken by those discussions.

Accordingly we have published here a collection of the formally presented papers, summaries of the discussions which arose from those papers, together with some linking material which the editors believe will be of help to the reader in seeing the significance of some of the ideas which were put forward during the workshop. This linking material has been prepared in Edinburgh. After the contributions were to hand, we came to believe that some of the potential readership might wish to have available introductions to the main sections, outlining areas not touched on in any of the particular papers and giving a few general references not quoted in these papers.

We must apologise to our colleagues and admit with regret to the readers of this book that some interesting points made during discussion have been lost. The recording quality of the tapes, in spite of preliminary testing, turned out to be defective in places. Summary outlines were prepared during the discussions by some speakers, yet participants often, in the heat of discussion, did not find the time to write these out. Apart from the unavoidable gaps we hope that this has not led us accidentally to misrepresent any of the participants.

PREFACE

The topic of the regulability of cells which have already undergone a degree of differentiation or, to put it another way, the stability of their differentiated state, has some interest to clinicians, especially to oncologists and pathologists, and it also relates to one of the most lively areas of current biology, namely the way in which the expression of genes is controlled both in normal and abnormal development. The rapid expansion of our knowledge of gene structure and the details of gene transcription and the translation of RNA to give rise to cellular proteins gives an excitement to this area of research, but the organisers believed in the importance of relating this molecular data to current concepts in cell biology and to ideas which have been with us from the earliest days of experimental embryology such as notions of competence and determination. The proceedings published here follow the structure of the conference, with an introductory session aimed at defining and classifying the problems to be discussed, followed by sections on the molecular basis of differentiation and competence; on reversible malignancy, transdifferentiation and related topics; and on strategies of regulation. The final session of the conference was a round-table discussion which pursued in detail a number of important issues which had arisen earlier, in particular the extent to which differentiated cells can modify their gene expression or, after cell division, give rise to progeny expressing genes characteristic of other cell types. The types of molecular mechanism which would explain the balance between stability and plasticity of gene expression were also discussed.

An international workshop of this type is only made possible by adequate funding and the conference organisers wish to express their gratitude to Tenovus-Scotland, the Cancer Research Campaign, the Wellcome Trust, the International Society of Developmental Biologists, Glaxo Holdings Ltd., Gibco Europe Ltd., Sigma (London) Chemical Company and to Flow Laboratories for their financial assistance. We also wish to record our gratitude to the companies who participated in the exhibition associated with the meeting.

The meeting itself and the preparation of the proceedings for publication owe much to the efforts of D.J. Bower, J. Cuthbert, L.H.E. Errington, C.E. Patek, B. Pollock, and M. Wilson.

The organisers' debt to Louise Dobbie and Muriel Alexander for their secretarial and organisational abilities and efforts is very great indeed and it is a pleasure to record it.

The task of converting authors' manuscripts to camera-ready copy for printing was carried out with great care by Muriel Alexander and the editors wish to express their admiration and gratitude for this work.

R.M. Clayton
D.E.S. Truman

CONTENTS

Preface v

Introduction 1
   R. M. Clayton and D. E. S. Truman

CLASSIFICATION OF THE PROBLEMS

Some Topics of Common Interest to Developmental
   Biologists and Oncologists 7
   M. F. A. Woodruff

Gene Switching and Cellular Differentiation 13
   J. Paul

The Molecular Basis for Competence, Determination,
   and Transdifferentiation: A Hypothesis 23
   R. M. Clayton

Problems of Regulation 39
   H. Bloemendal

Taxonomies of Differentiation 45
   D. E. S. Truman

Summary of Discussion 55

MOLECULAR BASIS OF DIFFERENTIATION AND COMPETENCE

Introductory Review: the Molecular Basis of
   Differentiation and Competence 61
   A. P. Bird, D. E. S. Truman and R. M. Clayton

The Structure and Expression of the Haemoglobin
    Genes 65
  F. Grosveld, M. Busslinger, G. Grosveld,
    J. Graffen, A. de Klein and R. Flavell

Analysis of Red Blood Cell Differentiation 81
  P. R. Harrison, N. Affara, P. S. Goldfarb,
    K. Kasturi, Q.-S. Yang, A. Lyons, J. O'Prey,
    J. Fleming, E. Black and R. Nicholas

Expression of the Human Epsilon Globin Gene in
    Mouse Erythroleukaemic Cells Transformed
    with Recombinants between it and a Bacterial
    Plasmic Vector 89
  J. Paul and D. A. Spandidos

Isolation of Immunoglobulin Class-switch and Variable-
    region Variants from Mouse Myeloma and Hybridoma
    Cell Lines 93
  A. Radbruch, M. Bruggemann, B. Liesegang
    and K. Rajewsky

Monoclonal Antibodies Recognising Surface Antigens
    on Subpopulation of Neurones 107
  R. Mirsky, T. Vulliamy and S. Rattray

Effects of Rous Sarcoma Virus on the Different-
    iation of Chick and Quail Neuroretina Cells
    in Culture 115
  P. Crisanti-Combes, A.-M. Lorinet, A. Girard,
    B. Pessac, M. Wasseff, and G. Calothy

Molecules that Define a Dorsal-Ventral Axis of
    Retina can be used to Identify Cell Position 123
  G. D. Trisler, M. D. Schneider, J. R. Moskal
    and M. Nirenberg

Analysis of Myogenesis with Recombinant DNA
    Techniques 127
  D. Yaffe, U. Nudel, H. Czosnek, R. Zakut,
    Y. Caromon and M. Shani

Summary of Discussion 139

    REVERSIBLE MALIGNANCY, TRANSDIFFERENTIATION AND RELATED TOPICS

Introductory Review: Tumours, Transposition and
    Transdifferentiation 143
  A. H. Wyllie, R. M. Clayton and D. E. S. Truman

CONTENTS

Ectopic Hormones 155
    J. G. Ratcliffe

Neuronal-Glial Differentiation of a Stem Cell Line
    from a Rat Neurotumor RT4 - Branch Determination 165
    N. Sueoka, M. Imada, Y. Tomozawa, K. Droms,
    T. Chow and T. Leighton

The Crystallins of Normal and Regenerated Newt
    Eye Lenses 177
    D. S. McDevitt

Formation of Lentoids from Neural Retina Cells:
    Glial Origin of the Transformed Cells 187
    A. A. Moscona and L. Degenstein

Serum Factors Affecting Transdifferentiation in
    Chick Embryo Neuro-Retinal Cultures 199
    D. I. de Pomerai and M. A. H. Gali

Microenvironments Controlling the Transdifferentiation
    of Vertebrate Pigmented Epithelial Cells in
    in _vitro_ culture 209
    G. Eguchi, A. Masuda, Y. Karasawa, R. Kodama
    and Y. Itoh

Can Neuronally Specified Cells Transdifferentiate
    into Lens? 223
    T. S. Okada, K. Yasuda, H. Kondoh, S. Takagi
    and K. Nomura

The Effects of N-Methyl-N'-Nitro-N-Nitrosoguanidine
    on the Differentiation of Chicken Lens
    Epithelium _in vitro_: The Occurence of Unusual
    Cell Types 229
    R. M. Clayton and C. E. Patek

Summary of Discussion 239

THE STRATEGIES OF REGULATION: EXTERNAL SIGNALS, RECEPTORS AND EFFECTOR SYSTEMS

Introductory Review: Strategies of Regulation:
    Signals, Receptors and Effectors 245
    D. E. S. Truman and R. M. Clayton

Commitment, DNA synthesis and Gene Expression in
    Erythroleukemia Cells 253
    R. A. Rifkind, E. Epner and P. A. Marks

Regulability of Gene Expression and Different-
    iation during Myogenesis 259
    H. A. John

The Murine Haemopoietic Stem Cell: Patterns of
    Proliferation and Differentiation 275
    B. I. Lord

Inhibitor and Stimulator in the Regulation of
    Haemopoietic Stem Cell Proliferation (Abstract) 285
    B. I. Lord

Lymphocyte Differentiation within the Thymus (Abstract) . . . 287
    J. J. T. Owen

The Role of a Growth Factor Derived from the Retina
    (EDGF) in Controlling the Differentiated
    Stages of Several Ocular and Non-ocular Tissues 289
    Y. Courtois, C. Arruti, D. Barritault, J. Courty,
    J. Tassin, M. Olivie, J. Plouet, M. Laurent
    and M. Perry

Tissue Culture of Chick Embryonic Choroidal Cells:
    Cell Aggregation and Pigment Accumulation 307
    G. J. Chader, E. Masterson and A. Goldman

Gene Expression in the Cell Surface of Interneurones
    and Astrocytes (Abstract) 321
    R. Balazs

Summary of Discussion 323

QUANTITATIVE REGULATION OF GENE EXPRESSION

Introductory Review: Quantitative Regulation
    of Gene Expression 327
    R. M. Clayton, D. E. S. Truman and A. P. Bird

Skeletal Muscle Myogenesis: The Expression of Actin
    and Myosin mRNAs 331
    M. E. Buckingham, M. Carvatti, A. Minty,
    B. Robert, S. Alonso, A. Cohen, P. Daubas
    and A. Weydert

Genes Coding for Vimentin and Actin in Mammals
    and Birds 349
    W. J. Quax, H. J. Dodemont, J. A. Lenstra,
    F. C. S. Ramaekers, P. Soriano, M. E. S. van Workum,
    G. Bernardi and H. Bloemendal

CONTENTS

The Unusually Long Rat Crystallin αA_2 mRNA is
    Monocistronic 359
  R. Moormann, H. Dodemont, J. G. G. Shoenmakers,
  and H. Bloemendal

Commitment: A Multi-Step Process during Induced
    MELC Differentiation 367
  R. A. Rifkind, Z.-X. Chen, J. Banks and P. A. Marks

Studies of the Relationship between DNA Methylation
    and Transcription of the Ribosomal RNA Genes 375
  A. P. Bird, D. Macleod and M. H. Taggart

Tissue Specific Expression of Mouse α-Amylase Genes 381
  U. Schibler, O. Hagenbüchle, R. A. Young, M. Tosi
  and P. K. Wellauer

Regulation of the Relative Abundances of mRNAs
    in Hepatoma and Liver 387
  G. D. Birnie, H. Jacobs, R. Shott and P. R. Wilkes

Summary of Discussion 395

THE DIFFERENTIATED STATE AND ITS REGULABILITY

Final Discussion 399
  J. Paul, B. I. Lord, R. M. Clayton, R. A. Rifkind,
  P. R. Harrison, A. A. Moscona, A. Wylie,
  D. I. de Pomerai, M. E. Buckingham, D. S. McDevitt,
  G. J. Chader, N. N. Iscove, G. D. Birnie, T. S. Okada,
  H. Bloemendal, N. Sueoka, B. Pessac, Y. Courtois,
  A. Radbruch, U. Schibler, R. Balazs, and A. Bird

DEMONSTRATIONS

High Levels of δ-Crystallin Precursor RNA in
    Early Development of Chick Lens Cells 445
  D. J. Bower, B. J. Pollock and R. M. Clayton

Hybridization Selection and Cell Free Translation
    of mRNA from a δ-crystallin Clone 451
  L. H. Errington, C. Sime and R. M. Clayton

The Dedifferentiation of Chick Lens Epithelium
    in Cell Culture and the Effects on this
    Process of Exposure to a Carcinogen in Vitro 455
  C. E. Patek and R. M. Clayton

Cell Deletion in Differentiation 461
    A. H. Wyllie, R. G. Morris and A. J. Strain

Cloning and Structure of δ-Crystallin Genes 467
    K. Yasuda, N. Nakajima, K. Agata, H. Kondoh,
      T. S. Okada and Y. Shimura

List of Contributors and Participants 473

Index 479

INTRODUCTION

    R. M. Clayton and D. E. S. Truman

    Dept. of Genetics
    University of Edinburgh
    Edinburgh EH9 3JN

    When biologists talk of a differentiated cell, or of different-
iation, do they all mean exactly the same thing in the same way as
they do when they talk of reticulocytes, astroglia, myoblasts, lens
fibre cells and so on? The original definition of cellular differentiation used by embryologists involved criteria dependent upon the
light microscope. When cells appeared different, differentiation had
taken place. When it was established that characteristic compositions
of proteins, and especially enzymes, provided the basis for the
structure, appearance and behaviour of the cell, then the stage of
earliest detection of a distinct protein composition could be used
to define the differentiation. The development of sensitive and dis-
criminating methods of protein analysis (O'Farrell et al., 1977) has
extended the use of protein composition as a yardstick of differentiation, and it is now possible to define a cell as being in a differentiated state earlier in the ontogeny of the system. It is now expected that the appearance of characteristic proteins would be accompanied or preceded by the accumulation of their mRNA species and recombinant DNA technology has made it possible to assay specific mRNA
accumulation with sensitivity and relative accuracy. Such accumulation may be considered to give an earlier indication that differentiation has occurred, if we restate the criteria for differentiation.

    However high resolution procedures for proteins and mRNA species
have also shown that there is considerable overlap in the content of
these gene products between cell types, and that some of this analysis
cannot be brushed aside as the products of so-called 'housekeeping
genes' since it may include low levels of products which are important
in some other cell type.

    The notion that the differentiated cells of an individual all
contain the entire genetic information in a potentially usable form
derives much of its support, at least in the case of higher animals,

from studies of transplantation of nuclei from differentiated cells into enucleated eggs (reviewed by Gurdon, 1977), but there is also dissent, centering on the low rate of successful development of such eggs (Briggs, 1977). Molecular techniques have now shown that some of the overlap in mRNA species between different tissues involves the population of nuclear mRNAs: whether some of these are processed at all or are processed at lower levels in some tissues than in others is an important problem. Antecedent to any nuclear mRNA population there are a whole range of changes in the chromatin, affecting transcription, but the state of chromatin found in actively transcribed genes appear to be a necessary but not a sufficient condition for active transcription (see reviews by Lilley and Pardon, 1979, and McGhee and Felsenfeld, 1980). Perhaps the problems with nuclear transplants are wholly technical - but perhaps not every aspect of nuclear heterogeneity is equally susceptible to reversal. We may understand better regulatory mechanisms, and also apparent discrepancies, from studies of conditions permitting transcription of a locus, including the duration of this availability during the cell cycle.

Determination defines the state of a cell or group of cells in an embryo which will have not acquired histiotypic characteristics, but which will in the normal course of events exhibit a limitation of differentiation potential following manipulation. However, the possibility of detecting very early signs of differentiation may push back the boundary between determination and differentiation, and new methods of destabilising cells may likewise undermine the concept of determination, as well as our concepts of stem cells.

If we look at determination as a restriction in potential, this might require that pathways be switched off, perhaps with some relatively irreversible inactivation of the genes involved in that particular pathway, as well as that some pathways be switched on. When we consider a particular course of events in determination and differentiation it seems probable that the formation of a particular specialised cell type involves both the stimulation of certain genetic activities above levels obtaining in the undifferentiated cells, and the restriction of the activity of other sets of genes to levels of activity much lower than those in undifferentiated cells. The search for a single molecular mechanism might, therefore, be fruitless.

Many morphologically and functionally distinctive cell types such as muscle cells, neurones, lens fibre cells, red cells, lymphocytes and so on, no longer divide (indeed two of these cell types have lost their nucleus). However there are other cell types whose function requires that they continue to divide, but their progeny are restricted in type. Such cells include basal cells of the skin, cells in the gut epithelium, lens epithelial cells. How should we define stem cells? Should our criteria have in mind properties and definitions appropriate to those cells, still unidentified, which are re-

INTRODUCTION

cognised by their progeny, – cells which can give rise to a specific range of different cell types, as in the haemopoietic system? Can we assume that the same definitions, and the same types of change in descendant cells, will be appropriate models for cells giving rise to a more restricted, or a single cell type, as in the lens? Should we seek for stem cells in every system, or is their existence characteristic of systems whose function includes a requirement for constant renewal or constant growth? If the number of cells is restricted after a certain stage, as in brain, there might be nothing equivalent to a stem cell.

Should we refuse to accept as truly differentiated a cell of restricted nature, with a range of specific biochemical parameters, which is discovered given appropriate provocation, to have a potential to enter into mitosis and produce progeny which may acquire a new apparently unrelated cell phenotype? Is it appropriate to apply stem cell definitions in such cases?

A major focus of interest of cell biology is the acquisition of differentiation and acceptance of its stability (with the exception of the formation of tumours). How do we accomodate changes in differentiation – which at present are regarded as comparatively rare events?

Any molecular theory of determination and differentiation must seek to explain its comparative stability. Various possible stabilizing mechanisms have been considered. Modifications of the DNA which might limit transcription are one explanation, but perhaps these can, under some circumstances (not involving nuclear transplantation), be reversed. Another possibility is that signals specifying the differentiated state accumulate in the cytoplasm, and following cell division, are available to both daughter cells to programme differentiation along the same path (Gurdon and Woodland, 1970). Yet another possibility is that extracellular materials may have a similar programming function. This concept is also touched on in the section on Strategies of Regulation and such a regulatory function would account for the necessity to dissociate cells from their microenvironment if transdifferentiation is to occur.

Interdisciplinary contacts permit the exchange of thought patterns as well as of data, and history shows how significant this has been in the development of science. The application of procedures or theoretical frameworks derived from one area of research may lead to radical reassessments in another, the rephrasing of questions hitherto unsolved, the discovery that a problem is no longer inaccessible or even that there is a problem previously unformulated. Examples also abound of accurate and acute observations which lay fallow for want of a context in which their value might become apparent, until technical advances in other fields led to the formulation of a body of theory which could accomodate, or make use of these 'premature' observations (since we define phenomena by what we can observe even our definitions are often limited by existing technology). There are many well documented examples of these various propositions. Wilstätter prepared haemoglobin crystals many years before they were studied again for the purpose of structural analyses under the impetus of finding that many genetic anaemias were due to changes in haemoglobin properties, which together with the techniques of crystallography developed by the Braggs and others, led, inevitably to these studies. Duchesnes' thesis on 'War among Microbes: moulds against bacteria', described the effects of penicillin in 1896, but its preparation awaited a rediscovery by Fleming and the chemical technologies used by Chain and Florey, some 50 years later. When Garrod reported, in the first decade of this century, on the genetics and biochemistry of several human conditions he discussed their significance with complete clarity. The techniques he used were readily avilable, but his approach did not seem very relevant to others, and biochemical genetics did not really take off until, following experiments of <u>Drosophila</u> eye colours - interpreted according to notions derived from the physiology of hormones - Beadle and Tatum turned in the '50s to experiments with <u>Neurospora</u>. Today the wheel has come full circle. McKusick's (1978) latest listing shows that the number of studies on biochemical genetics of man are of an order of magnitude higher than that of any other organism. The early application of immunolgoical techniques, which have great discrimination, to the identification of compatible blood transfusions provided data that the genetic theory of the '50s and '60s was ill equiped to explain: blood groups showed numerous multiple alleles, all of them detectable in the heterozygote, while morphological or functional characterisitics apparently were governed by two or at most a very few recessive alleles. Many did not realise that this did not point to a special peculiarity of the red cell, nor was it due to imcompatible theories, but to incommensurate techniques. Delightful accounts of similar historical events will be found in Synge (1962) and Stent (1972).

The close relationship between our mental framework and our perception of relevance, and between theory and the types of data permitted by available technology, is surely not merely an aberration of the past, but an essential function of the search for knowledge. In ten years time, how will we see the evaluations and definitions we are now using?

In planning the workshop we naturally spent a considerable time discussing and listing the problems concerning the nature and stability of cell differentiation and cell committment. We believed that these problems could best be approached by an interdisciplinary workshop in which scientists whose work was addressed to different specific aspects of these more general problems, using different cell systems and a range of methodological approaches, could share their data and engage in discussions. We hoped that we could, as a group, test whether findings in one field could be extrapolated to others, or were instead, in part defined by special requirements of a particular system and of the levels of regulation studied; we hoped that we might be able to ask what extent problems in the several systems could be mutually redefined in terms of the different types of controls found in particular systems.

The topics chosen for each of the sessions were based firmly on our preliminary discussions. However, this introduction, written after the workshop took place, is in part able to lean on the always lively and often exciting workshop sessions, and to present a view which is both wider and less specialized than that of any one or two people. We try here to indicate not only the general problem areas but also to touch on a few of the specific questions which emerged and which we all felt were of fundamental importance.

Finally, and certainly unfairly, since we are now all dispersed to our several laboratories, we have touched in the Introduction on some questions which in retrospect, we believe emerged from our discussions overtly or by implication, as those which we may hope to return to in similar fashion in the not too distant future.

We must absolve our colleagues of responsibility for any idiosyncracy in our choice of these last tentative questions, while acknowledging our debt to them for the pleasure of their company, the excellence of their new data, the generosity and freedom of their discussion, and the great stimulus of the enquiries we all set out to share.

One phrase, "it is tempting to speculate that..." turns up in the Discussion section of scientific publications with sufficient frequency to lead one to suspect that even the scientific community is susceptible - albeit always, one hopes, with explicit reservations to Algernon's Syndrome, a condition best described in Wilde's original words, "I can resist anything except temptation" (Wilde, 1899).

REFERENCES

Briggs, R., 1977, Genetics of cell type determination, in:"Cell Interactions in Differentiation," M. Karkinen-Jaaskelainen, L. Saxen, and L. Weiss, eds., Academic Press, London, p.3.
Gurdon, J. B., 1977, Egg cytoplasm and gene control in development, Proc. R. Soc. B., 198:211.
Gurdon, J. B., and Woodland, H. R., 1970, On the long-term control of nuclear activity during cell differentiation, Current Topics in Develop. Biol., 5:39.
Lilley, D. M., and Pardon, J. F., 1979, Structure and function of chromatin, Ann. Rev. Genet., 13:197.
McGhee, J. D., and Felsenfeld, G., 1980, Nucleosome structure, Ann. Rev. Biochem., 49:1115.
McKusick, V. A., 1978, "Mendelian Inheritance in Man: Catalogues of

autosomal dominant, autosomal recessive and X-linked phenotypes," 5th ed., John Hopkins University Press, Baltimore.

O'Farrell, P. Z., Goodman, H. M., and O'Farrell, P. H., 1977, High resolution two-dimensional electrophoresis of basic as well as acidic proteins, Cell, 12:1133.

Stent, G. S., 1972, Prematurity and Uniqueness in Scientific Discovery, Scientific American, 227(6):84.

Synge, R. L. M., 1962, Tsvet, Willstätter, and the use of adsorption for purification of proteins, Arch. Biochem. Biophys., Suppl. 1:1.

Wilde, O., 1899, "The Importance of Being Earnest," Leonard Smithers and Co., London.

SOME TOPICS OF COMMON INTEREST

TO DEVELOPMENTAL BIOLOGISTS AND ONCOLOGISTS

M. F. A. Woodruff

MRC Clinical and Population Cytogenetics Unit
Western General Hospital
Edinburgh EH4 2XU

For a retired surgeon to find himself opening the batting at a meeting of developmental biologists is the kind of nightmarish situation that I don't think even Walter Mitty would have thought up for himself. You know who to blame for this. Oncologists, and I label myself as such, have certainly much to learn from developmental biologists. Ruth Clayton assures me that the converse is true, and has persuaded me to give this talk about three topics which interest me and which she says will interest you. I only hope that she is right in this rash prognostication.

The topics I am going to say a few words about are:

1. The nature of malignant transformation.
2. Uniformity and diversity in tumour cell populations.
3. The regulability of tumours.

I have discussed all these questions at greater length elsewhere (Woodruff, 1980).

THE NATURE OF MALIGNANT TRANSFORMATION

It is widely accepted by oncologists at least that the primary event in the development of a malignant tumour is a heritable change in a cell or number of cells which they call transformation. We might note in passing that people have suggested that the change is not in the cells which eventually make the tumour but in cells which produce somewhat vaguely envisaged regulatory factors, but this to me is unconvincing. Transformation is variously attributed to mutation or to some kind of epigenetic change. By epigenetic change I mean something which oncologists believe that developmental biologists believe plays a major role in the generation of diversity during development, except perhaps in the development of cells concerned with immunological responses, and perhaps also, as Macfarlane Burnet has suggested, in the generation of diversity in receptors of olfactory epithelium, which has some analogies with the immune system.

I don't know if you still accept Waddington's term 'epigenetic', or indeed if you still accept that the distinction I have drawn is valid, because I am totally out of date with your subject. But I suspect we would agree that we can distinguish two categories of hypothesis about transformation: namely, those that imply that the process is irreversible (or that the chance of it being reversed is so remote as to be negligible), and those which do not carry this implication. You are doubtless familiar with the arguments that are bandied backwards and forwards; based on the one hand, for example, on the fact that most carcinogens are mutagens and vice versa, and that many tumours show chromosomal abnormalities; on the other hand, on the existence in or on tumour cells of stage-specific substances, including so-called oncofoetal antigens, that belong to a particular stage of development, the ectopic hormone syndromes which Dr Ratcliffe is to speak about tomorrow, and, perhaps most interesting of all, the identification of markers of mouse teratocarcinomas in the normal tissues of adult chimaeric mice which have been injected with teratocarcinoma cells at the blastocyst stage. I have three comments. Firstly, it seems surprising that developmental biologists have not shown more interest in whether oncofoetal antigens play an important role in normal development. Secondly, I can see no compelling reason for supposing that all carcinogenesis is necessarily brought about exclusively by one or other of the two kinds of process we have been considering. Thirdly, whether you attribute the results of injecting teratocarcinoma cells into early mouse embryos to the reversal of an epigenetic change or to the loss by a mutant cell of the phenotypic characters of malignancy, these observations do illustrate the reversal of a change which possesses the generally accepted hall-marks of malignant transformation. The important question therefore is whether the mouse teratocarcinomas, which are a product of the laboratory, are a valid model for all, or some, or none of the tumours which arise in animals and people outside the laboratory. I carefully avoided saying "tumours which arise spontaneously". The word 'spontaneous' is one of the weasel words of oncology that I must warn you against. It may imply either that the tumour was not induced by some deliberate laboratory procedure or that it is of unknown origin. This first use of the term is dangerous because, from a biological point of view, there is no reason to distinguish between, for example, inducing a mouse tumour by injecting a chemical carcinogen or exposing the animal to irradiation, and acquiring a tumour oneself by cigarette smoking or accidental exposure to irradiation. Of course valid distinctions may be drawn on the basis of the nature, dose and duration of administration of the carcinogen, but that is another question.

The second use of the term is redundant. What is worse, however, is that people sometimes talk as if the term spontaneous implies that the tumour has developed in the absence of any definable cause, by a process analogous to the disintegration of a radioactive atom. It is perhaps conceivable that some tumours arise in this way, but there is no proof of this and personally I rather doubt whether such tumours exist.

The development of apparently normal cells from mouse teratocarcinoma cells reminds me of a question I heard asked over 30 years ago by a particularly biologically-minded radiotherapist, Dr Ralston Paterson, concerning the origin of the new epithelium seen when an ulcerating squamous-cell carcinoma of the skin heals following radiotherapy. If the ulcer is large, re-epithelialization does not appear to occur by ingrowth from the edge of the ulcer. Does it then occur from invisible islets of non-transformed cells in the ulcer floor which have survived the irradiation, or do some of the carcinoma cells undergo a reversal of malignancy? This question might conceivably be answered experimentally if one had a marker, as for example one form of the enzyme G6PD, which was present in all the tumour cells but in only about fifty per cent of surrounding epithelial cells, since in this event, if healing was due to reversal of malignancy virtually all the cells at the site of the healed ulcer should carry the marker.

UNIFORMITY AND DIVERSITY IN TUMOUR CELL POPULATIONS

The mention of G6PD brings me to my second topic, the question of uniformity and diversity in tumour cell populations. Let me begin by reminding you that what both laboratory workers and clinicians call a tumour is not just a collection of transformed cells but a complex ecosystem which contains transformed cells, vascular connective tissue, lymphocytes, macrophages and various other types of cell; the common assertion that tumours in general are monoclonal is therefore absurd, except in the trivial sense that they arise in an individual derived from a single fertilized egg, unless you add the qualification that you are talking about the transformed cells only. Is this qualified statement true? The answer usually given is yes, and this is based principally (though not exclusively) on the grounds that the myelomas produce only one kind of immunoglobulin and that tumours derived from persons who are heterozygous for the G6PD locus, which, I need hardly remind you, is carried on the X-chromosome and should therefore be extinguished in a more or less random manner, possess only one form of the enzyme. I would, however, make four comments:

(1) The myelomas may be a very special case in view of the complex, and possibly to some extent unique, mechanisms involved in the generation of diversity among immunoglobulin producing cells.

(2) I think the isoenzyme studies are far from decisive. Some reported results seem too good to be true because unless great care is taken to eliminate non-transformed cells from the population of cells tested one should find a good deal of heterogeneity - more I suspect than many people have found hitherto. Secondly, there is the problem of patch size. If the tumour arises from more than one cell in the same small patch where the same X-chromosome is being extinguished then one may find homogeneity without the tumour being truly monoclonal. Thirdly, heterogeneity may be missed either because the resolving power of the method is insufficient or, more importantly, minority clones of transformed cells may be eliminated by precautions taken to eliminate non-transformed cells. We are doing some experiments with Spedding Micklem and John Ansell of the Department of Zoology at the University of Edinburgh using tumours raised in female mice heteroygous at the PGK locus, which is analogous to the G6PD locus in humans. It would be premature to go into detail about this but I can say that we are all convinced that the possibilities of error that I have mentioned are very real possibilities indeed. So I think you have to interpret the enzyme studies with considerable caution.

(3) If there are minority clones in a tumour it is conceivable that these may, on occasion, be the origin of metastatic deposits and also of recurrent tumours after inadequate surgical or chemotherapeutic treatment, because the characteristics that a cell requires either to move around in the blood stream or the lymphatics and form a little focus of a tumour somewhere else, or to form a recurrence after most of the tumour has been excised, may not correspond at all closely with the kind of characteristics required to form an expanding tumour at the primary site. There is no conclusive evidence to show that minority populations can function in this way but Pimm and Baldwin have begun to look at the antigenic structure of recurrent tumours and to compare this with the antigenic structure of the original tumour.

(4) My last comment on this particular problem is that in the lab it is easy by giving a large dose of a chemical carcinogen to ensure that virtually all the animals you treat develop a tumour. In this event the total process of carcinogenesis, whether it occurs in one stage or two stages or two thousand stages, constitutes an event which is not rare, and in consequence, one would expect it to happen more than once in any given mouse. We must, therefore, consider the possibility that the tumour begins by being polyclonal, even though it ends up apparently essentially monocolonal. This raises the question of how the other clones are eliminated. Many people seem to regard this as a trivial question and, if pressed for an explanation, mutter something about different growth rates and survival of the fittest. I think however that the nature of the competition and interaction between these multiple clones, if they exist, merits serious investigation.

DEVELOPMENTAL BIOLOGY AND ONCOLOGY

THE REGULABILITY OF TUMOURS

Tumours are often said to be autonomous. Obviously this does not mean that they wander about the place independently of their host, but it does seem to imply two propositions which I find almost equally untenable, namely that malignant cells are not influenced in any way by any of the factors that control growth of normal tissues, and, secondly, that no specific homeostatic mechanisms have developed either for eliminating cancer cells or for preventing them from developing. The first of these propositions is manifestly false because, as Charles Huggins first established, some tumours at some stage of their development are clearly influenced by endocrinological factors, i.e. by the presence or absence of hormones of one kind or another. There are other instances of tumours being influenced by normal mechanisms, but the phenomenon of hormone dependence is sufficient to establish the point. As regards the second proposition, it would I think be surprising if the metazoa had not evolved homeostatic mechanisms to get rid of what Michael Stoker has called <u>asocial cells</u> - cells which have a high capacity for survival as cells but which are disadvantageous to the organism of which they form a part. So far as cancer is concerned this is often disputed on the ground that most cancer develops in old age, and that there is therefore no evolutionary drive to eliminate something which is essentially a disease of post-reproductive life. I think that takes a very anthropomorphic view of evolution. Undoubtedly many species show a decline in reproductive activity with age, but the abrupt termination of reproduction that we see in the human female is, in the main, a human phenomenon, or at least restricted to some of the higher primates. To sustain the argument against homeostasis one would need to know a great deal about the relationship between birth rate and parental age not only for man but some of his forebears as well. There is moreover quite a bit of direct evidence in support of the hypothesis of homeostasis, notably the occasional complete spontaneous regression of tumours, the fact that some tumours remain latent for long periods of time and then suddenly burst into activity, the high rate of cell loss sometimes found in kinetic studies of tumour cell populations, and the immunological and para-immunological reactions to tumour antigens and other surface determinants on tumour cells. Clearly, if homeostatic mechanisms exist they often fail otherwise nobody would die of cancer, but that doesn't mean that they don't exist. If my view is correct it follows that one major objective of cancer research should be to identify the reasons for the failure of these mechnaisms; a second objective must be to find ways of manipulating the host's response so as to reduce the incidence of cancer or to eliminate tumours after they have appeared. But we are getting a long way from developmental biology and this, as they say, is another story.

REFERENCE

Woodruff, M. F. A., 1980, "The Interaction of Cancer and Host - Its Therapeutic Significance," Academic Press, London; Grune and Stratton, New York.

GENE SWITCHING AND CELLULAR DIFFERENTIATION

J. Paul

Beatson Institute for Cancer Research
Switchback Road
Glasgow G61 1BD

Haemoglobin is characteristic of erythrocytes, ovalbumin of oviduct epithelial cells and insulin of pancreatic islet cells because these proteins are present in the corresponding cells at high concentrations but are not detectable in others. It is a broad biochemical generalisation that the character of differentiated cells is a manifestation of the proteins of which they are composed. Since most cells have a full complement of genes this has led to the popular speculation that differentiation is brought about by "switching genes on and off", a proposition which has frequently been linked with evidence that "active" and "inactive" genes are in structurally different states. The assumption inherent in these ideas is that important differences in gene expression are qualitative rather than quantitative. Yet this is certainly not true of the great majority of genes and may not be true of any.

The development of two-dimensional electrophoretic methods has made it possible to resolve at least 1000 peptides in a single preparation and comparison of different cells or tissues (Garrels, 1979) has revealed that most detectable proteins are common to most cells, major differences being limited to a small number. However, even this technique reveals only proteins present in relatively high concentrations. The number of different peptides in a eukaryotic cell is thought to be of the order of 10 000 and, therefore, since the 2-D electrophoretic technique enables us to resolve only about 1000, we have no reliable quantitative information about the majority of peptides is different cell types.

Mammalian cells contain from 50-1000 pg of protein. In an erythrocyte this is mainly haemoglobin (about 30 pg) and amounts to 10^9 molecules. In a hepatocyte there are altogether about 10^{10} protein molecules. The most abundant may be present as $2-5 \times 10^8$ molecules; the least abundant are, on average, present as 5×10^5 molecules, some undoubtedly being at considerably lower levels at which most proteins are not measurable.

The concentration of a protein in a cell represents a steady state between simultaneous synthesis and degradation. Degradation is not passive as has emerged clearly from measurements of the half lives of a variety of proteins in rat liver (Schimke, 1975) which vary from minutes (ornithine decarboxylase and aminolaevulinic synthetase) to days (arginase and lactic dehydrogenase isozyme 5). Moreover, induction of some enzymes is due to stabilisation rather than changes in rates of synthesis, for example, the induction of tryptophan oxygenase by tryptophan (Schimke et al., 1965) results from an increased stability of the enzyme. Some proteins made at significant rates may not be measurable because rapid breakdown leads to low steady state values. Hence, the possibility has to be considered that nearly all the proteins characteristic of an organism may be present in all cells, the differences being due to variations in concentrations over many orders of magnitude. A corollary would be that nearly all functional genes may be active in all cells.

RNA COMPLEXITY AND TURNOVER

By studying RNA instead of protein, much more precise information can be obtained because it is possible to measure less than one molecule of specific mRNAs per cell. In general the relative rates of synthesis of proteins are governed by the concentrations of the corresponding messenger RNAs (although there are some well-established exceptions to this in which translational control is important, for example, in the masked mRNA in eggs and as a consequence of heat shock, notably in Drosophila cells: Storti et al., 1980). The steady state concentrations of RNA populations can readily be measured by hybridisation experiments, either by following the kinetics of hybridisation of RNA to cDNA or by measuring the saturation of unique sequences of genomic DNA by RNA, and information obtained from studies of RNA complexity has reinforced the idea that most genes may be transcribed in all cells. These experiments have led to the following conclusions (Getz et al., 1975; Hastie and Bishop, 1976; Hough et al., 1975; Young et al., 1976). First, the number of messenger RNAs, which may be taken to reflect the number of structural genes in an organism is of the order of 10-20 000. Secondly, most cells in a mature organism contain nearly all these messengers. Thirdly, the complexity of nuclear RNA is much higher than mRNA, by a factor of 5-10 in the more complex eukaryotes.

In most eukaryotes the proportion of DNA which is represented in messenger RNA is small but the proportion represented in nuclear RNA is quite large, frequently about 20%. Much of the difference in complexity between nuclear and cytoplasmic RNA can be accounted for by processing of mRNA precursors, particularly the excision of introns. However, some RNA species may be confined to the nucleus, particularly the small nuclear RNAs, and yet others may be confined to the nucleus in some cells but processed to yield messenger RNAs in others. There is quite good evidence that the latter mechanism exists during the early stages of sea urchin development (Hough et al., 1975). In mammalian cells qualitative differences of this kind between different cells are less convincing but there are examples of major quantitative differences in processing such that a given nuclear precursor in one cell may be very efficiently processed into messenger RNA, whereas in another cell the same precursor may be processed very inefficiently (Birnie et al., 1981).

ARE SOME STRUCTURAL GENES COMPLETELY TRANSCRIPTIONALLY INERT?

The major differences in mRNA between different mammalian cell types are marked variations in the concentrations of certain messengers and the question, as with proteins, is whether these differences are ever absolute. It seems likely that in the few instances, such as the immunoglobulin genes, in which genetic rearrangement occurs, differences are indeed absolute, but there is no evidence of rearrangement in most genes.

As with proteins, it has been established for some years that RNA turnover may be important in regulating the levels of RNAs, the half lives of which can vary greatly even within the same cell and can fluctuate from time to time (Kafatos and Gelinas, 1975).

Consider the globin genes: are they transcribed or not in non-erythroid cells? The number of globin messenger RNA molecules in a mature reticulocyte is about 150 000. Low concentrations have been detected in non-erythroid (but haemopoietic) cells and in transformed cells and claims have been made that the odd copy can be measured in normal non-erythroid cells (Humphries et al., 1976) although in some cases this must be less than one per cell. These enormous differences almost certainly imply great differences in transcriptional rates but do they imply that the globin genes are not transcribed at all in non-erythroid cells? The difficulty is in determining whether the genes may be transcribed but the transcript "instantaneously processed". Little is known about RNA turnover in the cell nucleus except that the turnover rate of nuclear RNAs is rather short and it is not easy to exclude the possibility that the failure to detect a transcript is due to the fact that it is degraded while being transcribed.

There is, therefore, ample evidence for regulation of individual protein concentrations within the cell at all possible post-transcriptional levels. The unresolved question concerns the relative importance of these post-transcriptional events and transcriptional control. In particular, is transcriptional control of structural genes ever absolute? There is little doubt that many sequences in eukaryotes are never transcribed and there is excellenct ultrastructural evidence that ribosomal genes can vary from being entirely inert to being highly active (Trendelenburg et al., 1974), but does this also apply to those genes which are transcribed by RNA polymerase II and which give rise to the proteins which characterise cells?

One of the classical observations which has encouraged the view that cell differentiation is mainly achieved by switching genes on and off is the behaviour of "puffs" in Dipteran giant chromosomes. When Dipteran cells are incubated with a labelled RNA precursor and then autoradiographed, the main accumulation of grains is over puffs. However, Pelling (1964) showed that if the pulses of radioactive precursor were made very short, then the differences between puffed regions and non-puffed interband regions were much less, which suggested that transcription might be occurring at almost all sites within the chromosome. Another observation which raises this possibility is the finding of Jamrich et al. (1977) that RNA polymerase is present in <u>all</u> interband regions. Moreover, measurements of RNA complexity in Drosophila cells carry the implication that some thousands of genes are being transcribed.

Consideration of some of the numbers relating to transcription emphasises the difficulties of distinguishing between active and inactive genes. The great majority of genes give rise to mRNAs which are present in from 5-50 copies per cell. Galau et al. (1977) have reviewed the evidence that these sparse messengers are functionally effective in giving rise to adequate amounts of the corresponding proteins. Take the case of a mRNA which is present at 10 molecules per cell in the steady state. For a messenger with a half life of 2.5 h (an average value) in a diploid cell only 1.5 transcripts per hour are required to maintain this level. If we assume the transcriptional unit to be about 5 kb then transcription will take less than 2 min (assuming an elongation rate of about 4×10^3 nucleotides per min). Hence, for nearly 97% of the time this "active" gene will be totally "inactive".

The one questionable assumption in the above calculation is that the elongation rate is always roughly the same. This is what the evidence suggests but in the case of those genes which give rise to few transcripts, in the absence of direct evidence it has to be borne in mind that some control over elongation may operate. In relation to most genes, it may well be better, therefore, to think not of "active" and "inactive" structural genes but of highly active and less active genes, falling on a continuous scale covering 3-4 orders of magnitude (from about 0.01 to 40 transcripts per gene per minute).

Nevertheless there is persuasive evidence for structural differences in the chromatin structure of highly active and less active genes.

STRUCTURAL CHARACTERISTICS OF ACTIVE GENES

During the past 15 years many attempts have been made to separate "active" from "inactive" chromatin or to distinguish between them. Very little success has attended attempts at separation and one has to conclude that physically the differences may not be striking. On the other hand, despite the artifacts which have dogged many experiments, persistent evidence has accumulated that the more active can be distinguished from the less active. Earlier studies on transcription of chromatin in vitro indicated that active genes were more accessible to exogenous RNA polymerases (Gilmour, 1978). More incontrovertible results have been obtained in studies on DNAse I sensitivity which have consistently revealed that actively transcribed chromatin is more rapidly digested by this enzyme than "inactive" chromatin. Particularly relevant is the evidence that globin genes are more readily degraded in nuclei from erythroid cells than nuclei from non-erythroid cells (Stalder et al., 1980). Weisbrod and Weintraub (1979) have shown that this is associated with the presence of the special non-histone proteins HMG14 and 17 and several authors have shown that genes from tissues in which they are actually expressed are under-methylated as compared with the same genes in other tissues (e.g. Van der Ploeg and Flavell, 1980). However, there are exceptions to this and it cannot be considered a general rule.

How are these observations related to the conclusion that functional genes may be completely trascriptionally inert 97% of the time? Paul et al. (1978) found that, so far as can be measured, most genes giving rise to transcripts were digested when 20% of the DNA chromatin was digested by DNAse I. This implies that even those genes which are infrequently transcribed are structurally distinguishable from the chromatin which contains inactive DNA. This could lead to the speculation that there is a clear-cut distinction between structural genes, i.e. potentially transcribable chromatin, and the truly inactive never-transcribable chromatin. However, studies by Weintraub and his colleagues (1981) cannot readily be reconciled with such a hypothesis. They have demonstrated different degrees of sensitivity of genes to DNAse I. In particular, they have shown that actively transcribed globin genes in the chick are exquisitely sensitive, whereas adjacent genes, including those which are expressed at earlier stages of ontogeny or will be expressed later, have an intermediate degree of sensitivity. But the ovalbumin gene which may not be expressed at all in erythrocytes is highly resistant to DNAse I. These differences in sensitivity to the nuclease are presumably related to different configurations of chromatin. Once again we gain the impression that genes may not simply fall into the categories of digestible and resistant but may form a continuum possibly correlated with transcriptional activity.

THE SEARCH FOR TRANSCRIPTIONAL CONTROL SEQUENCES IN DNA

A direct study of the mechanism of transcription is difficult in intact cells because of the complexity of HnRNA. In order to simplify the problem attempts have been made to devise techniques for the study of transcription in vitro using DNA as template or, alternatively, to study the behaviour of genes introduced into cells.

Success in demonstrating "faithful" transcription of eukaryotic genes has been achieved using crude preparations of RNA polymerase II or purified preparations with an added crude cellular extract (Luse and Roeder, 1980; Manley et al., 1979). The template in these experiments has usually been naked DNA derived from recombinants between eukaryotic genes and plasmids or bacteriophage. These experiments have attracted attention to sequences originally identified as consensus sequences located on the 5' side of the genes. The Goldberg-Hogness box, or TATA box, which occurs 15-20 base pairs upstream from the initiation site for RNA transcription is found in one form or another in association with many genes but it is not invariably present and is absent from some viral genes. It first attracted interest because of its similarity to the Pribnow box which is part of the promoter sequence in prokaryotes. Experiments designed to investigate the function of this sequence have led to conflicting results. In genes injected into oocytes removal of this box does not eliminate transcription, although it does reduce it, but it gives rise to several transcripts initiating at different points (Grosschedl and Birnstiel, 1980a). Hence, in this system, the evidence suggests that the TATA box has to do with precision of initiation. On the other hand, in a completely in vitro system, using RNA polymerase II, removal of the TATA box seems to eliminate initiation completely in both adenovirus DNA and chicken conalbumin genes (Benoist and Chambon,1980; Corden et al., 1980). Similar ambiguity applies to the CAAT box which occurs about 80 nucleotides upstream from the cap site of several different genes. This site seems to be unimportant in vitro, whereas its removal leads to a lower rate of initiation in oocytes. It is difficult to envisage a specific regulatory role for TATA and CAAT sequences by themselves as they occur in most genes, including those like the globin genes which seem to be under transcriptional control.

One can question whether studies using naked DNA as template have relevance to what occurs in eukaryotic cells where DNA is invariably associated with proteins in chromatin. Hence, studies on genes inserted into living cells may turn out to be more relevant and, in this connection, certain sequences which have no known analogy in prokaryotic cells may be particularly interesting. One has been demonstrated by Grosschedl and Birnstiel (1980b) and their colleagues in relation to the histone gene H2A in the sea urchin. A particular component (segment E) from 184 to 524 bases upstream of the cap site of the H2A gene has a dramatic influence on expression of this gene in the Xenopus oocyte. Deletion of the element reduces transcription of the H2A gene by 15-20 fold. Moreover, when the element is inverted the H2A gene is still expressed and at an even higher level than in normal DNA. The significance of this sequence is at present difficult to understand but it may be related to the LTR (Long Terminal Repeat) sequences in retrovirus proviral DNA which greatly enhance the transcription of adjacent genes.

Long terminal repeats (Dhar et al., 1980; Hayward et al., 1981; Neel et al., 1981) contain sequences corresponding to the 3' and 5' ends of the viral RNA and they occur both at the 5' and at the 3' ends of the provirus sequences. The 5' component contains putative promoter sequences and promotes transcription of viral sequences but, since it is also situated at the 3' end of the provirus, it can promote trascription of adjacent cellular sequences also. When cells are co-transformed with LTRs and other genes, they facilitate transformation by these latter genes probably by enhancing their transcription and this might be expected if they became ligated to the transforming genes in the correct polarity. However, what is much more difficult to explain is that in some instances the sequences seem to be effective even when integrated in the opposite orientation. Hence, it is possible that sequences of this kind, like fragment E from the histone genes, have a more general influence on the conformation of chromatin in the adjacent region.

Many questions are, therefore, completely unanswered with regard to transcriptional control in differentiation. Do these absolute differences reflect different structures between "active" and "inactive" structural genes or are the transcriptional differences all quantitative and under continous control? What distinguishes a gene actively transcribed only 5% of the time from DNA which is never transcribed? Are differences of polymerase and nuclease access and binding of HMG14 and 17 secondary to the transcriptional state or do they reflect a mandatory conditional state for transcription? Are "special" genes, like globin and ovalbumin, subject to a special class of regulatory mechanisms or are they extreme examples of a standard type of control?

REFERENCES

Benoist, C., and Chambon, P., 1980, Deletions covering the putative promoter region of early mRNAs of Simian Virus 40 do not abolish T-antigen expression, Proc. Natl. Acad. Sci. U.S.A., 77:3865.
Birnie, G. D., Balmain, A., Jacobs, H., Shott, R., Wilkes, P. R., and Paul, J., 1981, Post-transcriptional control of messenger abundance, Molec. Biol. Rep.,7:159.
Corden J., Wasylyk, B., Buchwalder, A., Sassone-Corsi, P., Kedinger, C., and Chambon, P., 1980, Promoter sequences of eukaryotic protein-coding genes, Science, 209:1406.

Dhar, R., McClements, W. L., Enquist, L. W., and Woude, G. F. V., 1980, Nucleotide sequences of integrated Moloney sarcoma provirus long terminal repeats and their host and viral junctions, Proc. Natl. Acad. Sci. U.S.A., 77:3937.

Galau, G. A., Klein, W. H., Britten, R. J., and Davidson, E. H., 1977, Significance of rare mRNA sequences in liver, Archs. Biochem. Biophys., 179:584.

Garrels, J. I., 1979, Two-dimensional gel electrophoresis and computer analysis of proteins synthesized by clonal cell lines, J. biol. Chem., 254:7961.

Getz, M. J., Birnie, G. D., Young, B. D., Macphail, E., and Paul, J., 1975, A kinetic estimation of base sequence complexity of nuclear poly(A)-containing RNA in mouse Friend cells, Cell, 4:121.

Gilmour, R. S., 1978, Structure and control of the globin gene, in: "The Cell Nucleus," H. Busch, ed., Academic Press, New York, Vol. 6:329.

Grosschedl, R., and Birnstiel, M. L., 1980a, Identification of regulatory sequences in the prelude sequences of an H2A histone gene by the study of specific deletion mutants in vivo, Proc. Natl. Acad. Sci. U.S.A., 77:1432.

Grosschedl, R., and Birnstiel, M. L., 1980b, Spacer DNA sequences upstream of the TATAAATA sequence are essential for promotion of H2A histone gene transcription in vivo, Proc. Natl. Acad. Sci. U.S.A., 77:7102.

Hastie, N. D., and Bishop, J. O., 1976, The expression of three abundance classes of messenger RNA in mouse tissue, Cell, 9:761.

Hayward, W. S., Neel, B. G., and Astrin, S. M., 1981, Activation of a cellular onc gene by promoter insertion in ALV-induced lymphoid leukosis, Nature, Lond., 290:475.

Hough, B. R., Smith, M. J., Britten, R. J., and Davidson, E. H., 1975, Sequence complexity of heterogeneous nuclear RNA in sea urchin embryos, Cell, 5:291.

Humphries, S., Windass, J., and Williamson, R., 1976, Mouse globin gene expression in erythroid and nonerythroid tissues, Cell, 7:267.

Jamrich, M., Greenleaf, A. L., Bautz, F. A., and Bautz, E. K. F., 1977, Functional organization of polytene chromosomes, Cold Spring Harb. Symp. quant. Biol., 42:389.

Kafatos, F. C., and Gelinas, R., 1975, mRNA stability and the control of specific protein synthesis in highly differentiated cells, Biochem. Cell Diffn., 9:223.

Luse, D. S., and Roeder, R. G., 1980, Accurate transcription initiation on a purified mouse β-globin fragment in a cell-free system, Cell, 20:691.

Manley, J. L., Sharp, P. A., and Gefter, M. L., 1979, DNA-dependent transcription of adenovirus genes in a soluble whole-cell extract, Proc. Natl. Acad. Sci. U.S.A., 76:160.

Neel, B. G., Hayward, W. S., Robinson, H. L., Fang, J., and Astrin, S. M., 1981, Avian leukosis virus-induced tumors have common proviral integration sites and synthesize discrete new

RNAs: oncogenesis by promoter insertion, Cell, 23:323.
Pelling, C., 1964, Ribonukleinsaure-synthese der Reisen-chromomen, Autoradiographische Untersuchungen an Chironomus tentans, Chromosoma, 15:71.
Van der Ploeg, L. H. T., and Flavell, R. A., 1980, DNA Methylation in the human γδβ-globin locus in erythroid and nonerythroid tissues, Cell, 19:947.
Schimke, R. T., 1975, Protein synthesis and degradation in animal tissue, Biochem. Cell Diffn., 9:183.
Schimke, R. T., 1975, Sweeney, E. W., and Berlin, C. M., 1965, The roles of synthesis and degradation in the control of rat liver tryptophan pyrrolase, J. biol. Chem., 240:322.
Stalder, J., Larsen, A., Engel, J. D., Dolan, M., Groudine, M., and Weintraub, H., 1980, Tissue specific DNA cleavages in the globin chromatin domain introduced by DNase I, Cell, 20:451.
Storti, R. V., Scott, M. P., Rich, A., and Pardue, M. L., 1980, Translational control of protein synthesis in response to heat shock in D. melanogaster cells, Cell, 22:825.
Trendelenburg, M. F., Spring, H., Scheer, U., and Franke, W. W., 1974, Morphology of nucleolar cystrons in a plant cell Acetabularia mediterranea, Proc. Natl. Acad. Sci. U.S.A., 71:3626.
Weintraub, H., Larsen, A., and Groudine, M., 1981, α-globin-gene switching during the development of chicken embryos: expression and chromosome structure, Cell, 24:333.
Weisbrod, S., and Weintraub, H., 1979, Isolation of a subclass of nuclear proteins responsible for conferring a DNase I sensitive structure on globin chromatin, Proc. Natl. Acad. Sci. U.S.A., 76:630.
Young, B. D., Birnie, G. D., and Paul, J., 1976, Complexity and specificity of polysomal poly(A)+RNA in mouse tissues, Biochemistry, N.Y., 15:2823.

THE MOLECULAR BASIS FOR COMPETENCE, DETERMINATION

AND TRANSDIFFERENTIATION: A HYPOTHESIS

R.M. Clayton

Department of Genetics
University of Edinburgh
Edinburgh   EH9 3JN

DIFFERENTIATION AND COMPETENCE

The major concepts of classical experimental embryology, competence, determination (or committment) and induction, are essentially descriptive: they spring from the observations following such experimental manipulations as excision, transplantation or explantation; and each of these terms is definable only in the context of the other two. Competence refers to the capacity of a multipotential embryonic tissue to respond to induction (normally contact with another embryonic tissue) by entering into certain pathways of differentiation other than those which would obtain in the absence of the induction (see for example Waddington, 1956; Saxen and Toivonen, 1962). During the course of development there is a succession of inductions and a temporal hierarchy of competences operating in the development of any structure (see for example Jacobson, 1966). The reciprocal term, determination, refers to the restriction of the possible future differentiation pathways to one. We may add a more recent concept: transdifferentiation: the exchange of one pattern of differentiation for another, which normally characterises a tissue of different developmental origin (Okada, 1976, 1980; Eguchi, 1976, these proceedings; Clayton, 1978, 1982). However, as with multi-potential competent cells, the possible exchanges in fate of transdifferentiating cells are restricted.

We must assume that all these phenomena are capable of definition in terms of patterns of gene expression, and regulatory changes in these patterns. Although no such molecular exegisis of these embryonic events is yet to hand, the available information from various lines of investigation points to the importance of the regulation of the levels of expression of gene products, which may vary over a wide range of quantitative values, during both differentiation and transdifferentiation. Assessments of the natural levels of a gene product which may be expressed in several tissues also points to the importance of mechanisms of quantitative regulation.

Recently attention has tended to be focussed on specific gene derepression, but the thesis tentatively advanced here is that a multipotential, competent embryonic tissue is so by virtue of the coexistence in it of a range of mRNA species at intermediate abundance levels, which when expressed at abundant levels, severally characterise different descendant tissue types. This proposition requires that cellular experience, including embryonic inductions, operates by selectively increasing the representation of some mRNA species, until they are relatively abundant, this corresponding to cellular commitment, while others decline in their representation, this process corresponding to a loss of cellular multipotentiality.

It is well established that areas of embryonic competence form a series of overlapping fields: that for the eye overlaps that for the ear and the nasal placode (Jacobson, 1966). The area which actually forms a rudiment is always smaller than the competent area, and the potential of a field declines at the periphery as the degree of commitment increases in the centre. Thus the probabilities of an embryonic area entering into various alternative options segregate out as development proceeds. That these changes may have molecular concomittants was suggested by studies of changes in the distribution of antigens during embryogenesis: in the formation of the chick heart (Ebert, 1953), the segregation of amphibian mesoderm and ectoderm, and later, neural plate and ectoderm (Clayton, 1953), during Xenopus development (Bravo and Knowland, 1979), and in the segregation of blastomeres (Monroy and Rosati, 1979).

Recent studies of the distribution and quantitation of mRNA species show similar patterns of developmental change: globin mRNA is found in a range of embryonic tissues but it persists and increases only in the haemopoietic lineages and is lost from others (Perlman et al., 1977). Similarly, 50% of the mRNA species in the limb rudiment are lost when histodifferentiation of muscle begins (Anderson et al., 1976).

TISSUE SPECIFICITY AND CELL DIFFERENTIATION

The concept of tissue specific proteins which uniquely defined a particular type of differentiated cell guided many studies of differentiating cell systems in which the earliest detectable appearance of organ specificity was sought for in the tissue or cell lineage leading to that organ. This orientation led to considerable efforts to follow the ontogeny of a single substance in a specific system and to eliminate 'non-specific' tissue cross reactive material (e.g.

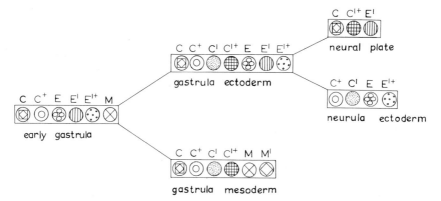

Figure 1. A model for the distribution of antigenic molecules in the
early newt embryo. The figure is derived from the data in
Clayton, 1953, by the deductive procedures often used in
analysing blood group antigens if their biochemistry is
unknown: it enumerates and identifies by elimination as
shown in the following table:

Tissue:			A	B	C
Antiserum	Anti A		+	+	+
"	"	absorbed B	+	−	+
"	"	" C	+	+	−
"	"	" B + C	+	−	−

A pattern of this kind shows that anti A must detect at
least 4 distinct antigens: 1) Common to A, B and C,
2) Common to A and C, 3) Common to A and B, 4) Specific
to A.

Takata et al., 1966; Ikeda and Zwaan, 1967). Few studies attempted
to search for complex patterns of change by seeking components found
in early embryos in the diverging descendant tissues, or by examining
the fate of components of the early embryonic stage of an organ as
that organ differentiated: but such studies suggested that loss of
some components accompanied the gain of others (Burke et al., 1944;
Clayton, 1953 and Fig. 1).

   The development of methods for the detection and quantitation
of specific proteins at very low levels have shown that many proteins
previously thought to be organ or tissue specific are actually superabundant in that organ or tissue but may be detected at low levels
in certain other organs or tissues (Table 1) (reviewed Clayton, 1982).
Similarly, the development of methods for the assessment of mRNA species and the identification of specific mRNAs have shown that the mRNAs
which are in the abundant class of one tissue may be in the intermediate abundance class of others (Table 1). An investigation with

Table 1. Superabundant gene products found at low levels in other tissues (see also Clayton, 1982).

Protein and major tissue source	Other tissues with low levels	Comments	Source
Crystallins – Lens	Crystallins reported in: reptile pineal, iris, retina, embryonic adenohypophysis, cornea. Low levels of crystallin reactivity in brain, liver, kidney, skin. Crystallin mRNA reported in embryo eye cup and in neural retina and pigmented retina layers.	Iris, neural retina, pigment retina, embryo adenohypophysis, cornea can all trans-differentiate into lens cells with high levels of lens proteins _in vitro_.	1, 2, 3, 4, 5, 6, 7.
Chick ovalbumin – Oviduct	Ovalbumin mRNA sequences found in nuclear RNA of liver, spleen, brain, heart, and on polysomes of liver, brain.	Responsive to oestrogen stimulation in all tissues.	8.
Insulin – Pancreas	Liver, brain, cultured lymphocytes, cultured fibroblasts. Produced in some non-endocrine neoplasms.	Levels unaffected by insulin depletion.	9.
ACTH – Pituitary	Carcinomata, mainly of lung (oat cell carcinoma), thymus, pancreas, phaeochromocytoma, some ovarian tumours.	Levels of ACTH and related peptides can be sufficient in these carcinomata to produce symptoms.	10, 11, 12.

| Chorionic gonadotrophin – Placenta | Normal tissues: pituitary, liver, colon, ovary, testis, cultured fibroblasts. Carcinomata of spleen, liver, pancreas, breast, bladder, lung, kidney teratomas, melanomas, myelomas. | Not glycosylated in normal non-placental tissues. | 13, 14, 15, 16, 17. |

1. Clayton et al., 1968.
2. Clayton, 1970.
3. Bours et al., 1972.
4. Jackson et al., 1978.
5. Thomson et al., 1979.
6. Thomson et al., 1981.
7. Clayton et al., 1979.
8. Tsai et al., 1979.
9. Rosenzweig et al., 1980.
10. Bertagna et al., 1979.
11. Jeffcoate and Rees, 1978.
12. Ratcliffe, 1982.
13. Kawamura et al., 1978.
14. Yoshimoto et al., 1977.
15. Yoshimoto et al., 1979.
16. Braunstein et al., 1979.
17. Rosen et al., 1979.

high titred antibody to total lens proteins (Clayton et al., 1968) detected crystallin antigenicities in several extra lenticular tissues at low levels. α, β and δ crystallins are heteropolymers, and each of those, like the monomeric γ crystallins are members of a gene family (Clayton, 1974; de Jong, 1981). In many tissues the extra-lenticular cross reactivity found was only partial, suggesting either that only some members of a gene family, or that material antigenically similar but not identical was found in extra-lenticular sites. The pattern of cross reactivities found led to the conclusion that "much tissue specificity is combinatorial and quantitative in nature" (Clayton et al., 1968). Subsequent studies (Bours and van Doorenmaalen, 1972; Bours, 1973) have confirmed the presence of extra-lenticular crystallins.

Factors regulating the high levels of a superabundant protein in the appropriate tissue are under study in several systems but it is evident that there are also regulatory responses in tissues in which the same protein remains at a low level (Tsai et al., 1979). Why should certain tissues maintain, at low levels, a product which is superabundant and has an important function in some other tissue? The generality of this situation might lead one to assume some evolutionary significance. One explanation might be that leakiness permitting some noise in the system and the possibility of regulation is the means by which developmental divergence and flexibility of response is maintained. Another possibility is that it is the properties of the protein itself which renders it useful in more than one situation, and that tissues which become specialised for this necessary function have then evolved. Contractile proteins, once thought of as confined to muscle, are not only found at low levels in many other cell types but exert vital contractile functions in those cells. As to the crystallins; their structure evidently enables them to enter into close orderly packing with each other and achieve particularly high protein concentrations (reviewed by de Jong, 1981). The consequences of such properties for evolutionary stability are discussed in Clayton, 1974, 1978, and de Jong, 1981. Perhaps they are found elsewhere because they can associate with the plasma membrane (Benedetti et al., 1976) and are therefore possibly useful for subcellular localised orderly water-excluding arrays in other types of cell.

Another possibility is that the evolution of differentiated cells with a range of responsive capacities, based on permutations of existing gene products, is probably faster - and more likely - than an evolution which depends on the appearance of wholly new gene products for each new function.

All these hypotheses avoid the general problem of the mechanisms for regulating the level of a gene product in a cell, and mechanisms for changing it: the duration of transcription, differential processing efficiencies, mRNA stability, availability for translation, and rates of translation are all possible candidates and the likelihood is that many mechanisms, each of them effective quantitatively rather than qualitatively, operate for any one product. There is evidence that crystallins can be regulated non-coordinately and that each crystallin may be regulated in more than one way (Clayton et al., 1976; reviews: Clayton, 1979a, 1982; Piatigorsky, 1980, 1981), but the problem of selective regulatory effects which have the effect of changing the representation of gene products, and thus effecting differentiation, remains to be investigated.

There are almost certainly many possible different regulatory mechanisms acting at each of several different cellular levels. In the transdifferentiating neural retina system, various specific agencies which may affect different metabolic pathways are highly effective in promoting or preventing the entry into the lens or pigment transdifferentiation pathways (e.g. Itoh, 1976; Clayton et al., 1977; Pritchard, 1981; Eguchi et al., these proceedings; de Pomerai and Gali, these proceedings).

Cell-cell interactions in culture, which may be facilitated by high cell density or by such manoeuvres as cell aggregation, can also act as regulators of differentiation pathways. In the chick embryo retina for example, cell aggregation renders the cells inducible for glutamine synthetase (Linser and Moscona, 1979) and promotes crystallin synthesis at an early stage of culture (Moscona and Degenstein, 1981, and these proceedings). Similarly, multilayering, whether promoted by high intial cell density, by folding of the cell sheet or occuring naturally in regions of overlap between neighbouring colonies, promotes transdifferentiation to lentoids: but in cultures set up at low cell densities, lentoid transdifferentiation is greatly delayed, until the cell density has risen sufficiently (Clayton et al., 1976). Another cell density mediated differentiation, is the differentiation of neuroblasts over the neuroepithelial cell sheet. This potential evidently must differ from that for lentoid formation, since it is available only in the early stages of culture. Neuroblasts are wholly absent from low density cultures, yet when, after about 40 days, the cultures are sufficiently dense for lentoid formation, no neuroblasts are ever seen (Clayton et al., 1977). We pointed out that cell selection is insufficient to account for all these phenomena and that some selection of alternative pathways of potential within the cells must also occur.

Transdifferentiation Potential

Three embryonic structures, the eye, the hypophysis and the pineal are topologically homologous: they have some potential for histiotypic interchange: transdifferentiation potential has been shown for two of them so far, and there is evidence of extralenticular expression of crystallin mRNAs in these structures (reviewed Clayton, 1982).

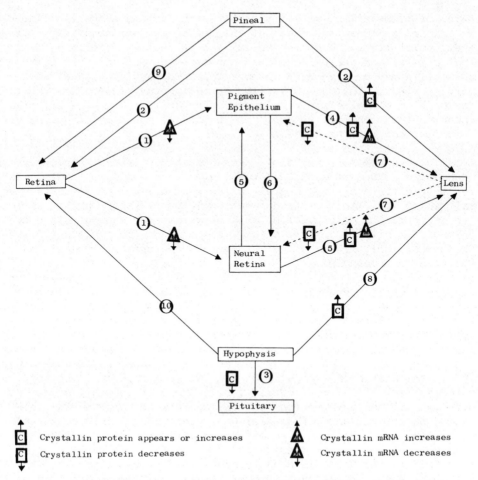

Figure 2. The transitions possible between retina, pineal, adenohypophysis, and lens. The conditions which permit any particular transposition are described in the text.
1. Normal development, all vertebrates.
2. Normal development, lizards, where lens like cells and retina like cells develop in the same structure, Eakin (1973); presence of crystallins in chameleon pineal, McDevitt (1972).
3. δ-crystallin in chick embryo hypophysis is lost by 7 days incubation, Barabanov (1977).
4. Cell culture, Eguchi and Okada (1973) (for other references see Clayton, 1981). Presence of crystallin mRNA in pigment epithelium, Jackson et al. (1978); Thomson et al. (1979).
5. Cell culture, Okada et al. (1975). Presence of crystallin mRNA in neural retina, Jackson et al. (1978); Thomson et al. (1979).

The retina, the pineal and the neurohypophysis form as outpushings of the embryo brain; the lens and the adenohypophysis form as associated invaginations of ectoderm (the head and the roof of the mouth respectively). In birds and mammals the pineal develops as an endocrine structure, but it is in some way responsive to diurnal changes in light, and a rhodopsin-like response has been detected in chick pineal (Deguchi, 1981). The pituitary is, of course, an endocrine organ: curiously the retina, in addition to being the visual effector also has some regulatory function on other tissues. Lens fibre elongation (Coulombre and Coulombre, 1965) and a change in crystallin representation (McAvoy, 1978) are normally regulated by proximity to the retina, while a growth regulating factor, EDGF, has been obtained from retinas (Arruti and Courtois, 1978; Barritault et al., 1979) which affects both the rate of mitosis and the differentiation of a wide range of tissues (Courtois et al., 1981, and these proceedings).

Although retina, pineal and adenohypophysis normally develop into three dissimilar structures, overlapping developmental potentials are suggested by such data as is available (Fig. 2).

The pineal forms both retina and lens cells in lizards (Eakin, 1973) and the pineal lens contains crystallins (McDevitt, 1972). A pinealoma in man with typical retinal histology has been reported (Stefanko and Manschot, 1979), suggesting that some oncogenic process was able to elicit a normally dormant retinotypic potential.

Both retina and lens potential appear to reside in the embryo hypophysis. Between 4 and 7 days of incubation the adenohypophysis of the chick expresses sufficient δ-crystallin protein to be readily detectable by immunological techniques (Barabanov, 1977; Clayton, 1979b) and the 4½ day embryo adenohypophysis can transdifferentiate into lens in cell culture (Clayton, 1979b). On the other hand, in the amphibian, the interposition of a mechanical barrier between brain

6. Organ culture, Tsunematsu, quoted in Okada (1980).
7. MNNG treatment in cell culture leads to production, in secondary culture of some 'pigment epithelium-like' and some 'neurone-like' cells, Clayton et al. (1980); Clayton and Patek, these proceedings.
8. Cell culture, Clayton (1979a).
9. Pinealoma, Stefanko and Manschot (1979).
10. Mechanical separation from brain, Sacerdote (1971).

and hypophysis in vivo led to the formation of retinal architecture (Sacerdote, 1971). It would be of interest to investigate further which conditions favour retina-like differentiation of the future pituitary and which favour lens cells.

Heterologous differentiation potentials have been most studied in the eye. In all vertebrates so far tested the iris, the retina, pigment epithelium, and the neural retina all have the potential to transdifferentiate into the other in suitable conditions in vitro (Okada, 1981; Clayton, 1981; Eguchi, 1979). In the chick, the embryo pigment epithelium and neural retina both express crystallin mRNAs in the intermediate abundance class (Jackson et al., 1978) and as transdifferentiation proceeds the crystallin mRNA increases until it is in the high abundance class (Thomson et al., 1979, 1981). Similar levels of crystallin mRNA are achieved in terminally transdifferentiated cultures of 3½ day and 8 day chick embryo NR cultures, but the 3½ day embryo cultures, with a higher initial level, achieve terminal levels in half the time required by the 8 day embryo cultures. We have argued (Thomson et al., 1981) that the data are more readily explained if both the numbers of cells expressing crystallin mRNA, and the levels in these cells, are higher at the earlier stage.

Our earlier studies of levels of hybridisable crystallin mRNA in retina were based on mRNA from homogenised tissues. We have preliminary evidence that the distribution of δ-crystallin mRNA in the 4½ day chick embryo retina is not homogeneous, as judged by in situ hybridisation to δ-crystallin cDNA. We therefore tentatively conclude that the true level in those retina cells which do express crystallin mRNA is actually higher than the previously estimated average values for the whole structure.

As histiotypic differentiation proceeds, the initial levels of crystallin mRNA and the potential for transdifferentiation both decline and the time required is increased (Clayton et al., 1979; Thomson et al., 1981; Clayton, 1982). A similar decline in extra lenticular crystallin mRNA must also occur in the developing pituitary, since after 7 days of incubation, Barabanov failed to detect δ-crystallin, which suggests that there too, available heterologous mRNAs are lost as histiotypic changes are intiated.

Mechanisms of Transdifferentiation

Evidence has been assembled that transdifferentiation is probably a multistep phenomenon (Clayton, 1979b, c, 1982), and the following requirements have been suggested.
Preconditions. So far, all tissues which have the potential for transdifferentiation into lens appear to express low levels of crystallin or crystallin mRNA in the intermediate abundance class. The rate of transdifferentiation appears to be affected both by the init-

ial values and by some other features of the tissue of origin and the culture conditions.

Permissive conditions. Removal from basement membrane, or disruption of cell-cell contacts or cell dissociation appears to be an essential requirement for transdifferentiation.

Given these two pre-requisites, transdifferentiation may still not take place, and the likelihood and direction of transdifferentiation is strongly affected by a range of directive conditions including growth conditions or culture media. Finally, even if lentoids form by transdifferentiation, the exact balance of crystallins expressed is modifiable by a range of modulatory conditions. The literature on which these hypotheses are based is reviewed elsewhere (Clayton, 1982).

It seems unlikely that transdifferentiation is a phenomenon unique to ocular tissues. Aberrant cell types, distinguished by the expression of an inappropriate gene product at high levels, which is however normal in some other tissue or stage of development, is reported in several malignancies including the ectopic hormone syndromes (see Ratcliffe, these proceedings). The possibility that some of these may be due to transdifferentiation-like events following some oncogenic event or events might be accessible to investigation (Clayton and Patek, 1982). Indeed a carcinogen, MNNG, facilitates some transdifferentiation events (Eguchi and Watanbe, 1973; Clayton et al., 1980; Clayton, 1982, and these proceedings).

Whatever the oncogenic event in the history of a tumour synthesising an ectopic product, some at least of these tend to arise in tissues in which those same products are normally always present but at very low levels. Cell immigration or contamination is not always a possible explanation. These data are reviewed by Baylin and Mendelsohn, 1980.

CONCLUSIONS

The expression of mRNA in the low abundance class of cells may be in large part simply a measure of the inescapable noise in the transcription mechanism. However the parallel between the potential for transdifferentiation to lens and the expression of crystallin mRNAs at intermediate abundance levels makes it likely that the relationship of this class of mRNAs to cell potential may be substantive, and if so a significant area of research for studies of the processes of embryonic determination and of oncogenic and other modifications of the epigenotype of a cell may depend on an understanding of the quantitative controls for levels of expression in the intermediate and abundant classes.

ACKNOWLEDGEMENTS

I am glad to acknowledge those many friends and colleagues with whom, over many years, I have shared experiments, arguments and discussions, or both together. I am greatly indebted to M. Alexander and L. Dobbie; without their painstaking and patient help this paper would not have been ready in time. I also thank L. Dobbie for the diagrams. I am grateful to the M.R.C. and the C.R.C. for their support.

REFERENCES

Anderson, D. M., Galau, C. A., Britten, R. J., and Davidson, E. H., 1976, Sequence complexity of the RNA accumulated in oocytes of Arbacia puctulata, Develop. Biol., 51:138.

Arruti, C., and Courtois, Y., 1978, Morphological changes and growth stimulation of bovine epithelial lens cells by a retina extract in vitro, Expl. Cell Res., 117:283.

Barabanov, V. M., 1977, Detection of δ-crystallin in the adenohypophysis of chick embryos, Doklady Akademii Nauk S.S.S.R., 234:181.

Barritault, D., Arruti, C., Whalen, R. G., and Courtois, Y., 1979, Adult bovine epithelial lens cells in culture. Electrophoretic pattern of total protein on longterm cultures and morphological changes induced by retinal extract, Ophthal. Res., 11:316.

Baylin, S. B., and Mendelsohn, G., 1980, Ectopic (inappropriate) hormone production by tumours: mechanisms involved and the biological and clinical implications, Endocr. Rev., 1:45.

Benedetti, E. L., Dunia, I., Bentzel, C. J., Vermorken, A. J. M., Kibbelaar, M., and Bloemendal, H., 1976, A portrait of plasma membrane specializations in eye lens epithelium and fibers, Biochim. biophys. Acta, 457:353.

Bertagna, X. Y., Nicholson, W. E., Tanaka, K., Mount, C. D., Sorenson, G. D., Pettengill, O. S., and Orth, D. N., 1979, Ectopic production of ACTH, lipotropin and β-endorphin by human cancer cells. Structurally related tumor markers, Rec. Res. Cancer Res., 67:16.

Bours, J., 1973, The presence of lens crystallins as well as albumin and other serum proteins in chick iris tissue extracts, Expl. Eye Res., 15:299.

Bours, J., and van Doorenmaalen, W. J., 1972, The presence of lens antigens in the intra-ocular tissues of the chick eye, Expl. Eye Res., 13:236.

Braunstein, G. C., Kamdar, V., Rasor, J., Swaminathan, N., and Wade, M., 1979, Widespread distribution of a chorionic gonadotropin-like substance in normal human tissues, J. Clin. Endocr. Metab., 49:917.

Bravo, R., and Knowland, J., 1979, Classes of proteins synthesized in oocytes, eggs, embryos, and differentiated tissues of Xenopus laevis, Differentiation, 13:101.

Burke, V., Sullivan, N. P., Petersen, H., and Weed, R., 1944, Ontogenic change in antigenic specificity of the organs of the chick, J. Infect. Dis., 74:225.

Clayton, R. M., 1970, Problems of differentiation in the vertebrate lens, in:"Current Topics in Developmental Biology, Vol. 5," A. Moscona and A. Monroy, eds., Academic Press, London, p.115.

Clayton, R. M., 1953, Distribution of antigens in the developing newt embryo, J. Embryol. exp Morph., 1:25.

Clayton, R. M., 1974, Comparative aspects of lens protiens, in:"The Eye, Vol. 5," H. Davson, ed., Academic Press, London, p.399.

Clayton, R. M., 1978, Divergence and convergence in lens cell differentiation: regulation of the formation and specific content of lens fiber cells, in:"Stem Cells and Tissue Homeostasis," B. I. Lord, C. S. Potten, and R. J. Cole, eds., Cambridge University Press, Cambridge, p.115.

Clayton, R. M., 1979a, Genetic regulation in the vertebrate lens cells, in:"Mechanisms of Cell Change," J. Ebert and T. S. Okada, eds., John Wiley, New York, p.129.

Clayton, R. M., 1979b, Regulatory factors for lens fibre formation in cell culture. I. Possible requirement for pre-existing levels of crystallin mRNA, Ophthal. Res., 11:324.

Clayton, R. M., 1979c, Regulatory factors for lens fibre formation in cell culture. II. The role of growth conditions and factors affecting cell cycle duration, Ophthal. Res., 11:329.

Clayton, R. M., 1982, Cellular and molecular aspects of differentiation and transdifferentiation of ocular tissues in vitro, in: "Differentiation In Vitro," M. M. Yeoman and D. E. S. Truman, eds., Cambridge University Press, Cambridge, p.83.

Clayton, R. M., Campbell, J. C., and Truman, D. E. S., 1968, A reexamination of the organ specificity of lens antigen, Expl. Eye Res., 7:11.

Clayton, R. M., Odeigah, P. G., de Pomerai, D. I., Pritchard, D. J., Thomson, I., and Truman, D. E. S., 1976, Experimental modifications of the quantitative pattern of crystallin synthesis in normal and hyperplastic lens epithelia, INSERM, 60:123.

Clayton, R. M., de Pomerai, D. I., and Pritchard, D. J., 1977, Experimental manipulation of alternative pathways of differentiation in cultures of embryonic chick neural retina, Develop. Growth and Differ., 19:319.

Clayton, R. M., Thomson, I., and de Pomerai, D. I., 1979, Relationship between crystallin mRNA expression in retina cells and their capacity to rc-differentiate into lens cells, Nature, 282:628.

Clayton, R. M., Bower, D. J., Clayton, P. R., Patek, C. E., Randall, F. E., Sime, C., Wainwright, N. R., and Zehir, A., 1980, Cell culture in the investigation of normal and abnormal differentiation of eye tissues in,"Tissue Culture in Medical Research (II)," R. J. Richards and K. T. Rajan, eds., Pergamon Press, Oxford, p.185.

Clayton, R. M., and Patek, C. E., 1982, Apparent redifferentiation of chicken lens epithelium by N-methyl-N'-nitro-N-nitrosoguanidine

in vitro, These proceedings.
Coulombre, J. L., and Coulombre, A. J., 1965, Regeneration of neural retina from the pigmented epithelium in the chick embryo, Develop. Biol., 12:79.
Courtois, Y., Arruti, C., Barritault, D., Tassin, J., Olivié, M., and Hughes, R. C., 1981, Modulation of the shape of epithelial lens cells in vitro detected by a retinal extract factor, Differentiation, 18:11.
Courtois, Y., Arruti, C., Barritault, D., Courty, J., Tassin, J., Olivié, M., Plouet, J., Laurent, M., and Perry, M., 1982, The role of a growth factor derived from the retina (EDGF) in controlling the differentiated stages of several ocular and nonocular tissues, These proceedings.
Deguchi, T., 1981, Rhodopsin-like photosensitivity of isolated chicken pineal gland, Nature, Lond., 290:706.
Eakin, R. M., 1973, "The Third Eye," University of California Press, Berkely.
Ebert, J. D., 1953, An analysis of the synthesis and distribution of the contractile protein, myosin, in the development of the heart, Proc. Natl. Acad. Sci. U.S.A., 39:333.
Eguchi, G., 1976, Transdifferentiation of vertebrate cells in cell culture, in:"Embryogenesis in Mammals," Ciba Foundation Symposium 40, Elsevier, Amsterdam, p.241.
Eguchi, G., 1979, Transdifferentiation in pigment epithelial cells of the vertebrate eye in vitro, in:"Mechanisms of Cell Change," J. D. Ebert and T. S. Okada, eds., John Wiley, New York, p.273.
Eguchi, G., Masuda, A., Karasawa, Y., and Kodama, R., 1982, Microenvironments controlling the transdifferentiation of vertebrate pigmented epithelial cells in in vitro culture, These proceedings.
Eguchi, G., and Okada, T. S., 1973, Differentiation of lens tissue from the progeny of chick retinal pigment cells cultured in vitro: a demonstration of a switch of cell types in clonal cell culture, Proc. Natl. Acad. Sci. U.S.A., 70:1495.
Eguchi, G., and Watanabe, K., 1973, Elicitation of lens formation from 'ventral iris' epithelium of the newt by a carcinogen, N-methyl-N'-nitro-N-nitrosoguanidine, J. Embryol. exp. Morph., 30:63.
Ikeda, A., and Zwaan, J., 1967, The changing cellular localization of α-crystallin in the lens of the chicken embryo, studied by immunofluorescence, Develop. Biol., 15:348.
Itoh, Y., 1976, Enhancement of differentiation of lens and pigment cells by ascorbic acid in cultures of neural retinal cells of chick embryo, Develop. Biol., 54:157.
Jackson, J. F., Clayton, R. M., Williamson, R., Thomson, I., Truman, D. E. S., and de Pomerai, D. I., 1978, Sequence complexity and tissue distribution of chick lens crystallin mRNAs, Develop. Biol., 65:383.
Jacobson, A. G., 1966, Inductive processes in embryonic development, Science, 152:25.
Jeffcoate, W. J., and Rees, L. H., 1978, Adrenocorticotropin and

related peptides in nonendocrine tumors, Current Topics in Exp. Endocr., 3:57.

de Jong, W. W., 1981, Evolution of lens and crystallins, in:"Molecular and Cellular Biology of the Eye Lens," H. Bloemendal, ed., John Wiley, New York, p.221.

Kawamura, J., Machida, S., Yoshida, O., Oseko, F., Imura, H., and Hattori, M., 1978, Bladder carcinoma associated with ectopic gonadotropin, Cancer, 42:2773.

Linser, P., and Moscona, A. A., 1979, Induction of glutamine synthetase in embryonic neural retina: localization in Müller fibers and dependence on cell interactions, Proc. Natl. Acad. Sci. U.S.A., 76:6476.

McAvoy, J. W., 1978, Cell division, cell elongation and the co-ordination of crystallin gene expression during lens morphogenesis in the rat, J. Emb. exp. Morph., 45:271.

McDevitt, D. S., 1972, Presence of lateral eye crystallins in the median eye of the American chameleon, Science, 175:763.

Monroy, A., and Rosati, F., 1979, Cell surface differentiations during early embryonic development, Current Topics in Develop. Biol., 13:45.

Moscona, A. A., and Degenstein, L., 1981, Lentoids in aggregates of embryonic neural retina cell, Cell Differentiation, 10:39.

Moscona, A. A., and Degenstein, L., 1982, Formation of lentoids from neural retina cells: glial origin of the transformed cells, These proceedings.

Okada, T. S., 1976, "Transdifferentiation" of cells of specialised eye tissues in cell culture, in:"Tests of Teratogenicity In Vitro," J. D. Ebert and M. Marois, eds., Elsevier, Amsterdam, p.91.

Okada, T. S., 1980, Cellular metaplasia or transdifferentiation as a model for retinal cell differentiation, Current Topics in Develop. Biol., 16:349.

Okada, T. S., 1981, Phenotypic expression of embryonic neural retinal cells in cell culture, Vis. Res., 21:83.

Okada, T. S., Itoh, Y., Watanabe, K., and Eguchi, G., 1975, Differentiation of lens in cultures of neural retinal cells of chick embryos, Develop. Biol., 45:318.

Perlman, S. M., Ford, P. J., and Rosbash, M. M., 1977, Presence of tadpole and adult globin RNA sequences in oocytes of Xenopus laevis, Proc. Natl. Acad. Sci. U.S.A., 74:3835.

Piatigorsky, J., 1980, Intracellular ions, protein metabolism and cataract formation in:"Current Topics in Eye Research, Vol. 3," Academic Press, London, p.1.

Piatigorsky, J., 1981, Lens differentiation in vertebrates. A review of cellular and molecular features, Differentiation, in press.

de Pomerai, D. I., and Gali, M. A. H., 1982, Serum factors affecting transdifferentiation in chick embryo neuro-retinal cultures, These proceedings.

Pritchard, D. J., 1981, Transdifferentation of chicken embryo neural retina into pigment epithelium: indications of its biochemical basis, J. Emb. exp. Morphol., 62:47.

Ratcliffe, J. G., 1982, Ectopic hormones, These proceedings.

Rosen, S. W., Kaminska, J., Calvert, I. S., and Aaronson, S. A., 1979, Human fibroblasts produce "pregnancy-specific" beta-1-glycoprotein in vitro, Am. J. Obs. Gynecol., 134:734.

Rosenzweig, J. L., Havrankova, J., Lesniak, M. A., Brownstein, M., and Roth, J., 1980, Insulin is ubiquitous in extrapancreatic tissues of rats and humans, Proc. Natl. Acad. Sci. U.S.A., 77:572.

Sacerdote, M., 1971, Differentiation of ectopic retinal structures in the hypothalamo-hypophysial area in the adult crested newt bearing a permanent hypothalamic lesion, Z. Anat. Entwickl. Gesch., 134:49.

Saxen, L., and Toivonen, S., 1962, "Primary Embryonic Induction," Logos Press, London.

Stefanko, S. Z., and Manschot, W. A., 1979, Pinealoblastoma with retinoblastomatous differentiation, Brain, 102:321.

Takata, C., Albright, J. F., and Yamada, T., 1966, Gamma crystallins in Wolffian lens regeneration demonstrated by immunofluorescence, Develop. Biol., 14:382.

Thomson, I., de Pomerai, D. I., Jackson, J. F., and Clayton, R. M., 1979, Lens-specific mRNA in cultures of embryonic chick neural retina and pigmented epithelium, Expl. Cell Res., 122:73.

Thomson, I., Yasuda, K., de Pomerai, D. I., Clayton, R. M., and Okada, T. S., 1981, The accumulation of lens-specific protein and mRNA in cultures of eye cup from 3½ day chick embryos, Expl. Cell Res., 135:445.

Tsai, S. Y., Tsai, M.-J., Lin, C.-T., and O'Malley, B. W., 1979, Effect of estrogen on ovalbumin gene-expression in differentiated non-target tissues, Biochemistry, 18:5726.

Waddington, C. H., 1956, "Principles of Embryology," Allen & Unwin, London.

Yoshimoto, Y., Wolfsen, A. R., and Odell, W. D., 1977, Human chorionic gonadotropin-like substance in normal human tissues, Science, 197:575

Yoshimoto, Y., Wolfsen, A. R., Hirose, F., and Odell, W. D., 1979, Human chorionic gonadotropin-like material: presence in normal human tissues, Am. J. Obs. Gynecol., 134:729.

PROBLEMS OF REGULATION

Hans Bloemendal

Department of Biochemistry
University of Nijmegen
Geert Grooteplein Noord 21
6525 EZ Nijmegen, The Netherlands

I am always rather embarrassed when I read the word "regulation" in connection with differentiation and only because Ruth Clayton told me that I am allowed to ask questions of the audience instead of explaining mechanisms of regulation of the differentiated state did I agree to give a short introduction to this topic.

Even without knowing too much about the real events involved in this regulatory mechanism one can think about it. One can for instance put a fundamental question: Are there a small number of regulatory steps or are we confronted with a multitude of very tiny effects which together generate regulation at the molecular level. If the latter assumption is true I am afraid the solution of the problem is, if not hopeless, at least tremendously complex. My question to you therefore is: What do you think? Are there indeed very few principal steps of regulation or do infinitesimal changes occurring in differentiation cause the end effect, or might it be that the situation is somewhat in between these extremes.

To talk about regulation and control to an audience like you is as difficult as it is easy in a course to students. In the latter case I usually give the diagram shown in Table 1. So in a cell you may have regulation everywhere; on the level of replication and transcription, on the level of translation and you may even have regulatory steps after translation in post-translational events, say by phosphorylation or acetylation of proteins, assembly of subunits or all of these modifications together. Anyway it is always better to design experimental approaches, to check whether an assumption concerning regulation is true rather than to have long speculative discussions without ever doing the experiment.

To start the present discussion I shall restrict myself to one

Table 1. Levels of regulation

Process	Macromolecular Level
replication	DNA
transcription	DNA, RNA
translation	mRNA, protein
post-translation	protein

type of possible regulation at the DNA level and to special cases. It has been observed that almost all genes studied so far occur in families. The formation and divergence of multigene families may represent a regulatory mechanism per se because they can ensure high rates of synthesis of the end products. The question of course has to be answered, is this a general rule? Does the occurrence of multigene families always enable a high rate of synthesis of the gene product? Examples which are in favour of this asumption are listed in Table 2. With interferon, however, we already run into difficulties, since in fibroblasts the situation seems to be different from that in leukocytes, where the interferon genes do occur as a family. The multigene concept is certainly true for actin and we shall hear more about it from Dr. Buckingham. Tubulin genes also fit the schedule. There are only rare instances known so far (Table 2) where it has been proven that certain genes exist in one copy. We have found a new example in the lens matrix.

We recently became interested in the eye lens cytoskeleton. In order to introduce the system I want to show a cross section through

Table 2. Gene organization

Multigene Families	Single Genes
histone	insulin
globin	conalbumin
interferon (human leukocyte)	interferon (fibroblast)
vitellogenin	vimentin (?)
actin	
immunoglobulins	

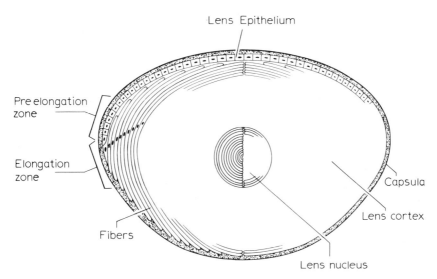

Figure 1. Cross section through a lens.

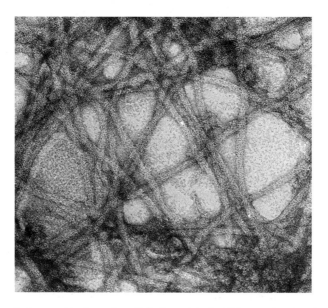

Figure 2. Intermediate-sized (vimentin) filaments in transformed hamster lens cells.

a mammalian lens, a classical tool for differentiation studies (Fig. 1). For those of you who are not familiar with this tissue: it arises from a monolayer of epithelial cells which at the stage of terminal differentiation result in a mass of very long anucleated fiber-like cells. The latter are not only characterized by a high concentration of the crystallins but contain also a well-defined cytoskeleton. The structural elements of this matrix include the so-called intermediate-sized filaments (IF). As a rule epithelial cells have prekeratin as IF protein subunit, whereas cells from mesenchymal origin produce vimentin-containing IF. The eye lens forms an exception to this rule in that it contains, surprisingly, the vimentin type of IF.

As soon as one transfers cells to culture conditions the expression of the IF protein gene changes and the cells express the vimentin gene irrespective of the source of the cells. You may ask the question: which regulation step induced by the environmental change controls the switching off, for instance, of the prekeratin gene and the switching on of the vimentin gene in culture.

Fig. 2 shows that IF appear in the electron microscope as naked filaments in contrast to actin filaments which are frequently decorated by protein particles. Some IF are running in the direction of the plasma membrane (Fig. 3) and meet the membrane at a recognition site. The mere fact of the existence of recognition sites already represents regulation, because the filaments will attach at that point and not at another one. This recognition site may only be "regulated" by virtue of the primary structure of the proteins involved in interaction between membrane and IF and which are encoded in the corresponding genes. As a matter of fact also the secondary and tertiary structures of the proteins involved are consequently "regulated" since the protein just cannot fold in another way and/or associate to multimeric entities.

Therefore it seems that in many cases structural conditions provide a simple and straight-forward means of regulation. Actually we all accept the slogan that the function of macromolecules is greatly determined by their structure. The experiment illustrated in Fig. 4 is suggestive of such specific recognition. If one isolates the total mRNA from hamster lens cells and incubates the translation products with isolated plasma membranes the latter interact preferentially with very few proteins, among which is vimentin. This experiment forms a basis of understanding of the micrograph (Fig. 3).

Our interest in vimentin arose also from other studies conducted in order to acquire some knowledge concerning regulation of lens cell differentiation. If one puts normal bovine lens cells into culture then after a while they will stop synthesizing their specific proteins, the crystallins. One can subculture these cells for months but they will not reinitiate crystallin synthesis. Which factors then regulate the synthesis of the specific proteins? Unfortunately I have

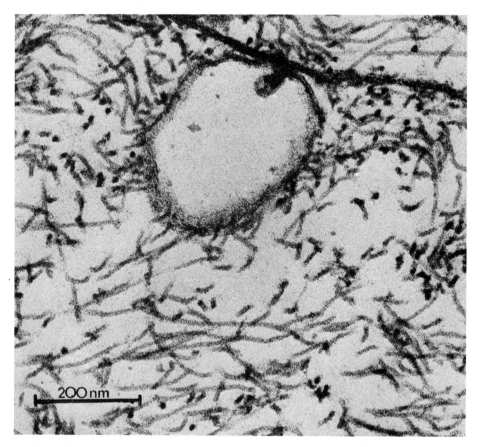

Figure 3. End on attachment of intermediate-sized filaments to the plasma membrane.

to confess that up till now we have not been able to solve the problem. As an experimental approach, we thought that if we transformed the lens cells they might synthesize crystallin again. The idea was that normal regulation in culture would then be disturbed and that the crystallin synthesis might start again. The result showed that it was not a brilliant idea because the (hamster) lens cells transformed with SV40 did not synthesize any crystallin. The spin off, however, from this study was that if we isolated the total messenger population from transformed cells we found that they direct mainly the synthesis of cytoskletal proteins among which are actin and vimentin. Subsequent studies of the organization of the cytoskeletal genes showed that the vimentin gene, in contrast to actin genes, exists as a single copy (details are given in the paper by Quax et al. in these proceedings).

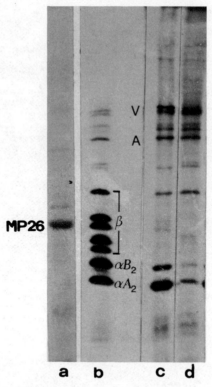

Figure 4. Isolated lens plasma membranes after incubation with the translation products of calf lens polyribosomes in a reticulocyte lysate.
a) The major lens membrane protein MP26 (for comparison).
b) The translation products in the reticulocyte lysate (no membranes added).
c) Some translation products among which preferentially vimentin (V) interact with the membranes.
d) Urea washing diminishes the uptake of α-crystallin chains but not of vimentin.
A = actins   β = β-crystallins.

When cultured lens cells grow older, actin synthesis increases (are more genes switched on?) but vimentin decreases. This phenomenon also must be regulated by some factor(s) but once again we do not know which one(s). In other words the external cause is ageing but what is the internal mechanism which slows down vimentin synthesis?

In conclusion then, if our discussion would result in some hints to elucidate such and similar problems many of us would be encouraged to continue our work devoted to unraveling of the still existing mystery of differentiation.

TAXONOMIES OF DIFFERENTIATION

D. E. S. Truman

Department of Genetics
University of Edinburgh
Edinburgh EH9 3JN

Studies on the regulability of differentiation often proceed without considering some concepts which are perhaps sufficiently elementary to be generally passed by, but on this occasion I would like to discuss some of these basic ideas in the hope that they might lead us to some insights into differentiation. The concepts I wish to pursue are:

1. That there is a limited number of types of cells in higher organisms.
2. That these cell types are discrete and relatively unchanged in most multicellular animals.
3. That though differentiated cells are generally stable, a cell may give rise to progeny of a different cell type. When this happens the change is generally into another discrete and recognisable cell type.

Given that there is a variety of cell types it is natural to a biologist to want to list, enumerate and classify them. I do not intend at present to do this but I would like to consider how we might begin to classify them and then to speculate on whether any system of classification might shed light on the control mechanism of differentiation.

THE DISCRETENESS OF CELL TYPES

The classical attempts to define cell differentiation have emphasised both the relative irreversibility of the process and also the discreteness of the cell types resulting from cytodifferentiation (e.g. Weiss, 1939; Waddington, 1956; Grobstein, 1959). More recent

```
                 Proximal optic cup ──→ Pigmented retinal epithelium
Optic vesicle ⤴
              ⤵
                 Distal optic cup ──→ Sensory retinal epithelium
```

Figure 1. A simple example of cell genealogy in eye tissues.

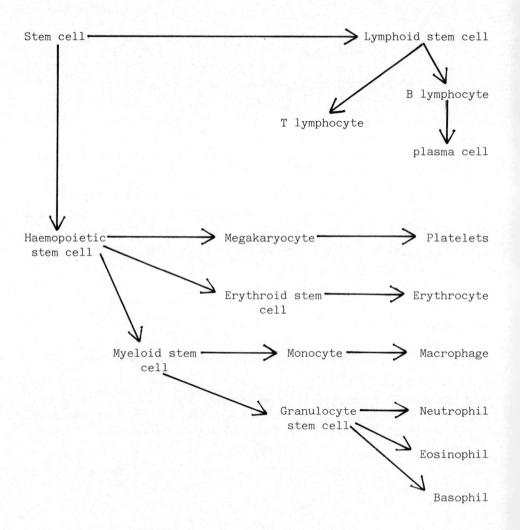

Figure 2. A genealogy of cell types in the blood.

observations have confirmed that the irreversibility of differentiation is only relative and that in certain circumstances there can
be quite dramatic changes in cell type - the transdifferentiations.
These are especially clearly seen in eye tissues in cell culture
(Eguchi and Okada, 1973; Okada et al., 1975; Eguchi, 1976, 1979;
Clayton, 1978, 1979, 1981). The proceedings of this workshop will
be much concerned with such transdifferentiations.

With such change of emphasis on the stability of differentiation,
what must we now believe about the discreteness of differentiation?
Is there really a limited number of cell types or is there a continuous gradation of cellular morphology and biochemistry within the
body without distinct boundaries between the cell types?

At the time when the 'classical' definitions of differentiation
referred to above were formulated an important feature of differentiation was the apparent existence of tissue-specific substances found
in one cell type but absent from another. The increasing sensitivity
of analytical techniques has gradually changed this viewpoint and we
frequently find that substances which are predominant in one cell type
may be found in very low concentrations in several other types. One
example of this is to be found in the crystallins of the lens of the
eye (Clayton et al., 1968; Clayton, 1981).

Such a change of ideas on tissue specificity might also lead to
an erosion of the idea of the discretenes of cell types. However,
it is quite possible to envisage a cell type defined, not in terms
of a specific substance but rather by the proportions of shared constituents (Truman, 1974, 1977; Paul, 1982). Our growing knowledge
of the mRNA populations of cell types also indicates that it is the
proportions of shared constituents which serve to characterise cell
types, e.g. Axel et al. (1976). However, if the characterisation is
so much in terms of quntitative rather than qualitative aspects then
it is easy to see that the boundaries between cell types might become
less distinct.

Many of the recent studies on transdifferentiation, such as those
of the eye tissues, together with earlier studies on cellular metaplasia, can be regarded as strengthening the concept of discreteness
of cell types since one of the notable features of these transdifferentiations is that the changes that occur are between distinct and
recognisable cell types.

THE CLASSIFICATION OF TYPES OF CELL

If then, there are discrete cell types, how many are there and
how may they be categorised? Many of the standard text-books of histology characterise their subject by the organs from which the tissues
are derived and cells are classified and enumerated according to the

tissues in which they are found, e.g. Maximow and Bloom (1953), Garven (1957). Ham (1965) classifies 41 types of cell on morphological grounds, grouped in a branching classification into what he calls the four basic tissues (epithelial, connective, muscular and nervous). Another text-book (Young, 1971) estimates that there are at least 1000 different cell types in the human body. In many cases we find that there is a confusion between classifying cell types and classifying tissues. The purpose of many classifications is to simplify the task of learning and recognition that faces the medical student and such schemes are what a taxonomist would recognise as artificial in that they are based upon a limited number of characters and do not attempt to establish a natural relationship between the subjects of the classification.

Is there a natural way of classifying the cells of the body? One way in which cells can be related is by descent, just as in human genealogies. We can, for example, see that the pigmented and the sensory epithelial cells of the retina are related as they are both derived from the optic vesicle (Fig. 1). A more complex instance would be the relationship of the various cell types in the blood, shown in Fig. 2.

Another way of classifying cells would be according to their biochemical functions. We can appreciate at once that skeletal and cardiac muscle have much in common. Is there a way in which we can quantify this? In a preliminary way a survey was made of five different tissues in which a wide range of enzyme activities have been measured, choosing fourteen enzymes in an arbitrary way. The tissues were placed in rank order of enzyme activity. The number of times pairs of tissues were adjacent in rank order was taken as an index of biochemical similarity of the tissues and we see the results in Table 1. Our prejudices are confirmed when we see that skeletal and cardiac muscle are rather similar and we may not be surprised by the strong similarity between kidney and liver, nor by the marked difference between liver and cardiac muscle. These are all tissues, rather than homogeneous cell types that are being compared. It is unlikely

Table 1. Index of enzymatic similarities based on measurements of enzyme activity.

	Kidney	Liver	Pancreas	Cardiac muscle
Liver	26	–	–	–
Pancreas	3	6	–	–
Cardiac Muscle	10	4	9	–
Skeletal Muscle	12	10	7	15

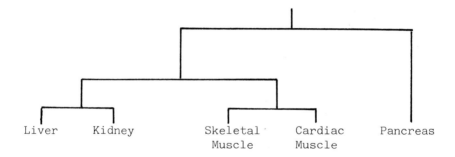

Figure 3. A taxonomy of tissues based on similarity of enzyme activities.

that many measurements have been made on pure cell types. The data in Table 1 can then be used to construct a hierarchical taxonomy of the tissues, as shown in Fig. 3.

An alternative approach to taxonomy comes when we have instances of metaplasia or transdifferentiation between cell types. One common example of metaplasia in vivo appears to be between muscle and bone and the relationship here is perhaps not surprising in view of the fact that in limb development, for example, the cell types share in their origin from the undifferentiated cells of the limb bud. Studies of metaplastic changes or transdifferentiations in cell culture provide us with a greater certainty that the changes that occur are between clearly differentiated cell types. The transdifferentiation of ocular tissues which are being studied by the groups of Okada, of Eguchi, of Clayton and of de Pomerai provide us with a range of interconvertions, summarised and simplified in Fig. 4. Here we

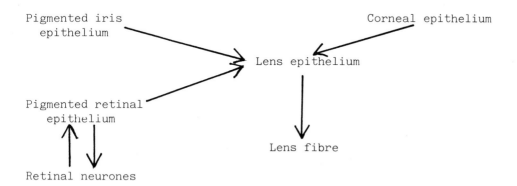

Figure 4. Summary of some transdifferentiations among eye tissues, based on Clayton (1981).

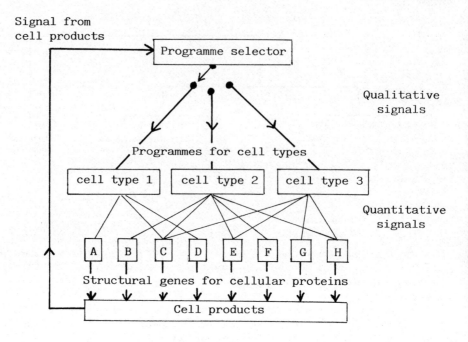

Figure 5. A model of information flow in the regulation of cell type.

have a group of cell types with a relationship characterised by their capacity for transdifferentiation. However, in some instances the embryonic origins of these cell types are rather diverse, the lens being derived during development from head ectoderm while the pigmented and neuronal retina cells, for example, are derived from the neural tube.

Sometimes the changes of cell type in transdifferentiation have only been observed in one direction while in other instances the interchangeability is mutual, such as that between the pigmented and neuronal cells of the retina, when these are grown in culture. Modifications to the culture medium can affect the extent and direction of changes in the cells (Itoh, 1976; Pritchard et al., 1978) and this suggests that with further experimentation a great range of transdifferentiations might occur. It is notable that it is ocular tissues which provide these examples and it is difficult to see a biological basis for this fact. Increasing studies of cultured cells may reveal more instances and treatment of cells with certain chemical agents may bring about changes that do not otherwise occur (R. M. Clayton personal communication). It may then be possible increasingly to use the extent of interconversion as a measure of the affinity of cell types.

TAXONOMIES OF DIFFERENTIATION

CONTROL MECHANISMS IN DIFFERENTIATION

If we accept that cells can be classified into discrete types and that there are varying degrees of affinity between these types, then I believe that we may be able to make some predictions about the types of control that may be involved. One way to bring about the differentiation of the discrete types would be by the existence of some kind of programme responsible for each type. This programme would bring about the quantitative regulation of the genes which are expressed in one particular cell type, and the programme itself would be subject to regulation of an 'on/off' type. (See Fig. 5.)

The affinity between tissue types would represent a measure of similarity between the programmes. If the programmes are in the form of genetic material, then the existence of a range of programmes of varying degrees of divergence might be expected if they had evolved by a process of duplication and divergence, such as we suppose has occurred in the structural genes for globins and crystallins, for example.

In the model shown in Fig. 5, transdifferentiation would involve a discrete change in the position of the switch mechanism of the programme selector. The examples of metaplasia and transdifferentiation described indicate that some changes in the switch are more likely than others: neural retina to lens is more likely than neural retina to muscle. Various methods of switching might entail such preferred changes. If the activation is brought about by the proximity of a control element to a programming element, then programmes which are close together in space might be interchanged more frequently. Alternatively, if the switch mechanism were to involve a sequence-dependent base-pairing between nucleic acids, then programmes activated by the most similar base sequences might be most likely to be interchanged.

Stability of Differentiation.

The model represented in Fig. 5 contains a feature that is intended to stabilise differentiation; a feedback loop, perhaps from the cell products to the programme selector so that once the switch has occurred it tends to remain in the same position.

Evolution of Differentiation.

The general features of a particular cell type are remarkably constant when we compare animals of quite disparate phylogeny, though it is evident that the number of different types of cells in the body has increased during evolution. The model proposed here (Fig. 5) could accomodate this observation: we would presume that there has been a gradual increase in the number of programmes for cell types but that the programmes themselves have been relatively conservative. In accordance with this we observe that the signals responsible for the selection of particular programmes are not species specific. For example the expression of mouse liver genes can be called forth by the presence of rat liver signals in rat hepatoma-mouse fibroblast hybrids (Paterson and Weiss, 1972).

Tumour formation.

The emphasis placed above on discrete cell types is intended to refer to normal cells. In the case of the formation of tumours it may be that the pattern of gene expression does not follow that usually generated by the programme. This might be expected if tumour formation were to arise from a defect in the programming element.

ACKNOWLEDGEMENTS

My ideas on differentiation are constantly developed by discussion with colleagues and students within the Department of Genetics of the University of Edinburgh but I would particularly wish to thank R. M. Clayton and J. C. Campbell for their stimulation and their willingness to listen.

REFERENCES

Axel, R., Feigelson, P., and Schutz, G., 1976, Analysis of the compplexity and diversity of mRNA from chicken liver and oviduct, Cell, 7:247.

Clayton, R. M., 1978, Divergence and convergence in lens cell differentiation: regulation of the formation and specific content of lens fibre cells, in: "Stem Cells and Tissue Homeostasis," B. I. Lord, C. S. Potten and R. J. Cole, eds., Cambridge University Press, p.115.

Clayton, R. M., 1979, Genetic regulation in the vertebrate lens cell, in: "Mechanisms of Cell Change," J. Ebert and T. Okada, eds., Wiley, New York, p.129.

Clayton, R. M., 1982, Cellular and molecular aspects of differeniation and transdifferentiation of ocular tissues in vitro, in: "Differentiation in vitro," M. M. Yeoman and D. E. S. Truman, eds., Cambridge University Press, p.83.

Clayton, R. M., Campbell, J. C., and Truman, D. E. S., 1968, A reexamination of the organ specificity of lens antigens, Exptl. Eye Res., 7:11.

Eguchi, G., 1976, "Transdifferentiation" of vertebrate cells in cell culture, in: Ciba Foundation Symposium 40: "Embryogenesis in Mammals," Elsevier, Amsterdam, p.241.

Eguchi, G., 1979, Transdifferentiation in pigment epithelia of the vertebrate eye in vitro, in: "Mechanisms of Cell Change," J. Ebert and T. Okada, eds., Wiley, New York, p.273.

Eguchi, G., and Okada, T., 1973, Differentiation of lens tissue from the progeny of chick retinal pigment cells cultured in vitro: a demonstration of a switch of cell types in clonal cell culture, Proc. Natl. Acad. Sci. U.S.A., 70:1495.

Garven, H. S. D., 1957, "A Student's Histology," Livingstone, Edinburgh and London.

Grobstein, C., 1959, Differentiation of vertebrate cells, in: "The Cell," J. Brachet and A. Mirsky, Eds., Academic Press, New York, Vol. 1:437.

Ham, A. W., 1965, "Histology," (5th ed.), Lippincott, Philadelphia and Montreal.

Itoh, Y., 1976, Enhancement of differentiation of lens and pigment cells by ascorbic acid in cultures of neural retinal cells of chick embryos, Develop. Biol., 54:157.

Maximow, A. A., and Bloom, W., 1953, "A Textbook of Histology," (6th ed.), Saunders, Philadelphia and London.

Okada, T. S., Itoh, Y., Watanabe, K., and Eguchi, G., 1975, Differentiation of lens in cultures of neural retinal cells of chick embryos, Develop. Biol., 45:318.

Paterson, J. A., and Weiss, M. C., 1972, Expression of differentiated functions in hepatoma cell hybrids: induction of mouse albumin production in rat hepatoma-mouse fibroblast hybrids, Proc. Natl. Acad. Sci. U.S.A., 69:571.

Paul, J., 1982, Gene switching and cellular differentiation, in this book.

Pritchard, D. J., Clayton, R. M., and de Pomerai, D. I., 1978, 'Transdifferentiation' of chicken neural retina into lens and pigment epithelium in culture: controlling influences, J. Embryol. exp. Morph., 48:1.

Truman, D. E. S., 1974, "The Biochemistry of Cytodifferentiation," Blackwell Scientific Publications, Oxford.

Truman, D. E. S., 1977, The Biochemistry of Cytodifferentiation: an outline of progress, in: "The Cultured Cell and Inherited Metabolic Disease," R. H. Harkness, and F. Cockburn, eds.; MTP Press, Lancaster, p.79.

Waddington, C. H., 1956, "Principles of Embryology," George Allen and Unwin, London.

Weiss, P., 1939, "Principles of Development," Holt, New York.

Young, J. Z., 1971, "An Introduction to the Study of Man," Clarendon Press, Oxford.

SUMMARY OF DISCUSSION ON CLASSIFICATION

OF THE PROBLEMS

The reversibility of malignant transformation was discussed by several speakers.

Bloemendal drew a distinction between two modes of obtaining a reversal of malignancy in a patient. One might be the loss of a tumour following the formation of cytotoxic antibodies, but a quite different mode of reversal would be a process whereby tumour cells gave rise to descendant cells which were not malignant. Woodruff cited experiments with teratocarcinoma cells which gave rise to cellular progeny indistinguishable from normal cells as evidence of true reversible malignancy. This was challenged by Buckingham, who emphasised that teratocarcinomata were derived from totipotent cells and so could be expected to give rise to normal progeny. Paul pointed out that a reversal of malignancy which was very rare would have no clinical significance but might be of great interest from the point of view of cellular control. He also distinguished between loss of malignancy on the one hand, and a reversion to normal differentiation on the other. Such behaviour was seen in neuroblastomas, and also virally transformed cells. It was suggested by Woodruff that the normal differentiation might arise from a minority cell population which eventually replaced the malignant cells. Clayton suggested that malignancy may be just one of the directions into which cells may be pushed by carcinogens, citing experiments in which substances generally regarded as carcinogenic could instead bring about a deflection of differentiation. In such cases it seemed likely that the substances were acting at an epigenetic level rather than by causing a mutation at the genomic level. The variety of causes of malignant transformation was raised by Yaffe who also described experiments which showed that changing the cellular environment of some types of cells could cause them to give rise to adenocarcinomas, but when they were transferred to an early embryo they reverted to normally regulated cells. Bloemendal and Schibler considered the evidence that malignant transformation involved triggering the same specific portions of the host genome by a variety of external mutagenic agencies, but Woodruff questioned whether we were yet in a position to consider a single, unified theory of the nature of malignant transformation.

The papers of Paul and of Clayton had both put forward evidence that differentiated cells might produce low concentrations of transcripts of many genes not previously regarded as typical of a particular cell type and this was taken up in discussion. Schibler suggested that one possible cause of the findings might be that transcription of an active gene was sometimes continued beyond the normal termination point giving rise to 'read-through' transcripts of portions of the DNA downstream of active genes. It was necessary to characterise transcripts in detail at the 5' and 3' termini in order to eliminate the possibility of such 'read-through' transcripts. Harrison also questioned the biological significance of transcripts formed at very low levels. They might be the inevitable consequences of the equilibrium constants of a whole series of reactions involved in DNA, RNA and protein synthesis and such leakiness of regulation may be tolerable for the cell, as the price of obtaining the transcripts which were functional. The significance of low level transcripts could only be established by showing a correlation with some developmental programme of the cell.

Experiments by Weintraub on the structure of the globin gene were cited by Buckingham as evidence against the view that low levels of transcription might be associated with the competence of a gene to become expressed following some developmental trigger. The evidence seemed to be that the globin gene was not accessible to DNaseI in the penultimate stage of differentiation of erythroid cells and that it only became accessible when the gene was expressed by abundant transcription. Clayton replied that it was not certain that the low levels of transcription and the competence of the cell were causally related but that, at least in the lens system, there seemed to be a coincidence between the two. As to the question of DNaseI sensitivity, the total time during which the chromatin would be actively transcribed to produce the low level transcripts could be so short that it would probably be difficult to demonstrate by methods such as DNaseI treatment.

The apparent presence of heterologous proteins or messenger must be interpreted with caution, according to Yaffe, because of the possibility that the transcripts found at low level in one tissue and characteristic, at high levels, of other tissues, might be due to the products of different genes in the two instances. The evidence from the actin and myosin gene families was that some of the genes were transcribed abundantly in certain tissues while other related but different genes were transcribed at lower levels in different tissues.

The importance of minority components during cell differentiation was emphasised by Buckingham. While muscle development requires large quantities of myosin, actin and tropomyosin, the presence of the acetylcholine receptor is equally essential, yet the acetylcholine receptor message falls into the class of very rare messages.

SUMMARY OF DISCUSSION

Sueoka reported experiments using nucleic acid hybridization to measure the proportion of the genome found transcribed among nuclear RNA in a number of rat tissues. In brain 30% of the DNA was expressed, in liver 20% and in kidney 10%. The 30% for brain included the 20% for liver, which included the 10% for kidney. The majority of the transcription products were present at low levels. The precise significance of the rare transcripts was uncertain and it remained possible that in every tissue all genes were transcribed but with very differing frequencies. The pattern of complexities of transcription reflected, to some extent, developmental processes, so that ectodermal tissues such as brain and mammary gland showed similar figures, and kidney and lymphocytes, of mesodermal origin, were also similar. Birnie enquired what degree of overlap there was between the sequences in nuclear RNA in different tissues. For example, were exactly the same genomic sequences transcribed in liver and lung tissue? Sueoka replied that this was not yet known, and also that an error of around 1-2% was to be expected in the figures. Paul agreed about the difficulty of making exact measurements and pointed to the problems arising when there was a very rapid rate of degradation of RNA. Moreover, some functional genes may be transcribed as infrequently as once in the cell cycle and if their half-life was as short as five minutes they would be undetectable. He too believed that all genes might be transcribed in all tissues.

With regard to the significance of infrequent messengers during development, McDevitt sought an explanation of how low levels of crystallin messages in tissue such as retina or iris might direct cells into the pathway of lens development during transdifferentiation. Clayton doubted whether they did have this directing influence: it seemed rather that they permitted the transdifferentiation and that extraneous events, such as the breaking of cell-cell connections served as a trigger which permitted mRNA molecules of low abundance to move into a high abundance class. Nevertheless the correlation between transdifferentiation to lens and the presence of crystallin messages appeared strong, and when neural retina becomes aged there is a fall both of capacity to transdifferentiate and also in the level of crystallin mRNA.

The question of transdifferentiation was widely discussed, and evidence was sought that there was indeed a switch of the pattern of gene expression in the progeny of a cell which was itself differentiated. Iscove enquired whether it was established beyond doubt that the observed change in cell population was not due to environmental selection, with the disappearance of one cell type and its replacement by a second cell type which had always been programmed to give rise to that cell type. Similar evidence was sought by Paul and Lord. By analogy with haemopoietic systems, it seemed possible that stem cells might be present among the differentiated cells and that such stem cells gave rise to the new cell type by multiplication and differentiation. Clayton pointed out that the experiments of Okada and

Eguchi, in which single differentiated cells of pigmented retina had been cloned and found to give rise to lens-like cells, provided the best evidence against the hypothesis that stem cells were involved. It was agreed that further discussion of this topic should be deferred until after the papers of Okada and Eguchi.

The nature of the triggers of directing influences in transdifferentiation was raised by Harrison, who wondered if the differences between the cell types which can give rise to lens arose as a result of their position in the embryo, and that when they were removed experimentally from their positional information it would be expected that they would become more similar to each other. Clayton agreed that pituitary, pineal and eyecups were all derived from head ectoderm and that they became different in response to inducers which varied from one position to another. When differentiated cells of such structures were deprived of cell-cell contacts or contact with the basement membrane, then these cells appeared to become more multipotential. Chader stressed the importance of position in cell development by reference to cells derived from the neural crest which, following migration would give rise to a wide range of cell types, an analogy with the effect of travel broadening the horizons. The wide range of gene expression found when cells were put into monolayer culture also illustrates the importance of cellular environment in regulating gene expression.

The role of cellular environment in regulating cellular differentiation was also discussed by Balazs. He cited as an example the selection of the neurotransmitter utilized by certain nerve cells derived from the neural crest. Factors in the cellular environment influence this selection and a protein has been isolated form heart conditioned medium which conveys cholinergic specificity to the cells.

Possible mechanisms whereby the cellular environment might regulate phenotypic expression were considered by Courtois. One mechanism would be through the interactions of the cytoskeleton and polysomes. There are many examples where cell shape or polarity can be manipulated while inducing a change in protein synthesis. Cell surface changes, involving receptor molecules or cell-cell contacts might lead to modification of the cytoskeletal framework. Citing recent experiments by Bloemendal using the lens in which polysomes associated with membrane protein synthesis were preferentially associated with cytoskeletal proteins while crystallins were not, Courtois asked whether it was possible that the organization of cytoskeletal elements might have an influence on the expression of mRNA already present in the cytoplasm. Bloemendal in reply expressed the view that an attempt to explain regulation in terms of control by cytoskeletal elements left the question of the control of the cytoskeleton open, and he believed that regulation of the DNA level should be seen as the ultimate level of control. In reply to Clayton's suggestion that this was a rather reductionist argument, Bloemendal accepted the importance of

SUMMARY OF DISCUSSION

post-transcriptional controls, some of which could respond to environmental influences.

Bloemendal's data on the relationship between the composition of cytoskeletal elements and the embryonic origins of cells received support from Balazs who reported that astroctyes, which are of mesenchymal origin, contain vimentin in the intermediate filaments. He believed that as more information is collected on the distribution of markers hitherto regarded as cell-specific it might become necessary to review our current classifications of cell types.

Eguchi reported some experiments at variance with those of Bloemendal who had found that in cultured lens cells crystallin' synthesis declined while cytoskeletal components accumulated. In Eguchi's experiments a small population of cells with a high growth potential continued to express crystallin genes, while a larger population of cells with lower growth potential produced large amounts of actin and tubulin. He believed that the synthesis of such cytoskeletal elements was an indication of senescence of the cells rather than the loss of differentiative potential. Bloemendal, in reply, pointed out that his cells also had a high growth potential.

The model of regulation of differentiation involving two distinct levels proposed by Truman was commented upon by Paul, who stressed the importance of cytoplasmic factors in selection of which portions of the genome were transcribed. He argued for the existence of a series of metastable states influenced by the cytoplasm as against regulation by supergenes.

Okada supported Truman's attempts to classify cell types, giving an example of the family of cells derived from neural crest. Despite the great diversity in their phenotypes, there were now several examples of transdifferentiation betweeen members of this family. However, the presence of multipotential progenitor cells has not yet been discovered in the neural crest of early embryos.

Owen stressed the difference between systems of cell classification based on cell lineage and those based on cell function. He believed there was no reason to expect any correlation between the two types of classification.

Chader asked Truman which of the two possibilities proposed by Bloemendal he favoured: a small number of regulatory factors or a large number. Truman accepted that there might be many regulators involved in differentiation of a cell, but believed that the discreteness of cell types suggested a relatively simple regulator for selecting between cell types.

INTRODUCTORY REVIEW:

THE MOLECULAR BASIS OF DIFFERENTIATION AND COMPETENCE

A. P. Bird, D. E. S. Truman*, and R. M. Clayton*

MRC Mammalian Genome Unit
Edinburgh   EH9 3JT

*Department of Genetics
University of Edinburgh
Edinburgh   EH9 3JN

The zygote of a developing organism contains all the information necessary for the specification of the mature form. As it divides by mitosis it gives rise to a population of cells which, in the type of regulative development occurring in vertebrates, are initially capable of a wide range of different developmental possibilities. As development proceeds, the variety of potentialities available to each cell may become restricted and it may be possible to predict the normal fate of a cell from a particular region of the embryo. Even so, if there is experimental interference with the embryo, a cell may manifest a different fate from that which would normally befall it. Thus the cell is competent to go in a number of directions depending in part on external stimuli. As development proceeds the range of competencies becomes limited, and when it ceases to be possible experimentally to change the cell from the path which it would normally follow, then the cell is said to be determined. It may not, at this stage, show any specific phenotypic features characteristic of a given cell type, but if it is determined, then it will eventually show these features and become differentiated. Once a cell has become differentiated it may cease cell division, in the case of terminally differentiated cells, or it may divide to produce progeny which will normally resemble it in the type of their differentiation. Only very rarely, and then usually in response to experimental manipulation, does a differentiated cell change its phenotype drastically and give rise to another distinctive cell type. When this does occur, it is known as transdifferentiation.

Each of the developmental phenomena mentioned above was origin-

ally identified at the macroscopic or microscopic level. It is now
a major task in biology to interpret them in molecular terms. Most
accessible is differentiation, because here alone there is a consensus
view about which molecular processes should be studied. It is reasonably certain that the structural and functional character of a cell
is a consequence of its distinctive protein compostition, and that
protein compostition is in turn the result of a distinctive pattern
of protein synthesis. Thus, in order to understand the molecular
basis of differentiation it is necessary to find out how patterns of
protein synthesis are set up. Protein synthesis is a multistep process, involving transcription, RNA processing and transport, and
translation. In theory, each step offers one or more control points
which could be used to affect the quality or quantity of protein synthesised. Although our knowledge is still primitive, it is already
apparent that many, and perhaps all, of the available control points
can be utilised in certain situations (Brown, 1981).

Cell differentiation has usually been thought to occur against
the background of a constant genome. Exceptions to this generalisation, such as chromosome diminution (Wilson, 1928), or ribosomal gene
amplification (Gall, 1968; Brown and Dawid, 1968), have long been recognised, but the available evidence suggested that these were atypical mechanisms. The β-globin gene, for example, is found at the same
level in erythrocytes, which express the gene at a high level, as in
liver or brain cells, which do not synthesise globin (Bishop and
Rosbach, 1973). Thus the globin gene is not deleted in non-erythropoietic cells, but nor is it amplified in erythrocytes. Evidence of
a different kind reinforces the view that irreversible alterations
to the genome do not accompany development. In a few cases the nucleus of a differentiated somatic cell has been shown to direct the
complete development of an enucleated Xenopus egg, thereby demonstrating that differentiation does not affect totipotency of the nucleus
(Gurdon, 1962). The implications of nuclear transplant experiments
have been critically reviewed by Briggs (1977) and Gurdon (1977).
The question of genome constancy has been reopened by recent results.
Thus, amplification of the chorion protein genes has been detected
in the follicle cells of Drosophila (Spradling and Mahowald, 1980),
and the rearrangement and somatic mutation of immunoglobulin genes
in specialised cells of the immune system is now well documented
(Seidman and Leder, 1978; Sakano et al., 1979; Radbruch et al., these
proceedings). These results emphasise that the genomes of somatic
cells can be altered structurally during development. In this connection, it is striking that all animals segregate germ cells from
somatic cells very early in development, and this raises the possibility that somatic cells are normally subject to changes at the genomic level from which germ cells must be isolated.

Apart from the possibility of dramatic control of protein synthesis by changes in the genome itself, there is clear evidence for
control at the transcriptional level. The sequences for ovalbumin

mRNA, for example, are only found at significant levels in the nuclei of oviduct cells after hormonal induction (Tsai et al., 1979). Clearly the genes are either "switched on" or substantially increased in activity as differentiation proceeds. The mechanism for transcriptional control of this kind is not yet understood, though several molecular events can be correlated with transcriptional activity. For example, DNA sequence involved in transcription are more accessible to the non-specific nuclease DNAse I than are non-transcribed sequences (Weintraub and Groudine, 1976). This suggests that a "loosening" of chromatin structure accompanies transcription, but it is not yet clear whether the change is a cause or a consequence of RNA synthesis. Reduced levels of DNA methylation also correlate with vertebrate gene expression in many cases (reviewed by Razin and Riggs, 1981; Doerfler, 1981), though many animals regulate gene expression in the absence of detectable DNA methylation. Present evidence suggests that unmethylated sites in the vicinity of a gene are necessary, but not sufficient, for transcription to occur.

Although transcriptional control has been demonstrated for a number of genes, it is not clear that gene expression is primarily regulated at this level. Another feasible control point is post-transcriptional processing of RNA. By differential processing it would be possible to discriminate between those sequences that are to be released from the nucleus as mRNA, and sequences that are to be destroyed in the nucleus. Since the great majority of transcribed sequences turn over in the nucleus without ever reaching the cytoplasm, differential processing may be occurring on a large scale. Several processing events involving newly synthesised RNA are known (e.g. capping, splicing, polyadenylation), but the regulatory value of any of these events has yet to be demonstrated (reviewed by Lewin, 1980).

Our knowledge of differentiation at the molecular level is obviously still at a preliminary stage, but already there are indications that the search for a single mechanism controlling differential gene activity will prove to be fruitless. The variety and inconsistency of mechanisms that are found suggest that evolution has been opportunistic in arriving at mechanisms for controlling gene activity. In different systems we find gene loss, gene amplification, gene rearrangement, and differential transcription, and it is likely that other regulatory phenomena have yet to be added to this list. Knowledge in this field is likely to grow rapidly in the next few years, but it may take longer to approach more fundamental questions about development. In particular, can we understand determination and competence by studying molecular gentics? Put another way, is differential gene expression a fundamental causative influence on development, or is it the final stage in a process whose origins lie with other cellular or supracellular processes?

REFERENCES

Bishop, J. O., and Rosbach, M., 1973, Reiteration frequency of duck hemoglobin genes, Nature New Biol., 241:204.

Briggs, R., 1977, Genetics of cell type determination, in:"Cell Interactions in Differentiation," M. Karkinen-Jaaskelainen, L. Saxen, and L. Weiss, eds., Academic Press, London, p.23.

Brown, D. D., 1981, Gene expression in eukaryotes, Science, 211:667.

Brown, D. D., and David, I. B., 1968, Specific gene amplification in oocytes, Science, 160:272.

Doerfler, W., 1981, DNA methylation - A regulatory signal in eukaryotic gene expression, J. Gen. Virol., 57:1.

Gall, J. G., 1968, Differential synthesis of the gene for ribosomal RNA during amphibian oogenesis, Proc. Natl. Acad. Sci. U.S.A., 60:553.

Gurdon, J. B., 1962, Adult frogs derived from the nuclei of single somatic cells, Develop. Biol., 256:73.

Gurdon, J. B., 1977, Egg cytoplasm and gene control in development, Proc. R. Soc. B., 198:211.

Lewin, B., 1980, "Gene Expression 2," 2nd edition, John Wiley, London, chapters 23 and 26.

Sakano, H., Rogers, J. H., Huppi, K., Brack, C., Trannecker, A., Maki, R., Wall, R., and Tonegawa, S., 1979, Domains and the hinge region of an immunoglobulin heavy chain are encoded in separate DNA segments, Nature, Lond., 277:627.

Seidman, J. G., and Leder, P., 1968, The arrangement and rearrangement of antibody genes, Nature, Lond., 276:790.

Spradling, A. C., and Mahowald, A. P., 1980, Amplification of genes for chorion proteins during oogensis in Drosophila melanogaster, Proc. Natl. Acad. Sci. U.S.A., 77:1096.

Tsai, S. Y., Tsai, M.-J., Lin, C.-T., and O'Malley, B. W., 1979, Effect of estrogen on ovalbumin gene expression in differentiated nontarget tissues, Biochemistry, 18:5726.

Weintraub, H., and Groudine, M., 1976, Chromosomal subunits in active genes have an altered conformation, Science, 193:848.

Wilson, E. B., 1928, "The Cell in Development and Heredity," 3rd edition, MacMillan, New York.

THE STRUCTURE AND EXPRESSION OF THE HAEMOGLOBIN GENES

F. Grosveld, M. Busslinger, G. Grosveld,
J. Groffen, A. DeKleine, and R. A. Flavell

Laboratory of Gene Structure and Expression
National Institute for Medical Research
Mill Hill
London NW7 1AA

INTRODUCTION

The haemoglobin genes provide a good model system for the study of the different aspects of gene expression in higher eukaryotes for the following reasons. (The problems have been reviewed by Maniatis et al., 1980.)
- Most of the protein products have been characterized in terms both of primary and of higher order structure.
- The primary DNA sequence of the genes has been elucidated.
- The changes in the expression of the different genes during development are known. In the case of the human β-related genes, the ε gene is expressed in the early embryo, the γ-genes are expressed in the foetus and the δ- and β-genes in the adult.
- The general arrangement of these genes has been elucidated (Fig. 1). They are present in a cluster on chromosome 11 in the same order as they are expressed during development.
- Several defects in the functioning of the globin genes have been defined.
i) abnormal globin protein can be produced, usually because of point mutations (e.g. sickle cell anaemia) and less commonly because of deletions and fusions (e.g. hemoglobin Lepore).
ii) the amount of the globin proteins can be reduced to a lower level, for example in the β+-thalassaemias, or can even be absent, as in the β°-thalassaemias. A number of these are due to gross alterations in the structure of the β-globin locus (Fig. 1).
We have recently studied a number of aspects of the β-globin gene expression at both the structural and functional level. Some of these data are described below.

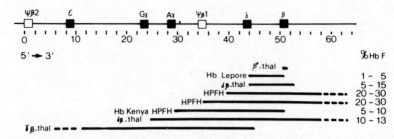

Figure 1. The structure of the εγδβ globin region in normal DNA and in a number of hereditary diseases. The DNA deleted in the hereditary diseases indicated is depicted by the horizontal line. A dashed line indicates that the precise location of the end point of the deletion in that area is not known. The deletion in Hb Kenya has not been established at the DNA level. See also Maniatis et al. (1981) and Flavell et al. (1981).

STRUCTURAL ANALYSIS OF INTERGENIC DNA SEQUENCES

We are interested in structural aspects of the DNA sequences which may regulate gene expression. In the β-gene region there are known β-related genes and the recently discovered pseudogenes, but there is also a tenfold excess of intergenic DNA. The function of this DNA is not clear, but there are several reasons to study this DNA in more detail. There is some evidence that certain features of the DNA sequence may act to regulate globin gene expression. Firstly, differences occur in the degree of methylation of this DNA in different tissues, and the degree of methylation appears to correlate with globin gene expression in those tissues (Van der Pleog and Flavell, 1980). Secondly, deletion of the β-globin intergene regions appear to affect gene function (Maniatis et al., 1980). Finally, a number of repeated sequences have been located in the β-globin region and several types of these have been localised. In the case of the rabbit β-globin cluster several different types of repeated sequences have been mapped (Hoeijmakers-van Dommelen et al., 1980; Shen and Maniatis, 1980), while in the case of the human β-globin cluster, two types of repeated sequences have been localized (Fritsch et al., 1980; Coggins et al., 1980; Kaufman et al., 1980). It would be of interest to compare the nature and location of several classes of repeated sequences between different species, since it might point towards a possible function of repeated sequences in relation to gene expression. To this end we have investigated the localization of a large number of repeated sequences in the human β-globin locus (Fig. 2; Groffen et al., 1981) and compared these with similar sequences present in the rabbit β-globin locus. This comparison shows that they have several classes of sequences in common (de Kleine et al., 1981).

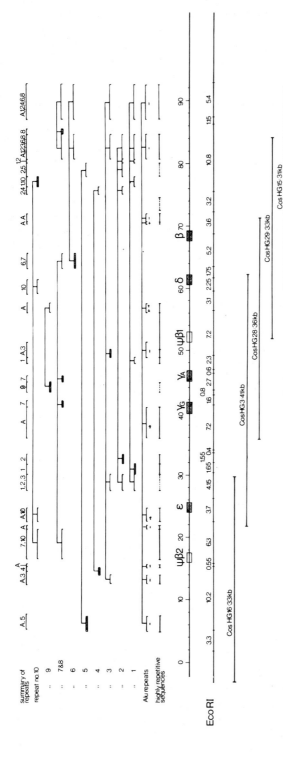

Figure 2. Classification of intergenic repeat sequences in the human β-globin locus. Repeat sequences were classified in eleven types (Alu and 1 through 10) on the basis of one and two dimensional electrophoresis-hybridization experiments (Groffen et al., 1981). The size of each fragment containing a repeat sequence is illustrated by a bracket as compared to the EcoRI map illustrated below the gene map (indicated by hatched boxes). Fragments containing homologous repeat sequences are connected by a straight line. The cosmid recombinants that were used for this analysis are shown on the bottom.

The first common factor is that both clusters contain a number of homologous so-called Alu repeats. These are each about 200 base pairs in length and occur about 300 000 times per genome (Jelinek et al., 1980). At least some of these repeats are transcribed in vivo and they contain a sequence that is homologous to a sequence found near the origin of replication in several eukaryotic viral DNAs. It has been speculated that these sequences may be the sites of initiation of DNA replication in different cells and that expression of a specific gene is regulated by the use of specific (Alu) replication origins (Smithies, pers. comm.). Two other classes of repeats, termed B and E in rabbit and 1 and 2 in man, show homology. While very little is known about the properties of these repetitive elements, it is interesting to notice that preliminary data suggest that the human type 1 sequence is homologous to RNA sequences in γ-globin expressing tissues (de Boer et al., 1981). Currently, further experiments are under way to confirm these observations and test the relationship between the presence of certain repeated elements and gene expression.

ANALYSIS OF SEQUENCES IN OR IMMEDIATELY FLANKING β-GLOBIN GENES

We have recently studied two aspects concerning the level of β-globin gene expression, by the introduction of cloned β-globin in a β^+-thalassaemia, and, the sequences 5' of the gene involved in the intiation of transcription. For the latter study the rabbit β-globin gene was chosen as the model system.

β^+-thalassaemia

In β^+-thalassaemia the level of β-globin chains is reduced relative to the α-globin chains and this reduction correlates with the level of the respective mRNAs in the cytoplasm. However, it has been shown that the ratio of β- to α-globin RNA is higher in the nucleus than in the cytoplasm of bone marrow cells and that precursors to the β-globin mRNA seem to accumulate in the nucleus (Nienhuis et al., 1977; Maquat et al., 1980; Kantor et al., 1980). The DNA sequence of two β-globin genes from two β^+-thalassaemic patients has been determined and in both cases, comparison with the normal β-globin DNA sequence shows a single G $\rightarrow$ A base substitution at 21 nucleotides from the end of the small intron (Spritz et al., 1981; Westaway et al., 1981).

In our study one of these β-globin genes and the normal β-globin gene were cloned in the expression vector SV40-pBR328 and introduced into Hela cells by calcium phosphate co-precipitation (Banerji et al., 1981) (Fig. 3). RNA from these transformed cells was then analyzed by S1 nuclease mapping and primer extension DNA sequence analysis (Figs. 4 and 5). These and additional results (Busslinger et al.,

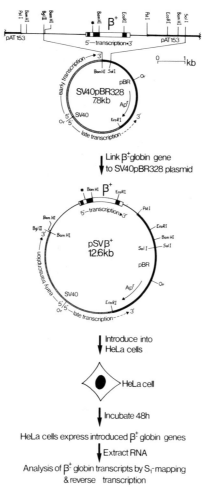

Figure 3. Procedure to test a β^+-thalassaemic globin gene for its functional defect. The β^+-globin gene was originally cloned into the plasmid pAT153 by Westaway and Williamson (1981) from the DNA of a Turkish-Cypriot patient diagnosed as doubly heterozygous for β^+-thalassaemia and Hb Lepore. The construction of the recombinant plasmid pSVβ^+ is described in detail in Busslinger et al. (1981). pSVβ^+ DNA was introduced into Hela cells by the calcium phosphate coprecipitation method (Wigler et al., 1977). Hatched lines denote SV40 DNA, solid lines correspond to plasmid sequences and thin lines to human DNA. The β-globin-coding sequences are represented by filled blocks, introns and sequences coding for the untranslated regions of the β-globin mRNA are shown by open blocks. The star above the first intron indicates the position, where the normal and the β^+-thalassaemic globin genes differ by a single base-change (GC$\leftrightarrow$AT; Westaway and Williamson, 1981).

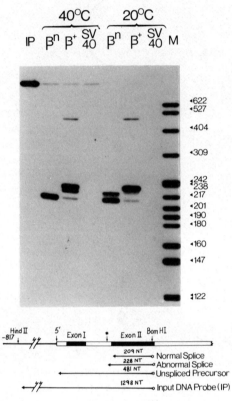

Figure 4. S_1-mapping of the intron 1/exon II splice junction: S_1 nuclease digestion of hybrids at 20°C and 40°C. The Hind II/Bam HI DNA probe was hybridized to RNA from HeLa cells transformed with either the β+-thalassaemic (pSVβ+ DNA) or the normal β-globin gene (pSVβn DNA). As a control, RNA from HeLa cells transformed with the SV40-pBR328 vector alone was used. Hybrids were digested with S_1 nuclease either at 40°C for 1 hour or at 20°C for 2 hours. The protected DNA fragments (lane β+, βn, SV40) were separated on an 8% polyacrylamide sequencing gel. IP denotes input DNA probe. 5' end-labelled pBR322 DNA x Hpa II was used as a size marker (M). The position of the input DNA probe as well as of the S_1 nuclease resistant DNA fragments are depicted as arrows below the map of the β+-globin gene.

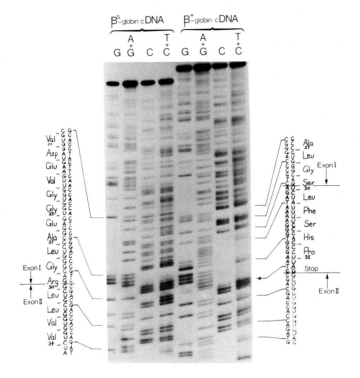

Figure 5. cDNA sequence over the junction of exon I and exon II of
β⁺-globin mRNA. cDNA was synthesized by hybridizing an
end-labelled 87 NT Cvn I/Cvn I DNA primer to β⁺-globin
mRNA obtained from transformed HeLa cells and subsequent
primer-extension by reverse transcriptase. Full-length
cDNA synthesized up to the 5' end of β-globin mRNA was
sequenced according to Maxam and Gilbert (1980) and then
separated on a 10% polyacrylamide sequencing gel. β^S-
globin mRNA isolated from reticulocytes of a patient with
sickle cell anaemia was used in the control experiment.
The cDNA and its complementary mRNA sequence together with
the deduced amino acid sequence are shown next to the
sequence lanes. Amino acids are numbered starting at the
N-terminus of the β-globin protein. The shaded area indicates the additional 19 nucleotides of the β⁺-globin mNRA
that are inserted between exon I and exon II of the normal
β-globin mRNA. The term exon II is used for both β⁺- and
β-globin mRNA to describe the sequences encoding amino
acids 31 to 104 of the normal β-globin sequence. Formally,
exon II of the abnormal β⁺-globin mRNA begins 19 nucleotides further upstream. A small contamination of the β⁺-
globin cNDA by unreacted primer gives rise to a faint band
in all sequencing lanes at the position indicated by the
arrow.

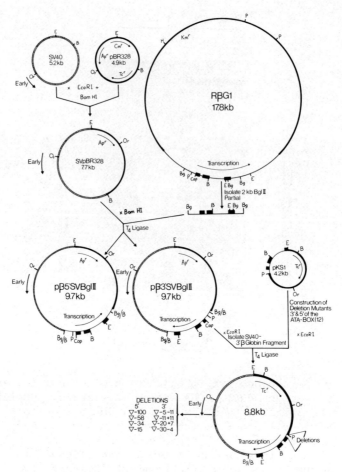

Figure 6. Construction of SV40 derivatives of β-globin DNAs deleted in the promotor region. To prepare SVpBR328, SV40 DNA and pBR328 DNA (each 5 μg) were digested with EcoRI and BamHI, both large Bam-Eco fragments were cut out from a 0.5% low melting agarose gel, purified on a DEAE 52 column (20) and ligated 4 hours at 15°C at a DNA concentration of 20 μg/ml with T4 ligase and transfected into E. coli HB101. From the Ampicillin resistant colonies an SV40 pBR328 recombinant was isolated. To clone the 2kb BglII fragment (which contains the rabbit β-globin gene) into this vector, 50 μg R G-1 DNA was partially digested with BglII and loaded on a 0.5% low melting agarose gel. The 2070 bp β-globin BglII partial band was cut out and the DNA was recovered by DEAE cellulose chromatography. This BglII globin fragment was ligated into the BamHI linearized pBSV328 DNA at a final concentration of 20 μg DNA/ml. From the Ampicillin resistant colonies both orientations of the globin fragment were

HAEMOGLOBIN GENE STRUCTURE AND EXPRESSION

1981) show that about 90% of the nuclear mRNA is spliced incorrectly due to the creation of a 3' acceptor splice site in the first intron by the point mutation. This mRNA could possibly give rise to a partial globin protein of 35 amino acids. In addition, there is an accumulation of unspliced globin RNA which suggests that the removal of the first intron is retarded and that this in turn influences the removal of the second intron.

Sequences involved in the initiation of transcription

Comparison of DNA sequences located immediately to the 5' side of the site of initiation of transcription (the CAP site, Ziff and Evans, 1978; Contreras and Fiers, 1981) of a number of genes shows two conserved regions which might have a promotor function. The first of these sequences, the ATA box, is located about 30 base pairs upstream from the CAP site (Proudfoot, 1979; Gannon et al., 1979). The second sequence, the CCAAT box, is located about 80 bp upstream from the CAP site (Benoist et al., 1980; Efstradiatis et al., 1980).

In vitro transcription systems used on a variety of gene templates indicate that the ATA box is required for transcription in vitro, but that there is little or no effect after deletion of the CCAAT box (Corden et al., 1980; Hu and Manley, 1981; Wasylyk et al.,

⬅︎ ─────────

isolated with the globin 5' end facing SV40 (p β5'SV BalII) and with the globin 3' end facing SV40 (pβ3' SV BglII) respectively. pKS1 DNA, a plasmid containing the Pst1-EcoRI rabbit β-globin gene fragment containing the globin sequences from -100 to +1120) inserted into PstI-EcoRI of PAT153 (12), was used to produce deletions in the globin promotor region (12) resulting in deletions with end points to the 5' side of the ATA box (-100, -50, -34, -15) and the 3' side of the ATA box (-11 -5, -11 +11, -20 +7, -30 +4). To clone these DNAs into an SV40 containing vector 25 μg 5' p 5'SV BglII DNA was digested with EcoRI and the 4.8 kb SV40-globin band which contains Bam-Eco SV40 and 3' end of the rabbit β-globin gene recovered from a 0.5% low melting agarose gel. Then EcoRI linearized pKS1 DNA or DNA from the derivative deletion clones was ligated (20 μg/ml) into this fragment. Tetracycline resistant colonies that hybridized to SV40 DNA were picked and screened by restriction analysis.
Hatched area = SV40 DNA; solid lines = plasmid sequences; thin lines = rabbit DNA; filled boxes represent the β-globin exons.

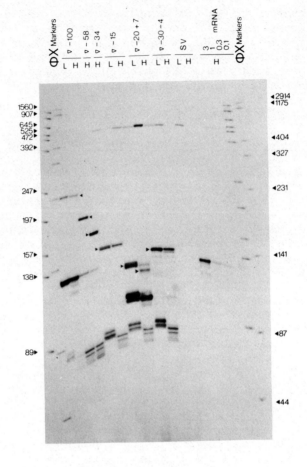
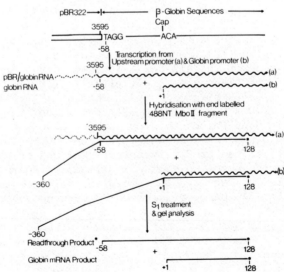

1980; Grosveld et al., 1981). These results, however, do not agree with the results obtained with in vivo systems, where it was shown that removal of the ATA box affects the specificity of initiation and that other sequences further upstream are required for initiation (Benoist and Chambon, 1981; Grosschedl and Birnstiel, 1980a, b).

To determine which sequences are required for the expression of the β-globin gene, we prepared a number of mutant recombinants of the rabbit β-globin gene linked to an expression vector and introduced these into Hela cells. The RNA from these cells was then analyzed for β-globin RNA sequences by S1 mapping (Grosveld et al., 1981). The construction of these mutants is illustrated in Fig. 6 and they consist of two classes; 5' deletion mutants which start at the insertion site in the plasmid (100 bp upstream from the CAP site) and 3' deletion mutants which start in between the ATA box and the CAP site. The S1 analysis of the RNA transcripts (Figs. 7 and 8) shows the following results:
a) Deletion of the ATA box (mutant -30 to -4) reduces the level of correct initiation considerably, indicating that this sequence is required for specific initiation of transcription in vivo.

Figure 7. Transcription of the rabbit β-globin gene from mutant templates lacking DNA sequences at the 5' and 3' boundaries of the promotor. HeLa cells were transformed with 5' deletion clones ▽ -100, ▽ -58, ▽ -34 and ▽ -15 and the 3' deletion clones ▽ -20 + 7 and ▽ -30 -4, as described in Figure 8. The RNA products from these cells were analyzed by S_1 mapping, using a 488NT 5' end labelled MboII probe that contains 360 bp upstream sequences and the first 128 bp of the 5' end of the rabbit β-globin gene (-360 to +128). With a number of RNAs the S_1 reaction was done at 20°C (L) or at 40°C (H) using 3 000 units S_1 nuclease per sample (0.3 ml). The bands labelled with ▶ are derived from RNA starting at an upstream promotor that is picked up by the MboII probe as is explained in the lower part of the figure. The bands at +42 to +47 shift at 40°C to a lower position in the gel because S_1 nuclease attacks the AT splice junction of these RNAs. To calibrate the hybridization natural rabbit β-globin mRNA (3.1, 0.3 and 0.1 ng) was hybridized and treated with S_1 nuclease at 40°C. The relative amounts of rabbit β-globin mRNAs were determined by scanning the bands on the X-ray film. The numbers are given in Figure 8. The labelled marker bands on each side of the gel are Ø X 174 X Taq 1 (left) and Ø X 174 X Rsa I (right), respectively.

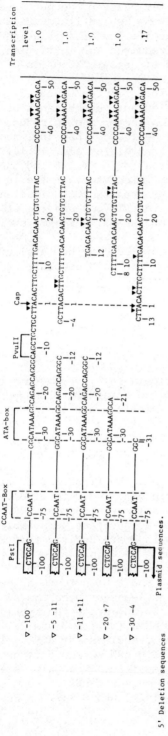
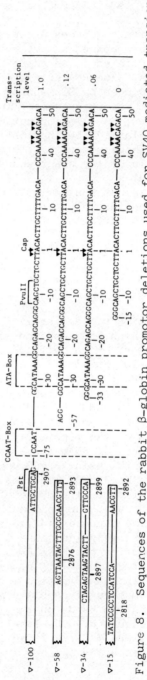

Figure 8. Sequences of the rabbit β-globin promotor deletions used for SV40 mediated transient expression in HeLa cells. The upper part of the figure gives the relevant areas of DNA (vector-globin junction, CCAAT-box, ATA-box, cap sequences and start sequences within the gene) of the β-globin clones that have deletions 3' of the ATA box. The size of the deletions are represented by the gap which is left open. Since the deletions were constructed from pKS1 (Fig. 2) all clones have 5' upstream sequences present to -100 (PstI site). The junction of vector DNA (pAT153) and rabbit β-globin sequences are indicated by the boxed nucleotides. The lower part of the figure gives the important sequences of the β-globin clones that have been deleted 5' to the ATA box. The size of the deletions is again represented by a gap. The clones have all lost the rabbit sequences 5' to the deletion

since the construction of these deletions was started from the BalI site at -75. The vector seqeunces (pAT 153 DNA) present in the clones are indicated by the sequences in the box. The numbers indicated within the plasmid sequences represent the nucleotide which is now at position -75 from the globin cap nucleotide. The tables behind the sequences represent relative levels of globin mRNA in the HeLa cells transformed with these DNAs. To determine this the autoradiogram of Figure 7 was scanned with a Joyce Lobel densitometer. The mRNA signals were compared to each other by assuming that the signal of readthrough RNA starting at an upstream promotor is not influenced by the various deletions in the globin promotor region and was therefore taken as an internal standard signal. ▶ indicates transcription starts seen in the various clones as concluded from the A + C sequence marker lanes in Figure 3, and the splice sites located at +44 to +48.

b) Deletion of sequences 3' of the ATA box (mutants -11 to -5, -11 to +11 and -20 to +7, does not affect the level of transcription. However, the 5' ends of these transcripts are displaced to the 3' side by the same number of nucleotides as the deletion. This would indicate that the RNA polymerase is bound to the ATA box during initiation and starts transcription about 30 bp downstream.
c) Deletion of sequences comprising the CCAAT box (5' mutants -57 and -34) affects the level of transcription, but not correct initiation, while in a mutant lacking both the CCAAT and ATA box (mutant -15) globin transcription is eliminated completely. These results agree with results obtained by Dierks et al., 1981 and seem to indicate that the CCAAT box is involved in the efficiency, but not in the specificity, of initiation of transcription although final proof awaits finer deletion and point mutation mapping.

Although the data described above give an insight into the mechanism and control of transcription of a gene when it is "open" for transcription, it should be noted that they do not provide an understanding of why a particular globin gene is transcribed at a particular stage of development. Probably other sequences much further removed from the gene are involved in this process, since distant deletions can affect globin gene expression in man (Bernards and Flavell, 1980). Elucidation of these mechanisms awaits further experimentation on gene clusters in systems that can mimic the in vivo developmental pathway and specificity of gene expression.

ACKNOWLEDGEMENTS

We thank Cora O'Carroll for help in preparing this manuscript. M.B. is a recipient of a fellowship from Schweizerischer Nationalfonds. This work was also supported by the Medical Research Council.

REFERENCES

Banerji, J., Rusconi, S., and Schaffner, W., 1981, Expression of a genomic segment of simian virus 40 DNA, Cell, 27:299.
Benoist, C., O'Hare, K., Breatnach, R., Chambon, P., 1980, The ovalbumin gene sequence of putative control regions, Nucleic Acids Res., 8:127.
Benoist, C., and Chambon, P., 1981, The ovalbumin gene-sequence of putative control elements, Nature, 290:304.
Bernards, R., and Flavell, R. A., 1980, Physical mapping of the globin gene deletion in hereditary persistence of foetal haemoglobin (HPFH) Nucleic Acids Res., 8:1521.
Coggins, L. W., Grindley, G. J., Voss, J. K., Slater, A.-A., Montagu, P., Stinson, M. A., and Paul, J., 1980, Repetitive DNA sequences near three human β-type globin genes, Nucleic Acids Res., 8:3319.

Contreras, R., and Fiers, W., 1981, Initiation of transcription by RNA polymerase II in permeable, SV40-infected or non-infected, CV1 cells, evidence for multiple promotors of SV40 late transcription, Nucleic Acids Res., 9:215.

Corden, J., Wasylyk, B., Buchwalder, A., Sassone-Corsi, P., Kedinger, C., and Chambon, P., 1980, Promotor sequences of eukaryotic protein-coding genes, Science, 209:1406.

Dierks, P., van Ooyen, A., Mantei, N., and Weissmann, C., 1981, DNA sequences preceding the rabbit β-globin gene are required for formation in mouse L cells of β-globin RNA with correct 5' terminus, Proc. Natl. Acad. Sci. U.S.A., 78:1411.

Efstratiadis, A., Posankony, J. W., Maniatis, T., Lawn, R. M., O'Connell, C., Spritz, R. A., DeRiel, J. K., Forget, B., Weissman, S. M., Slightom, J. L., Blechl, A. E., Smithies, O., Baralle, F. E., Shoulders, C. C., and Proudfoot, N. J., 1980, The structure and evolution of the human β-globin gene family, Cell, 21:653.

Flavell, R. A., Bud, H., Bullman, H., Dahl, H., deBoer, E., deLange, T., Groffen, J., Grosveld, F., Grosveld, G., Kioussis, D., Moschonas, N., and Shewmaker, C., 1981, Globin gene expression in vivo and in vitro, in:"Organization and Expression of Globin Genes," (2nd Conference on Hemoglobin Switching, Airlie, Virginia, U.S.A.), G. Stamatoyannopoulos and A. W. Nienhus, eds., Alan R. Liss, Inc., New York, p.119.

Grosschedl, R., and Birnstiel, M. L., 1980a, Identification of regulatory sequences in the prelude sequences of an H_2A histone gene by the study of specific deletion mutants in vivo, Proc. Natl. Acad. Sci. U.S.A., 77:1432.

Grosschedl, R., and Birnstiel, M. L., 1980b, Spacer DNA sequences upstream of the T-A-T-A-A-A-T-A sequence are essential for promotion of H_2A histone gene transcription in vivo, Proc. Natl. Acad. Sci. U.S.A., 77:7102.

Hu, S. L., and Manley, J. L., 1981, Identification of the DNA sequence required for the initiation of transcription in vitro from the major late promotor of adenovirus 2, Proc. Natl. Acad. Sci. U.S.A., 78:820.

Maniatis, T., Fritsch, E. F., Lauer, J., and Lawn, R. M., 1980, The molecular genetics of human haemoglobins, Ann. Rev. Genetics, 14:145.

Maxam, A. M., and Gilbert, W., 1980, Sequencing end-labelled DNA with base-specific chemical cleavages, in:"Methods in Enzymology," L. Grossman and K. Moldave, eds., Academic Press, New York, Vol. 65, p.499.

Van der Ploeg, L. H. T., Konings, A., Oort M., Roos, D., Bernini, L. F., and Flavell, R. A., 1980, γ-β-thalassaemia: deletion of the γ- and δ-genes influences β-globin gene expression in man, Nature, 283:637.

Wasylyk, B., Kedinger, C., Corde, J., Brison, O., and Chambon, P., 1980, Specific in vivo initiation of transcription on conalbumin and ovalbumin genes and comparison with adenovirus-2 early and late genes, Nature, 285:367.

Ziff, E., and Evans, R., 1978, Coincidence of the promotor and capped 5' terminus of RNA from the adenovirus 2 major late transcription unit, Cell, 15:1463.

ANALYSIS OF RED BLOOD CELL DIFFERENTIATION

P. R. Harrison, N. Affara, P. S. Goldfarb, K. Kasturi,
Q.-S. Yang, A. Lyons, J. O'Prey, J. Fleming, E. Black
and R. Nichols

Beatson Institute for Cancer Research
Switchback Road
Glasgow G61 1BD

Cell and molecular biologists have been fascinated for many years by the process of blood cell development, both on account of its basic interest as a multipotential differentiating system and its relevance to understanding leukemias and the ocogenic process. Experimentally, haemopoiesis (and erythropoiesis in particular) is characterised by a series of classical morphological, biochemical and antigenic markers and the ontogeny of individual types of blood cells is now fairly well understood (Fig. 1). Two areas of particular interest are how multipotential stem cells become committed (probably stochastically, Korn et al., 1973) to specific blood cell pathways and then how the proliferation and/or differentiation of these precursor cells is regulated by growth factors released by other blood cell types (see Till and McCulloch, 1980 for recent review). Thus the regulation of haemopoiesis is controlled by complex inter-acting networks operating at various stages of differentiation. In the case of erythropoiesis in particular, the earliest erythroid precursor cells (BFU-E) are regulated by growth factors released by T-lymphocytes whereas more mature erythroid precursors (CFU-E) are dependent on erythropoietin (Fig. 1).

A further boost to studies of haemopoiesis is the existence of developmentally competent cell lines expressing specific differentiation markers or antigens characteristic of particular normal mature blood cell types or their precursors. The discovery of the Friend cell (Friend et al., 1971) proved to be of particular importance in studying the maturation of early erythroblasts in vitro (see Harrison (1977) for review); the more recent isolation by Dexter et al. (1979) of cell lines (such as the 416 cell line) retaining bi- or multi-

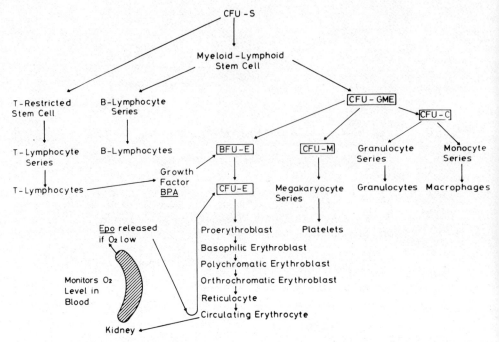

Figure 1. Scheme of haemopoiesis. See Till and McCulloch, 1980 for further details. The growth factor, BPA, is produced in association with the T-lymphocyte response but not necessarily by T-lymphocytes themselves.

potential stem cell properties when transplanted in vivo is also an important advance. Thus, together with the myelomas, and lymphomas and megakaryocyte and granulocyte-derived cell lines, cellular tools are available to test hypotheses concerning the mechanisms of gene expression during haemopoiesis. Of course, results based solely on studies with established (and often virus-transformed) cell lines must be interpreted with caution in terms of their relevance to the normal process, which is often much more difficult to study directly.

As noted earlier, cell differentiation involves the regulation of a battery of (usually unlinked) genes responsible for creating the specialised cell's phenotype and the synthesis of a series of characteristic proteins either coordinately or in a defined temporal sequence. Clearly, to elucidate the molecular basis of such control mechanisms we require probes with which to study the expression of the genes involved. Our research has therefore adopted two approaches: first the isolation of cDNA recombinants containing transcripts of non-globin mRNAs expressed at various stages of erythropoiesis as well as genomic DNA recombinants containing the genes coding for these mRNAs; secondly, the isolation of monoclonal antibodies raised

ANALYSIS OF RED BLOOD CELL DIFFERENTIATION

against specific red blood cell membrane components expressed during development, especially on early precursor cells.

ISOLATION OF NON-GLOBIN cDNA RECOMBINANTS

Total poly(A)-mRNAs from foetal liver erythroblasts or reticulocytes were transcribed into double-stranded cDNAs and then cloned in plasmid pAT153 using the blunt-end ligation method, as described previously (Affara et al., 1981a). Ampicillin-resistant colonies were isolated and screened by the Grunstein-Hogness technique using foetal liver cDNA or primary fibroblast cDNA to define recombinants containing transcripts of mRNAs expressed at high levels in erythroblasts but low in fibroblasts. Adult α- and β-globin cDNA recombinants were eliminated at this first screening by comparing the filter hybridisation of colonies hybridised with foetal liver or reticulocyte cDNA either before or after a pre-hybridisation with excess cloned α- or β-globin cDNA recombinant plasmid DNAs. Subsequently non-globin cDNA recombinants were characterised both by measurement of the sizes of the corresponding mRNAs for which the cloned cDNAs code (using the Northern transfer technique), and by a translation assay of the mRNAs selected from the total erythroblast mRNA population after hybridisation with the non-globin cDNA recombinant DNAs. These experiments are presented in detail elsewhere (Affara et al., 1981b), but the results are summarised in Table 1.

The identity of the proteins for which these non-globin mRNAs code remains elusive. In all probability, pFC5 and pFA6 cDNA recombinants are derived from transcripts of the same mRNA whereas pFA4 is derived from a mRNA coding for a bigger protein (Table 1). Based on the size of the protein for which the pFC5/pFA6 mRNA(s) code(s), we considered that either might represent a carbonic anhydrase mRNA. However, sequencing of the cDNA insert in recombinant pFA6 yields a 170 b.p. sequence which does not seem to correspond (in any reading frame read in either direction) with any of the extensively conserved regions of the carbonic anhydrase sequences published (results not shown). Nor does the sequence correspond to the amino acid sequence of spectrin, (by comparison to the sequence of human spectrin) although this is an unlikely candidate in view of the size of pFA6 protein (Table 1). We are currently trying to determine whether pFC5 and pFA6 recombinants contain transcripts of the mRNA coding for the chromosomal protein IP25 (Keppel, et al., 1977) which increases in differentiating tissue.

ISOLATION OF GENOMIC DNA RECOMBINANTS CONTAINING GENES CODING FOR NON-GLOBIN mRNAs.

Two approaches have been used to screen a Charon 4A-Balb/c mouse genomic DNA library (kindly provided by Dr. P. Leder, N.I.H., U.S.A.)

Table 1. Pattern of expression of non-globin mRNAs.

Percentage total poly(A)-mRNA represented by recombinant sequence

Recombinant	β-globin	pFC5	pFA6	pFD6	pFA4	pFD12
Friend cell uninduced	0.07	0.5	0.3	1.2	0.7	0.3
Friend cell induced	3.0	0.07	0.07	0.25	0.2	0.12
Reticulocyte	45.0	n.d.	n.d.	n.d.	n.d.	n.d.
Stem cell	n.d.	n.d.	n.d.	0.1	0.1	0.5
Fibroblast/ Lymphoma	n.d.	n.d.	n.d.	n.d.	n.d.	n.d.
Size mRNA coded	9 S	12 S	12 S	8 S	13 S	?
Size protein coded (kd)	14.0	30.0	30.0	?	34.0	?

0-30 µg of poly(A)-RNA from the various cell types was bound to DBM-filters and then hybridised with a fixed amount of nick-translated probe. The percentage of the specific mRNA was then calculated from the relative amounts of filter-bound RNA required to convert half the hybridisable probe into hybrid, using the value of 45% β-globin mRNA in reticulocyte poly(A)-RNA as a standard. n.d. indicates a very low level of hybridisation above the pAt hybridisation control (which may or may not be significant) representing less than 10^{-4}% of the total mRNA population. (We have found the alternative approach of hybridising labelled total cDNAs to cloned cDNA recombinants on filters to be less reliable since the extent of hybridisation is influenced greatly by the size and position of the cloned sequence within the overall mRNA sequence if oligo(dT)-primed cDNAs are used and it is subject to other problems if randomly primed cDNAs are employed.) The sizes of the mRNAs and proteins encoded by the recombinants was determined as described elsewhere (Affara et al., 1981a, b).

to isolate genomic DNA recombinants coding for non-globin mRNAs expressed in erythroblasts. First, the library was screened to detect phage recombinants containing genomic DNA sequences hybridising strongly to foetal liver erythroblast cDNA but not to primary fibroblast cDNA. Ten phage recombinants have shown such an "erythroid-specific" hybridisation pattern reproducibly on subsequent replaquing. Alternatively, the genomic DNA library was screened with characterised non-globin cDNA recombinants after excision of the cDNA sequence and nick-translation. In this way, ten genomic DNA recombinants have been isolated which show positive hybridisation with cDNA recombinant pFA6. We have also confirmed that the same genomic DNA clones contain genes expressed in foetal liver erythroblasts but not primary fibroblast mRNAs by hybridisation with the corresponding cDNAs.

PATTERN OF EXPRESSION OF NON-GLOBIN mRNAs

The pattern of expression of selected non-globin mRNAs during haemopoiesis has proved to be very interesting. The method we have used involves binding poly(A)-RNA from various cell types covalently to DMB-filters and then hybridising the filters with nick-translated cDNA-recombinant probes (Le Meur et al., 1981). This is thus essentially the same principle as the Northern transfer technique. One major advantage of this method is that it is useful whichever part of the mRNA sequence has been cloned in the cDNA-recombinant. The results of experiments carried out to date are summarised in Table 1. All the non-globin cDNA recombinants, obtained by cloning foetal liver erythroblast mRNAs, code for mRNAs relatively high in abundance in early erythroblasts (i.e. uninduced Friend cells) which then decline in abundance as the erythroblast matures and synthesises haemoglobin (i.e. induced Friend cells) and finally disappear by the reticulocyte stage. None of these non-globin mRNAs accumulate to significant extents in lymphoma or primary fibroblast cells. However, it is very interesting that three of these non-globin mRNAs accumulate in Dexter's 416B stem cell line (to similar levels to uninduced Friend cells) whereas two are entirely restricted to the erythroid lineage. One interesting hypothesis we are currently testing is whether the pFA4, pFD6 and pFD12 sequences are present in other haemopoietic cell lines more closely related to the erythroid lineage, such as macrophage, granulocyte or myeloma (B-lymphocyte) cells, than the lymphoma (T-lymphocyte) lineage which diverges from the common stem cell for the other blood cell types at a very early stage (see Fig. 1). We also wish to explore the structure of the corresponding genes in the chromatin of these related cell types, using the DNase 1 digestion method to probe chromatin structure.

ISOLATION OF MONOCLONAL ANTIBODIES AGAINST ERYTHROID CELL MEMBRANE PROTEINS

Membrane proteins

Since spectrin and glycophorin are the two major red blood cell-specific membrane proteins, they were chosen initially for production of monoclonal antibodies. Spectrin was isolated from red blood cell ghosts by EDTA extraction and then freed from contaminating membrane lipids, actin and small amounts of other proteins by gel chromatography in deoxycholate and dithiothreitol. Total glycoproteins were similarly extracted from red blood cell ghosts with lithium di-iodo-salicylate/phenol or using affinity chromatography on a wheat-germ-lectin-Sepharose column in the presence of SDS.

DA-strain rats were then immunised with purified spectrin and glycoproteins emulsified in Freund's complete adjuvant and their sera tested for antibody titres by radioimmunoassay using iodinated rabbit $F(ab')_2$ antibody to rat IgG prepared by ourselves. Spleens from hyper-immune rats boosted with antigen were fused with rat myeloma cells (strain Y3Ag1-2-3 obtained by courtesy of Dr. Secher, M.R.C. Laboratory for Molecular Biology, Cambridge) using polyethylene glycol. Growing hybridoma cells were selected in HAT medium using 3T3 cell feeder layers, and hybridomas producing antibodies against spectrin or glycoproteins identified by radioimmunoassay using antigen-coated plates and ^{125}I-labelled $F(ab')_2$ anti-rat IgG as second antibody.

After subcloning selected hybridomas by limiting dilution and by solid agar cloning, nine independent anti-spectrin antibody-producing subclones and 24 anti-glycoprotein clones have been isolated and these are being grown up in ascites form in pristane-primed DA/LOU-C F1 hybrids rats. By transferring spectrin chains separated by electrophoresis onto nitrocellulose and then performing the usual radioimmune assay, two antispectrin monoclonal antibodies have been shown to react with the alpha chain of spectrin primarily. Monoclonal antibodies against mouse red blood cell membrane glycoproteins are being analysed by similar procedures. Four anti-glycoprotein monoclonals react with antigens exposed on the red blood cell surface whereas the others react with internal determinants. Of these four monoclonal anti-glycoprotein antibodies, one reacts most readily with immature erythroblasts and two react primarily with mature red blood cells. However, the fourth reacts more readily with thymocytes than with erythroid cells of either type.

These specific antibodies are now being used to identify, using translation assays, recombinant cDNA clones coding for the appropriate mRNAs. They are also being exploited to isolate specific mRNAs directly by immune adsorption of polysomes on which the mRNAs in question are being translated.

Early antigens

In an attempt to extend this approach to define antigens expressed on the earliest erythroid precursors, fractions rich in early erythroid precursor cells have been isolated from foetal liver populations using density separations in Percoll. CFU-E-rich fractions banding at a density of about 1.075 were then used to raise monoclonal antibodies against their cell surface antigens using the techniques described above. It is planned to screen such antibodies for specificity against red blood cell precursors and a variety of other cell types by radioimmunoassays and functional tests. In this way we hope to use such monoclonal antibodies to define antigens characteristic of the earliest stages of erythropoiesis and then use immunological methods to sort the cells involved.

ACKNOWLEDGEMENTS

This work is supported by grants from the Medical Research Council and Cancer Research Campaign.

REFERENCES

Affara, N. A., Goldfarb, P. S., Vass, K., Lyons, A., and Harrison, P. R., 1981a, Cloning of a new mouse foetal β-globin mRNA sequence, Nucleic Acids Res., 9:3061.

Affara, N. A., Goldfarb, P. S., Lyons, A., Yang, Q.-S., and Harrison, P. R., 1981b, Cloning of non-globin mRNAs expressed during red blood cell differentiation, manuscript submitted.

Dexter, T. M., Allen, T. D., Scott, D., and Teich, N. M., 1979, Isolation and characterisation of a bi-potential haematopoietic cell line, Nature, Lond., 277:471.

Friend, C., Scher, W., Holland, J. G., and Sato, T., 1971, Hemoglobin synthesis in murine virus-induced leukemic cells in vitro: stimulation of erythroid differentiation by dimethyl sulphoxide, Proc. Natl. Acad. Sci. U.S.A., 68:378.

Harrison, P. R., 1977, The Biology of the Friend Cell, in "International Review of Biochemistry, Biochemistry of Cell Differentiation II," J. Paul, ed., University Park Press, Baltimore.

Keppel, F., Allet, B., and Eisen, H., 1977, Appearance of a chromatin protein during the erythroid differentiation of Friend virus-transformed cells, Proc. Natl. Acad. Sci. U.S.A., 74:653.

Korn, A. P., Henkelman, R. M., Ottensmeyer, F. P. and Till, J. E., 1973, Investigations of a stochastic model of haemopoiesis, Expl. Hemat., 1:362.

Le Meur, M., Glanville, N., Mandel, J. L., Gerlinger, P.,

Palmiter, R., and Chambon, P., 1981, The ovalbumin gene family: hormonal control of X and Y gene transcription and mRNA accumulation, Cell, 23:561.

Till, J. E., and McCulloch, E. A., 1980, Hemopoietic stem cell differentiation, Biochim. biophys. Acta, 605:431.

EXPRESSION OF THE HUMAN EPSILON GLOBIN GENE IN MOUSE
ERYTHROLEUKAEMIC CELLS TRANSFORMED WITH RECOMBINANTS
BETWEEN IT AND A BACTERIAL PLASMIC VECTOR

John Paul and Demetrios A. Spandidos

Beatson Institute for Cancer Research
Switchback Road
Glasgow   G61 1BD

INTRODUCTION

The human alpha and beta globin genes have been isolated and their nucleotide sequences determined but we still have little information about their regulation during growth and development. Different haemoglobins are synthesised at different stages. For example, the epsilon chain is the predominant beta-type chain present during the first few weeks of embryonic development. It is succeeded by gamma chains during the major part of foetal life and they are in turn replaced by the adult beta and delta chains around birth. The switching of the beta-type globin chains, therefore, offers a model system for the study of changing gene expression during development. However, we know very little of the mechanisms involved. Much of the evidence we have derives from studies of thalassaemia and suggests that cis-type controls are important (Jones et al., 1981).

Recently studies have been undertaken in a number of laboratories, including our own (Diecks et al., 1981; Mantei and Weissmann, 1979; Mulligan et al., 1979; Wold et al., 1979), on the introduction of rabbit or human globin genes into mouse L-cells and into oocytes. In all these cases the genes are transcribed and the question arises whether their regulation can be studied if they are not in their normal configuration within the chromosome. Moreover, the recipient cells used are not erythroid and, in testing for trans-type controls, it is likely to be more interesting to introduce human globin genes into erythroid cells. To this end we have studied the behaviour of human globin genes inserted into Friend erythroleukaemic cells. These mouse erythroleukaemic cells can be caused to enter a maturation pathway rather similar to that of normal erythroid cells and to accumulate haemoglobin on treatment with an inducer such as hexamethylene bis-acetamide (HMBA) and the behaviour of the transformed genes in response to induction of maturation in the cells studied. This paper describes preliminary results of investigations of this kind.

MATERIALS AND METHODS

The major difficulty in undertaking these experiments is in preparing a suitable vector and a suitable host cell. We required an inducible Friend cell carrying a stable selective marker and permitting a reasonably high efficiency of transformation. Several thymidine kinase less (tk-) Friend cells were avilable to us from Dr Wolfram Ostertag, Dr Paul Harrison and Dr David Conkie and these were first screened for reversion frequencies which were found to vary from 10^{-5} to 10^{-9}. Reversion frequencies of greater than 10^{-7} create serious background problems so attention was focussed on cell line 707B10/1 and cell line F4-12B2 which gave reversion rates of this order or less. Initial transformation studies were undertaken with the plasmid pTK-1 containing the HSV thymidine kinase gene (Wilkie et al., 1979). On this basis F4-12B2 was selected as the best cell for the experiments as it exhibited a much higher rate of transformation than 707B10/1.

As vectors a number of recombinants were constructed between pTK-1 and several human globin gene fragments. The recombinant used in the experiment to be described was made by inserting a Hind III fragment, 8 kilobase (kb) in length, of recombinant 5788 (Proudfoot and Baralle, 1979) into the single Hind III site of pTK1. The 8 kb fragment contains the epsilon globin gene and about 5 kb of flanking DNA. The recombinants obtained with the genomic DNA in both orientations were named pTKH G1 and pTKH G2. These plasmids were used without carrier to transform F412B2 cells employing modifications of existing techniques (to be published elsewhere).

RESULTS

Transformed cells resistant to HAT were investigated for the presence of the human epsilon gene using a spot hybridisation test with a nick translated ^{32}P-labelled epsilon gene and positive colonies were grown up. The presence of full length plasmids in these cultures was demonstrated by Southern blot hybridisation and the number of copies was found to vary from 3 to 11 per cell.

The presence of transcripts from the plasmid was studied by Northern blot hybridisation. Polyadenylated RNA was isolated from 200 micrograms of total RNA from each culture. These RNAs were electrophoresed and transferred to nitrocellulose paper in the presence of appropriate markers. Following hybridisation with a probe for the human epsilon gene, transcripts at 9S could be demonstrated.

Table 1. Accumulation of human epsilon globin mRNA in induced and uninduced Friend cells transformed with pTKH G-1 DNA.

	Total K652 RNA		Untransformed	Transformed cells (F4-12BTk$^+$)		
	10 µg	5 µg	F4-12BTk	F305	F307	F310
Non-induced	10	5.2	0.2	1.5	0.7	2.6
Induced	–	–	0.5	7.4	6.6	7.0

PolyA$^+$ RNA was prepared from 200 µg of total RNA from each Friend cell line, electrophoresed and blotted on to nitrocellulose paper. This "Northern blot" was probed with a ^{32}P-labelled DNA fragment from the 3' end of the gene. Autoradiographs were prepared and scanned. Densitometry measurements were made over the 9S region and the amounts of RNA are given in arbitrary numbers (10 µg total K562 DNA contains approximately 10 ng epsilon mRNA).

To test whether the level of human epsilon mRNA was affected by treating these cells with inducing substances, three independently isolated transformed Friend cell lines were used. A sample of each was treated with inducer (3 mM HMBA in HAT for 6 days), another sample being maintained in HAT for a similar period. RNA was extracted from all cells and run on Northern blots. The three lines responded differently but in all cases more human epsilon globin 9S mRNA was found in induced than in uninduced cells (Table 1).

This result suggests that the factors which promote accumulation of mouse globin messenger RNA in induced Friend cells can also promote accumulation of messenger RNA from globin genes inserted as free plasmids. This could be due to trans-acting regulators of transcription, factors affecting processing of an RNA precursor or stabilisation of the messenger RNA. These possibilities are currently being investigated.

ACKNOWLEDGEMENTS

This research was supported by grants from M.R.C. and C.R.C.

REFERENCES

Diecks, P., Van Ooyen, A., Mantei, N., and Weissmann, C., 1981, DNA sequences preceding the rabbit β-globin gene are required for formation in mouse L cells of β-globin RNA with the correct 5' terminus, Proc. Natl. Acad. Sci. U.S.A., 78:1411.

Jones, R. W., Old, J. M., Trent, R. J., Clegg, J. B., and
 Weatherall, D. J., 1981, Major rearrangement in the human
 β-globin gene cluster, Nature, Lond., 291:39.
Mantei, N., Boll, W., and Weissmann, C., 1979, Rabbit β-globin mRNA
 production in mouse L cells transformed with cloned rabbit
 β-globin chromosomal DNA, Nature, Lond., 281:40.
Mulligan, R. C., Howard, B. H., and Berg, P., 1979, Synthesis of
 rabbit β-globin in cultured monkey kidney cells following infection with SV40 β-globin recombinant genome, Nature, Lond.,
 277:108.
Proudfoot, N. J., and Baralle, F. E., 1979, Molecular cloning of human
 ε-globin gene, Proc. Natl. Acad. Sci. U.S.A., 76:5435.
Wilkie, N. M., Clements, J. B., Boll, W., Mantei, N., Lonsdale, D.,
 and Weissmann, C., 1979, Hybrid plasmids containing an active
 thymidine kinase gene of Herpes simplex 1, Nucl. Acids Res.,
 7:859.
Wold, B., Wigler, M., Lacy, E., Maniatis, T., Silverstein, S., and
 Axel, R., 1979, Introduction and expression of a rabbit β-globin
 gene in mouse fibroblasts, Proc. Natl. Acad. Sci. U.S.A.,
 76:5684.

ISOLATION OF IMMUNOGLOBULIN CLASS-SWITCH AND VARIABLE-REGION VARIANTS
FROM MOUSE MYELOMA AND HYBRIDOMA CELL LINES[1]

A. Radbruch, M. Brüggemann, B. Liesegang,
and K. Rajewsky

Institute for Genetics
University of Cologne
Weyertal 121
D-5000 Cologne 41

The immune system of vertebrates enables the organism to respond specifically to a great variety of antigenic structures, i.e. determinants not commonly occuring in the organism. The specificity is due to two cell types, namely B and T cells. While T cells use antigen receptors of unknown structure (and are thought mainly to regulate immune responses in either a suppressive or enhancing fashion), the B cell receptor and the genes encoding for it are known quite well. B cell antigen receptors - antibodies or immunoglobulins (Ig) - are found on the surface of B cells but also in large amounts in the body fluids. A schematic diagram of an immunoglobulin molecule is shown in Fig. 1. The molecule is composed of two identical heavy and two identical light chains. The organisation of the heavy chain gene locus is shown in Fig. 2. The κ light chain genes are organized similarly. During the ontogeny of a B cell the immunoglobulin genes are rearranged extensively (Fig. 2 b+c). B cells arise from the haemopoietic system by a process that is not very well understood. The first cell type that can be identified unambiguously is the "pre-B" cell that expresses Ig heavy chains in the cytoplasm (Burrows et al., 1979; Lewitt and Cooper, 1980). Expression of Ig heavy chain genes is the result of DNA rearrangement of the genes encoding one out of many V_H-, one out of several D- and one of 4 J_H-genes, to a contiguous VDJ-gene (Early et al., 1980; Fig. 2a, b) that is transcribed and expressed together with the Cμ-gene which is separated from the VDJ-gene by a large intron (Davis et al., 1980a; Maki et al., 1980). The next step in B cell differentiation is the corresponding rearrangement of the

---

[1] This work was supported by the Deutsche Forschumgsgemeinschaft through Sonderforschungsbereich 74.

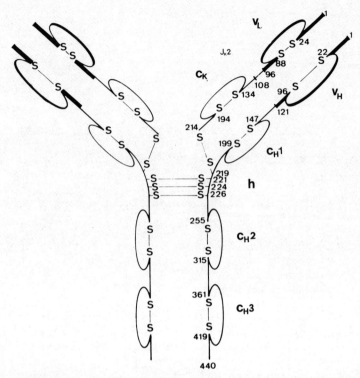

Figure 1. Structure of mouse IgG1 produced by the plasmacytoma MOPC 21. V_L-variable region of light chain; $C\kappa$-constant region of κ-light chain; V_H-variable region of heavy chain; C_H1- to C_H3- domains of heavy chain constant region; h-hinge region; $J\kappa2$ - J-region of κ chain; numbers designate amino acid positions starting from the amino terminus; s-s - disulfide bridges (according to Svasti and Milstein, 1972, Adetugbo et al., 1977 and Sakano et al., 1979).

Ig light chain genes leading to a $V_L J_L$-gene expressed together with either a $C\lambda$- or $C\kappa$-gene. The acquisition of light chain expression apparently enables a pre-B cell to put the Ig molecule on the cell surface where it may function as antigen receptor. B cells express surface Ig only and can be stimulated to clonal proliferation and differentiation into plasma cells. Plasma cells secrete high amounts of Ig and may or may not express surface Ig. Most of the plasma cells die off but it is not clear whether they cannot also revert to B cells.

Work of our group is mainly concerned with two particular aspects of this pathway of differentiation:
   I. Ig heavy chain class switching
   II. V_H-gene diversification.

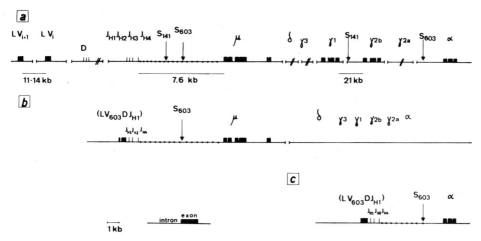

Figure 2. Ig heavy chain gene locus (Igh-locus).
L - protein leader sequence; V, D, J, μ, δ, $\gamma3$, $\gamma1$, $\gamma2b$, $\gamma2a$, α - genes for corresponding heavy chain regions; V 603 - gene for MOPC 603 V-region; S 603 - switch site in MOPC 603 according to Davis et al., 1980; (a) - embryonic state; (b) - IgM producing cell (hypothetically); (c) - IgA producing MOPC 603.

In an immune response to many antigens the class of the specific antibodies is IgM first (μ heavy chains) but later the specific antibodies are predominantly IgG ($\gamma3$, $\gamma1$, $\gamma2b$, $\gamma2a$ heavy chains) or IgA (α heavy chains). This corresponds to the differentiation state of many B cells that first produce IgM and then switch to the production of IgG or IgA. Both IgM and IgG/IgA have the same heavy chain V-region (Fig. 2b, c).

Ig HEAVY CHAIN CLASS SWITCHING

Ig heavy chain class switching was described first in 1964 (Nossal et al., 1964) but until recently no molecular basis could be provided for this phenomenon. Class switching results in expression of a C_H- gene other than $C\mu$, in conjunction with the $V_H DJ_H$-gene. In the mouse 8 different C_H genes have been described and their order on chromosome 12 is 5'-VDJ-Cμ-Cδ-Cγ3-Cγ1-Cγ2b-Cγ2a-Cϵ-Cα-3' (Fig. 2) (Liu et al., 1980; Roeder et al., 1981; Shimizu et al., 1981; Nishida et al., 1981). It is still not clear at what stage of B cell differentiation class switching occurs: Pre-B cells have only been found to contain cytoplasmic μ-chains (Burrows et al., 1979; Levitt and Cooper, 1980) although an Abelson virus-transformed cell line, with all the characteristics of pre-B cells, segregates cells that express cytoplasmic γ2b chains (Burrows et al., 1981). Most of the B-cells express IgM or IgM and IgD but a small percentage express IgG, IgA, or IgE. These later cells are either the result of class switching in B cells, or they are reverted plasma cells. Certain lymphocyte cell lines can undergo class switching, e.g. the line I29 switching from IgM to IgA in vivo (Sitia et al., 1981, personal communication). Evidence for class switching in B cell blasts is mainly derived from in vitro LPS-stimulation (bacterial lipopolysaccharide). LPS stimulation gives rise to clones of B cell blasts containing cells expressing IgM and cells expressing IgG or IgA (Kearney and Lawton, 1975).

We have looked at class switching in transformed plasma cells, (i.e. myelomas and hybridomas) in vitro, by separating cells that had undergone spontaneous class switching from the bulk of the cell population by fluorescent cell sorting with a fluorescence activated cell sorter (FACS) and class-specific antisera (Liesegang et al., 1978). It had been shown previously that class switching at high frequency can be induced by various mutagens in the myeloma cell line MPC11 (approx. 10^{-2}/cell/generation - Koskimies and Birshtein, 1976). We estimated the frequencies of spontaneous class switch variants by screening myeloma or hybridoma populations for such variants under the fluorescence microscope. For this purpose the cytoplasmic Ig of the cells was stained with fluoresceinated antibodies against one class and rhodaminated antibodies against another class, one of the classes being that expressed by the majority of cells in that cell line. An example is shown in Fig. 3, which identifies a γ2b producing cell in a γ1 producing cell line. Such cells were isolated by surface staining the cell population with antisera specific for the variant class and selecting positive cells with a cell sorter. After several rounds of cell sorting variants were sufficiently enriched and could be isolated by cloning. Details of the isolation and characterisation are published elsewhere (Liesegang et al., 1978; Radbruch et al., 1980; Neuberger and Rajewsky, 1981; Beyreuther et al., 1981; Baumhäckel et al., in press). Class switch variants isolated by our group are listed in Table 1. They are derived from the myeloma lines MPC11 and P3 X63 and from the hybridoma lines S24 and B1-8. The characterisation of the variant cell lines and Igs can be summarized as follows:

1) Spontaneous class switching in these cell lines is complete. We have partially sequenced the variant X63, MPC11 and B1-8 Ig heavy chains (Beyreuther et al., 1981). They carry "wild type" V-regions while the "wild type" C-regions are replaced by the complete C-region of the new class.

2) Class switching in myeloma and hybridoma cells is sequential. As is obvious from Table 1 class switch variants always express the next C_H-gene, following the order of C_H-genes on chromosome 12: μ to δ, γ3 to γ1, γ1 to γ2b to γ2a. Switching to other classes occurs at lower frequencies or does not occur at all. We never observed such switches

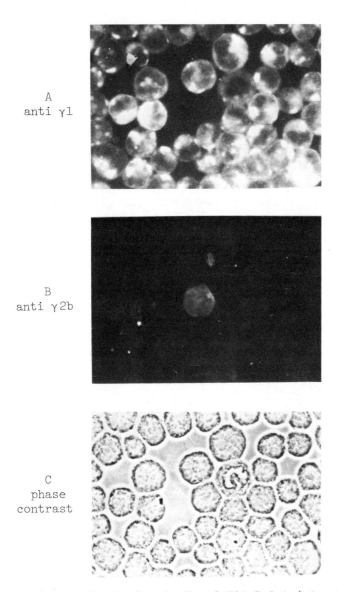

Figure 3. Analysis of cytoplasmic Ig of X63.5.3.1 (γ1, κ) cells by double-fluorescence staining with class-specific antibodies. a) staining with rhodamine-coupled anti-γ1 b) staining with fluorescein-coupled anti-γ2b c) phase contrast.

Table 1. From cloned cell lines indicated in the first row we have isolated clones of variant cells expressing Ig of another isotype (rows 3 and 4) than the wild type Ig (row 2). The detailed characterisation of variant Igs is published in the references listed in row 5: 1 - Neuberger and Rajewsky, 1981; 2 - Baumhackel et al., submitted; 3 - Liesegang et al., 1978; 4 - Radbruch et al., 1980. Frequencies of spontaneous class switching per cell per generation are given in brackets.

Cell line	Wild Type Ig	Ig of class switch variants isolated		Refs.
B1-8	μ, λ (secreted)	$\longrightarrow$ δ, λ (surface, secreted)		1
S24	$\gamma 3$, λ	$\longrightarrow$ $\gamma 1$		2
MPC11	$\gamma 2b$, κ	$\longrightarrow$ $\gamma 2a$, κ (2×10^{-7})		3
X63	$\gamma 1$, κ	$\longrightarrow$ $\gamma 2b$, κ (2×10^{-7})	$\longrightarrow$ $\gamma 2a$, κ (5×10^{-6})	4
X63		$\gamma 2b$, κ	$\longrightarrow$ $\gamma 1$, κ (10^{-5})	4
X63		2a, κ	$\longrightarrow$ $\gamma 2a$ and $\gamma 1$, κ	4

under the microscope nor could we enrich them by fluorescence-activated cell sorting (Radbruch et al., 1980).

3) The spontaneous class switching events described occur at frequencies of 10^{-7}-10^{-6}/cell/generation.

An apparent exception of this rule are the revertants isolated from P3-X63 variant cell lines. They occur at about tenfold higher frequency than other switch variants, and in one case proceed directly from $\gamma 2a$ to $\gamma 1$.

4) Class switching in hybridoma cell lines is accompanied by deletion of the wild type C_H gene (Neuberger and Rajewsky, 1981; Beyreuther et al., 1981). Such a deletion is difficult to demonstrate in the myeloma line X63 since these cells contain more than one C_H gene of the class which is expressed. Nevertheless, we could demonstrate in our P3-X63 cell lines changes in Ig gene rearrangement associated with class switching (Beyreuther et al., 1981). We hybridized a probe detecting a sequence of 1.2 kb downstream J_H (Maki et al., 1980) to fragments derived by digesting DNA from the various X63 lines with the restriction enzyme Kpn. One J_H-Kpn fragment changes its size upon switching from $\gamma 1$ to $\gamma 2b$. Upon reversion from $\gamma 2b$ to $\gamma 1$ two J_H

fragments disappear and a new J_H fragment is generated whose size is identical to that of the J_H fragment which was lost upon $\gamma 1 \rightarrow \gamma 2b$ switching. A likely explanation of this result is the involvement of two Igh-gene loci in the reversion.

Several features of class switching in transformed plasma cell lines as described above are clearly distinct from class switching as observed in vivo. Class switching in vivo occurs at higher frequencies and may proceed from μ to any class directly (Kearney et al., 1975; Gearhart et al., 1980). Revertants have never been demonstrated unambiguously in vivo although certain myeloma lines are probably themselves revertants (Davis et al., 1980b). Some of these differences, however, may be due to the fact that we cannot provide physiological stimuli to the myeloma and hybridoma cells that will enhance switch frequencies. All our attempts in this direction have failed so far. On the other hand these differences enable us to postulate specific switch-promoting factors that act in vivo.

V_H-GENE DIVERSIFICATION

V region diversity can be attributed to at least two different mechanisms. Sequence diversity is clustered in three hypervariable regions (Wu and Kabat, 1970) that are located at positions 31-35, 50-66 and 99-102 of the heavy chain and analogous positions of light chains. Diversity of the third hypervariable region originates from recombination of V-, D- and J-genes, the recombination itself generating additional diversity at the recombination sites (Sakano et al., 1979; Schilling et al., 1980). Diversity in the first and second hypervariable regions and in framework sequences cannot be accounted for by this mechanism. However, the observation that this type of diversity is much more pronounced in IgG and IgA than in IgM when comparing several hybridomas using the same V_H-gene has led to the suggestion that class switching may generate diversity in the VDJ-gene (Crews et al., 1981; Bothwell et al., 1981). However, we could not demonstrate any sequence difference in V-regions from our myeloma and hybridoma lines that underwent class switching. This result shows that class switching is not necessarily combined with V region modification. The more pronounced diversity of IgG and IgA as compared to that of IgM antibodies may in fact result from the selection of cells expressing Igs with variant V-regions by regulatory processes in the immune system. Class switching may accompany this selection as an independent step. Selection of lymphocytes on the basis of the structure of the variable region of their surface receptors may well occur in the immune system and several specific models have been developed along these lines (Burnet, 1966; Jerne, 1974). One of them, the "network" hypothesis (Jerne, 1974), postulates regulation of the immune system by cell surface V-regions recognizing V-regions on other cell surfaces i.e. anti-idiotypic interactions. Any particular monoclonal anti-idiotypic antibody will recognize a single determinant,

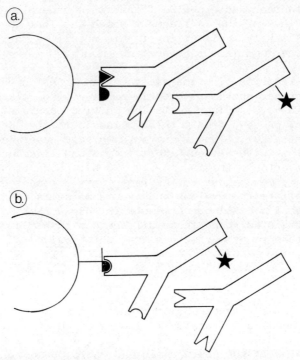

Figure 4. Principle of variant selection. a) the labelling of cells carrying wild type B1-8 IgD with fluorescnet Ac38 Ig is inhibited by excess (100 fold) of unlabelled Ac146 Ig.
b) cells expressing B1-8 IgD that has lost the Ac146 idiotope can no longer bind Ac146 Ig and are now stained by Ac38 Ig. (The figure is taken from Holtkamp et al., 1981.)

the idiotope[1], of the target antibody. The cell producing target antibody is no longer accessible to regulation by those anti-idiotope antibodies, if the V-region looses the idiotope determinants by a mutation of the VDJ-gene. To answer the question whether idiotypic regulation and idiotype variation are correlated we need estimates on the number of anti-idiotopes an organism can express against a given target idiotype, and estimates on the frequencies of the mutations changing an idiotope. Furthermore, we want to know whether those mutations can account for at least part of idiotype diversity ovserved in vivo.

In our laboratory the idiotypic regulation of target antibodies responding to the antigen NIP (4-hydroxy-5-iodo-3-nitro-phenyl-acetyl)

---

[1] idiotype - set of antigenic determinants specific for a V region
idiotope - one such determinant as recognized by a monoclonal antibody.

is analysed using the hybridoma technology (Köhler and Milstein, 1975; Reth et al., 1978). The monoclonal anti-NIP antibody cell line B1-8 (see Table 2) expresses a germline V_H gene (Bothwell et al., 1981). More than 10 iso- and allogeneic monoclonal anti-idiotope antibodies against the B1-8 Ig variable region have been obtained (Reth et al., 1979; Rajewsky et al., 1981) and several of them have been analysed for regulatory effects in vivo (Kelsoe et al., 1981; Reth et al., 1981). The anti-idiotope antibodies can be divided into those whose reaction with B1-8 Ig is hapten inhibitable (group 1) and those whose binding is only inhibitable by a hapten-carrier molecule (group 2) (Reth et al., 1979). A tentative assignment of the corresponding idiotypic determinants (idiotopes) to certain parts of the B1-8.δ1 V region has been possible by comparing its primary structure with that of V regions from related, but idiotope-negative myeloma and hybridoma proteins (Reth et al., 1981).

We have tried to isolate variant B1-8.δ cells producing IgD that no longer bears either of these two idiotopes in order to learn about frequencies and kinds of mutations affecting idiotope expression (Bruggemann et al., in preparation). The selection principle is schematically shown in Fig. 4. We separated cells that could be stained with a fluorescein-labelled anti-idiotope A in the presence of large excess of another anti-idiotope B that completely inhibits binding of the labelled anti-idiotope A to wild type B1-8 IgD. Using the fluorescein-labelled group 2 anti-idiotope Ac38 and the group 1 anti-idiotope Ac146 as inhibitor we have isolated variant B1-8.δ cells whose antibody product has selectively lost the Ac146 determinant. Preliminary characterization of several variant clones showed the following interesting data (Table 2):

1) B1-8 IgD produced by the variant clones does not bind Ac146 and As79 antibodies. However, it does bind two other anti-idiotope antibodies that also belong to group 1 (hapten-inhibitable), e.g. Ac22. Thus, the variants allow us to discriminate between previously indistinguishable idiotopes.

2) The frequency of spontaneous loss of the Ac146 idiotope is higher than approximately 10^{-7} per cell per generation as calculated from the number of cells involved in the selection. The frequency is probably lower than 10^{-6} per cell per generation because we could not detect a variant cell in 10^6 B1-8 cells that we screened for binding of fluoresceine-labelled Ac38 antibody but not rhodamine-labelled Ac146 by cytoplasmic Ig.

3) The variant heavy and light chains have the same size as wild type B1-8δ - and λ-chains. Preliminary data show no difference in number and size of CNBr-fragments of variant and wild type δ- and λ-chains (R. Dildrop, personal communication). Amino acid sequencing of the variant chains will reveal the nature of the mutation and tell us whether we have isolated one or several different variants in this

Table 2. Preliminary characterisation of a B1-8δ Ac146-idiotope loss variant (B1-8.δ18.1) and comparison to B1-8.δ wild type IgD. The molecular weight (Mr) was determined by sodium dodedyl-sulfate-polyacrylamide-gel electrophoresis of biosynthetically labelled Ig from tunicamycin-treated cells (Neuberger and Rajewsky, 1981). Binding of antigen and anti-idiotopes was determined in a solid-phase radioimmunoassay.

Cells	Mr		Binding of:			
	δ-chain	λ-chain	NIP	Ac38 (group 2)	Ac146 (group 1)	Ac22 (group 1)
B1-8.δ wild type	44 000	22 500	+	+	+	+
B1-8.δ 18.1	44 000	22 500	−	+	−	+

selection.

The Ac146 idiotope-negative variants demonstrate that profound changes in antigen- and anti-idiotope binding can result from presumably single-step mutations. Variants of this type arise spontaneously at not very low frequency although exact data are not yet available. We will continue to select variants that have lost idiotopes and characterize the mutations in detail to learn whether idiotypic interactions could serve as a "selector of diversity" in the immune system. We will also learn about the kind of mutations affecting structural determinants involved in antigen- or idiotope-binding. We will also learn about the kind of mutations affecting structural determinants involved in antigen- or idiotope binding.

In general we think that our approach of isolating cell lines showing distinct variations of Ig-expression that can be defined at the molecular level will teach us about the mechanisms that operate during B-cell differentiation. This applies to the mechanisms of DNA rearrangement as well as principles of selection of B-cell clones in the immune sytem, i.e. the differentiation of the immune system.

ACKNOWLEDGEMENTS

H. Baumhäckel, K. Beyreuther, J. Bovens, M. Bruggemann, R. Dildrop, R. Geske, G. Giels, W. Hülseneck, B. Liesegang, S. Irlenbusch, C. Muller, M. S. Neuberger, A. Radbruch, K. Rajewsky, F. Sablitzky, P. Schreier, K. Stackhouse, S. Zaiss and G. Zimmer form the group whose work is summarized in this article. We thank Å. Bohm for expert help in the preparation of the manuscript.

REFERENCES

Adetugbo, K., Milstein, C., and Secher, D. S., 1977, Molecular analysis of spontaneous somatic mutants, Nature, Lond., 265:299.

Baumhäckel, H., Liesegang, B., Radbruch, A., Rajewsky, K., and Sablitzky, F., Switch from NP specific IgG3 to IgG1 in the mouse hybridoma cell line S24/63/63, J. Immunol., in press.

Beyreuther, K., Bovens, J., Dildrop, R., Dorff, H., Geske, T., Liesegang, B., Müller, C., Neuberger, M. S., Radbruch, A., Rajewsky, K., Sablitzky, F., Schreier, P. H., and Zaiss, S., 1981, Isolation and characterisation of class switch variants of myeloma and hybridoma cells, in:"Immunoglobulin Idiotypes and Their Expression," (ICN-UCLA Symposia on Molecular and Cellular Biology) Vol. XX, E. E. Sercarz, H. Wigzell and C. F. Fox, eds., Academic Press, New York, in press.

Bothwell, A. L. M., Paskind, M., Reth, M., Imanishi-Kari, T., Rajewsky, K., and Baltimore, D., 1981, Heavy chain variable region contribution to the NP^b family of antibodies: Somatic mutation evident in a γ2a variable region, Cell, 24:625.

Burnet, M., 1966, A possible genetic basis for specific pattern in antibody, Nature, Lond., 210:1308.

Burrows, P., LeJeune, M., Kearney, J. F., 1979, Evidence that murine pre-B cells synthesize μ heavy chains but no light chains, Nature, Lond., 280:838.

Burrows, P., Beck, G. B., and Wabl, M. R., 1981, Expression of μ and γ immunoglobulin heavy chains in different cells for a cloned mouse lymphoid line, Proc. Natl. Acad. Sci. U.S.A., 78:569.

Crews, S., Griffin, J., Huang, H., Calame, K., and Hood, L., 1981, A single V_H gene segment encodes the immune response to phosphorylcholine: Somatic mutation is correlated with the class of the antibody, Cell, 25:59.

Davis, M. M., Calame, E., Early, P. W., Livant, D. K., Joho, R., Weissman, I. L., and Hood, L., 1980a, An immunoglobulin heavy-chain gene is formed by at least two recombinatorial events, Nature, Lond., 283:733.

Davis, M. M., Kimd, S. K., and Hood, L., 1980b, DNA sequences mediating class switching in α-immunoglobulins, Science, 209:1360.

Early, P., Huang, H., Davis, M., Calame, K., and Hood, L., 1980, An immunoglobulin heavy chain variable region gene is generated from three segments of DNA: V_H, D and J_H, Cell, 19:981.

Gearhart, P. J., Hurwitz, J. L., and Cebra, J. J., 1980, Successive switching of antibody isotypes expressed within the lines of a B-cell clone, Proc. Natl. Acad. Sci. U.S.A., 77:5424.

Holtkamp, B., Cramer, M., Lemke, H., and Rajewsky, K., 1981, Isolation of a cloned cell line expressing variant $H-2K^k$ using fluorescence activated cell sorting, Nature, Lond., 289:66.

Jerne, N. K., 1974, Towards a network theory of the immune system, Ann. Immunol (Inst. Pasteur), 125C:373.

Kearney, J. F., and Lawton, A. R., 1975, B lymphocyte differentiation induced by lipopolysaccharide, J. Immunol., 115:671.

Kelsoe, G., Reth, M., and Rajewsky, K., 1981, Controls of idiotope expresion by monoclonal anti-idiotope and idiotope-bearing antibody, Eur. J. Immunol., 11:418.

Köhler, G., and Milstein, C., 1975, Continuous cultures of fused cells secreting antibody of predefined specificity, Nature, Lond., 256:495.

Koskimies, S., and Birshtein, B. K., 1976, Primary and secondary variants in immunoglobulin heavy chain production, Nature, Lond., 264:480.

Levitt, D., and Cooper, M., 1980, Mouse pre-B cells synthesize and secrete μ-heavy chains but not light chains, Cell, 19:617.

Liesegang, B., Radbruch, A., and Rajewsky, K., 1978, Isolation of myeloma variants with predefined variant surface immunoglobulin by cell sorting, Proc. Natl. Acad. Sci. U.S.A., 75:3901.

Liu, C. P., Tucker, P., Mushinski, J., and Blattner, F., 1980, Mapping of heavy chain genes for mouse immunoglobulins M and D, Science, 209:1348.

Maki, R., Traunecker, A., Sakano, H., Roeder, W., and Tonegawa, S., 1980, Exon shuffling generates an immunoglobulin heavy chain gene, Proc. Natl. Acad. Sci. U.S.A., 77:2138.

Neuberger, M. S., and Rajewsky, K., 1981, Switch from hapten- specific immunoglobulin M to immunoglobulin D secretion in a hybrid mouse cell line, Proc. Natl. Acad. Sci. U.S.A., 78:1138.

Nishida, Y., Kataoka, T., Ishida, N., Nakai, S., Kishimoto, T., Böttcher, I., and Honjo, T., 1981, Cloning of mouse immunoglobulin ε gene and its location within the heavy chain gene cluster, Proc. Natl. Acad. Sci. U.S.A., 78:1581.

Nossal, G. J. V., Szenberg, A., Ada, G. L., and Austin, C. M., 1964, Single cell studies on 19 S antibody production, J. Exp. Med., 119:485.

Radbruch, A., Liesegang, B., and Rajewsky, J., 1980, Isolation of variants of mouse myeloma X63 that express changed immunoglobulin class, Proc. Natl. Acad. Sci. U.S.A., 77:2909.

Rajewsky, K., Takemori, T., and Reth, M., 1981, Analysis and regulation of V gene expression by monoclonal antibodies, in: "Monoclonal Antibodies and T Cell Hybridomas," J. F. Kearney, U. Hämmerling, and G. J. Hämmerling, eds., Elsevier/North Holland, in press.

Reth, M., Hämmerling, G. J., and Rajewsky, K., 1978, Analysis of the repertoire of anti-NP antibodies in C57BL/6 mice by cell fusion. I. Characterization of antibody families in the primary and hyperimmune response, Eur. J. Immunol., 8:393.

Reth, M., Imanishi-Kari, T., and Rajewsky, K., 1979, Analysis of the repertoire of anti-NP antibodies in C57BL/6 mice by cell fusion. II. Characterization of idiotopes by monoclonal anti-idiotope antibodies, Eur. J. Immunol., 9:1004.

Reth, M., Bothwell, A. L. M., and Rajewsky, K., 1981, Structural properties of the hapten binding site and of idiotopes in the NP^b antibody family, in: "Immunoglobulin Idiotypes and Their Expression," (ICN-UCLA Symposia on Molecular and Cellular

Biology) Vol. XX, E. E. Sercarz, H. Wigzell, and C. F. Fox, eds., Academic Pres, New York, in press.

Roeder, W., Maki, R., Traunecker, A., and Tonegawa, S., 1981, Linkage of the four γ subclass heavy chain genes, Proc. Natl. Acad. Sci. U.S.A., 78:474.

Sakano, H., Hüppi, K., Heinrich, G., and Tonegawa, S., 1979, Sequences at the somatic recombination sites of immunoglobulin light-chain genes, Nature, Lond., 280:288.

Schilling, J., Clevinger, B., Davie, J. M., Hood, L., 1980, Amino acid sequence of homogeneous antibodies to dextran and DNA rearrangements in heavy chain V-region gene segments, Nature, Lond., 283:35.

Shimizu, A., Takahashi, M., Yamawaki-Kataoka, Y., Nishida, Y., Kataoka, T., and Honjo, T., 1981, Ordering of mouse immunoglobulin heavy chain genes by molecular cloning, Nature, Lond., 289:149.

Sitia, R., Rubartelli, A., and Hämmerling, U., 1981, Expression of 2 immunoglobulin isotypes, IgM and IgA, with identical idiotype in the B cell lymphoma I.29, J. Immunol., 127:1388.

Svasti, J., and Milstein, C., 1982, The complete amino acid sequence of a mouse κ light chain, Biochem. J., 128:427.

Wu, T. T., and Kabat, E. A., 1970, Analysis of the sequences of the variable regions of Bence Jones proteins and myeloma light chains and their implications for antibody complementarity, J. Exp. Med., 132:211.

MONOCLONAL ANTIBODIES RECOGNISING SURFACE ANTIGENS

ON SUBPOPULATIONS OF NEURONES

Rhona Mirsky, Tom Vulliamy and Stephanie Rattray

MRC Neuroimmunology Group
University College London
London   WC1E 6BT

INTRODUCTION

Changes in surface characteristics of different cell types during maturation of the developing mammalian nervous system can now be studied at a molecular level by the use of monoclonal antibodies of defined specificity which recognize molecules confined to particular cell populations or subpopulations (Kohler and Milstein, 1975; Eisenbarth et al., 1979; Barnstable, 1980; Zipser and McKay, 1981).

We have made two hybridomas which produce monoclonal antibodies recognizing surface antigens on subpopulations of rat neurones. The first, 38/D7, defines an antigen present on the surface of all rat neurones with cell bodies situated in the peripheral nervous system (PNS) (Vulliamy et al., 1981) and the second, TR2, defines an antigen present on the surface of a subpopulation of PNS neurones. Both antigens are absent from central nervous system (CNS) neurones. This system should permit the study of the distribution and ontogeny of the antigens in the developing nervous system.

EXPERIMENTAL

Hybridoma 38/D7

Dorsal root ganglion (DRG) cultures from 5 day old Wistar/Furth or Sprague-Dawley rats were prepared as described previously (Fields et al., 1978). Dissociated cells were cultured at a density of $2 \times 10^5/cm^2$ in Dulbecco's modified Eagle's medium (DMEM) containing 10% foetal calf serum (FCS) and nerve growth factor, on collagen-coated Falcon plastic culture flasks and maintained in 95% air/5% CO_2. Cultures were enriched for neurones by treating with 10^{-5} M cytosine arabinoside after 1 day in culture for a period of 48 hours. After two weeks, cells for immunization were washed with DMEM, scraped off the culture flask and 7.5×10^5 neurones per mouse injected intraperitoneally into Balb/c mice. One week later the mice were boosted with 2.25×10^6 neurones per mouse. 3 days after the boost, serum was tested for the presence of antibodies to DRG cells by indirect immunofluorescence on living DRG cultures.

Hybridoma TR2

DRG from 15 day old rat embryos were gently dissociated by trituration. DRG from 6 rats were injected into a Balb/c mouse. The mouse was given three further immunizations, at weekly intervals and then given a final boost with DRG taken from twelve 15 day old embryo rats. The serum was tested for the presence of antibodies to DRG cells by indirect immunofluorescence of living cultures.

Cell Fusion

Spleen cells were fused with non-immunoglobulin secreting myeloma P3-NS1/1-Ag4-1 (NS1) (38/D7) or SP2 (TR2) using polyethylene glycol (Galfre et al., 1977). After each fusion cells were distributed in 24-well Linbro trays containing HAT medium. Cells were fed twice weekly with fresh HAT medium for two weeks, followed by two changes in HT medium and were thereafter grown in RPM1/FCS (38/D7) or DMEM/FCS (TR2). One hybrid from each fusion which produced antibodies that bound to DRG neurones was selected and cloned twice by limiting dilution on layers of peritoneal macrophages. Using class-specific immunoglobulins and indirect immunofluorescence we have shown that both antibodies belong to the IgM subclass.

RESULTS

The distribution of the antigen recognized by 38/D7 antibodies is shown in Table 1. It is expressed on the surface of all neurones in a variety of cultures derived from the rat peripheral nervous system. In addition to PNS neurones it is also present on the PC12 pheochromocytoma, a cell line derived from an adrenal medullary tumour and on chromaffin cells in adrenal medulla cultures. It is not present on neurones in early embryos, being first detectable on the surface of some dorsal root ganglion (DRG) neurones from 16 day old embryos. Interestingly it also develops in culture in a way which parallels its appearance in vivo. Neurones in dissociated cell cultures of DRG from 13 day old rat embryos first develop the antigen after 3 days in vitro, and over the next few days the number of neurones expressing the antigen increases until all neurones in the culture bind the antibody as they do in dissociated cell cultures from newborn, 5 day old and adult rat DRG.

Table 1. The distribution of antigens recognised by 38/D7 and TR2 antibodies in cultures of rat nervous tissue. Cultures were tested over the period 1-30 days in vitro.

		Antigen present in labelled neurones	
Culture	Age	Antiserum 38/D7	TR2
Dorsal root ganglion	13 day embryo	NT	+/-
"	14 day embryo	-	NT
"	15 day embryo	+	NT
"	Newborn	NT	+/-
"	5-6 days	+	+/-
"	Adult	+	+/-
Superior cervical ganglion	Newborn	+	+/-
Nodose ganglion	Newborn	+	+/-
Myenteric plexus	7-8 days	+	+/-
Adrenal medulla	5-6 days	NT	-
Cerebellum	5-6 days	-	-
"	4 weeks	-	-
Cerebrum	10 day embryo	-	-
Spinal cord	15 day embryo	-	-
Retina	Newborn	-	-

+ antigen present
- antigen absent
+/- antigen present in a sub-population of neurones
NT not tested

The antigen does not develop in culture on the surface of neurones originating in the central nervous system (CNS) and is absent from both peripheral and central glial cells, melanocytes and fibroblasts in a variety of cultures. It is also absent from a number of ethyl-nitrosourea-induced tumour lines derived from the CNS and from the rat-mouse neuroblastoma-glioma hybrid NG108. It can be visualised, though with difficulty, in frozen sections of paraformaldehyde-perfused adult rat DRG, using the peroxidase method. It is not seen on neurones in frozen sections of paraformaldehyde-perfused adult cerebellum.

TR2

The second monoclonal antibody, TR2, recognizes an antigen present on a subpopulation of PNS neurones in cultures from the superior cervical ganglion (SCG), nodose ganglion and myenteric plexus from newborn to 7 day old rats (see Table 1). The antibody does not bind to any cells in cultures from the adrenal medulla and is absent from glial cells and fibroblasts in the peripheral and central nervous system. Like 38/D7 it binds to no neurones in cultures from cerebellum, spinal cord, retina or cortex, all of which are derived from the central nervous system. It is, however, present on a very small population of cells with the morphology of epithelial cells in heart cultures and on occasional flat cells in DRG cultures.

A similar proportion of neurones, 50-60% of the total number of neurones in the culture, express the antigen in cultures made from DRG taken from newborn, 5 day old and adult rats, suggesting that all the neurones destined to acquire the antigen have done so by the time of birth.

It appears earlier in development than the antigen recognized by 38/D7, being detectable on about 10% of neurones in cultures from the DRG of 13 day old rat embryos after only one day in vitro and on a similar proportion of neurones in cultures from the DRG of 16 day old rat embryos. However, in contrast to 38/D7, neurones in culture do not develop this antigen in a way which parallels appearance in vivo, and the number of TR2 positive neurones relative to the total number of neurones declines in all cultures with the time in vitro. However, even after 20 days in culture, about 20% of all neurones from a 5-6 day old rat are still positive.

DISCUSSION

Both the monoclonal antibodies described here recognize antigens which are present on the surface of peripheral nervous system (PNS) but not central nervous system (CNS) neurones. The PNS neurones from the dorsal root ganglion (DRG), superior cervical ganglion (SCG) and myenteric plexus arise during development from the neural crest, but those from the nodose ganglion, (by analogy with chick) are probably of placodal origin. Crest-derived cells in the adrenal medulla and the pheochromocytoma PC12 also express the 38/D7 antigen whereas satellite cells and Schwann cells in various ganglia and melanocytes which are also crest derived, do not.

The TR2 antigen shows a more restricted distribution than 38/D7 but like 38/D7 is confined to neurones which originate from the neural crest or placodes rather than from the neural tube. Interestingly, an antigen showing reciprocal neuronal distribution which is present on the surface of neurones derived from the CNS, but absent from neurones derived from the PNS, has been defined by a monclonal antibody, A4, described by Cohen and Selvendran (1981). Thus three monclonal antibodies recognizing molecules present on the surface of neuronal subpopulations are restricted in expression either to neural crest of placodally derived cells or conversely to neural tube derived cells. This suggests that there may be a significant number of neuronal surface molecules which differentiate PNS from CNS neurones.

All the neurones destined to acquire both the 38/D7 and TR2 antigens in the DRG have done so by the time of birth. The 38/D7 antigen develops with the same time course in vivo and in vitro, indicating that the appearance of the antigen is not directly regulated by its immediate in vivo environment. However, the synthesis of the antigen detected by TR2 seems to be regulated in a more complex manner.

The development of the 38/D7 antigen in situ and in culture is reminiscent of the development of glial fibrillary acidic protein (GFAP) by astrocytes and galactocerebroside by oligodendrocytes. Both of these molecules develop in cells in cultures made from 10-day embryo rat brain with a time course of appearance equivalent to that observed in situ (Abney et al., 1981). It seems, therefore, that by the stage which cells from these embryos are put into culture (13 days for DRG, 10 days for embryo brain) no further information is needed for the expression later in development of these three antigens in neurones, astrocytes and oligodendrocytes respectively.

A second type of behaviour commonly seen in cultures is the failure to synthesize molecules which would normally be made by the cell type in situ. An example of this is the disappearance of the myelin-associated lipid galactocerebroside and proteins P_0 and P_1 from Schwann cells in culture.

When first dissociated and put into culture Schwann cells which in situ have been associated with axons which normally myelinate express these molecules whereas Schwann cells which have been associated with non-myelinating axons do not. However, in sciatic nerve or DRG cultures these molecules are no longer synthesized and therefore gradually disappear from the cells. This indicates the importance of the appropriate signal, in this case an axon capable of myelination, for the synthesis of both myelin lipids and proteins in the supporting Schwann cell.

A third type of behaviour, exemplified by the antigen TR2, can also be distinguished, exhibiting a pattern of expression which is more complex than that of the two extremes typified by 38/D7, GFAP, and galactocerebroside in oligodendrocytes on the one hand and by the behaviour of galactocerebroside and the myelin proteins in Schwann cells on the other hand. Conditions in DRG culture derived from early embryos are clearly insufficient to allow development of the TR2 antigen on neurones not expressing the antigen at the time of plating. Further, we have preliminary evidence that the antigen is lost over a period of several days from some neurones and not others in cultures made from DRG of all ages from 13 day old rat embryo onwards. However, a significant proportion of neurones are able to maintain synthesis over periods as long as 21 days in culture. This suggests that the regulation of the synthesis of this antigen is governed by more than one factor and that this pattern may be representative of a large number of molecules in the nervous system.

REFERENCES

Abney, E. R., Bartlett, P. P., and Raff, M. C., 1981, Astrocytes, ependymal cells and oligodendrocytes develop on schedule in dissociated cell cultures of embryonic rat brain, Develop. Biol., 83:301.
Barnstable, C. J., 1980, Monoclonal antibodies which recognize different cell types in the rat retina, Nature, Lond., 286:231.
Cohen, J., and Selvendran, S., 1981, A neuronal cell-surface antigen is found in the CNS but not in peripheral neurones, Nature, Lond., 291:421.
Eisenbarth, G. S., Walsh, F., and Nirenberg, M., 1979, Monoclonal antibody to a plasma membrane antigen of neurons, Proc. Natl. Acad. Sci. U.S.A., 76:4913.
Fields, K. L., Brockes, J. P., Mirsky, R., and Wendon, L. M. B., 1978, Cell surface markers for distinguishing different types of rat dorsal root ganglion cells in culture, Cell, 14:43.
Galfre, G., Howe, S. C., Milstein, C., Butcher, G. W., and Howard, J. C., 1977, Antibodies to major histocompatibility antigens produced by hybrid cell lines, Nature, Lond., 266:550.
Kohler, G., and Milstein, C., 1975, Continuous culture of fused cells secreting antibody of predefined specificity, Nature, Lond., 256:495.
Mirsky, R., Winter, J., Abney, E. R., Pruss, R. M., Gavrilovic, J., and Raff, M. C., 1980, Myelin-specific proteins and glycolipids in rat Schwann cells and oligodendrocytes in culture, J. Cell Biol., 84:483.

Vulliamy, T., Rattray, S., and Mirsky, R., 1981, Cell-surface antigen distinguishes sensory and autonomic peripheral neurones from central neurones, Nature, Lond., 291:418.

Zipser, B., and McKay, R., 1981, Monoclonal antibodies distinguish indentifiable neurones in the leech, Nature, Lond., 289:549.

EFFECTS OF ROUS SARCOMA VIRUS ON THE DIFFERENTIATION OF CHICK

AND QUAIL NEURORETINA CELLS IN CULTURE

>Patricia Crisanti-Combes, Anne-Marie Lorinet,
>Arlette Girard, Bernard Pessac[1], Marion Wasseff*
>and Georges Calothy**
>
>CNRS ER 231 & INSERM U178, Hôpital Broussais
>96 rue Didot, 75674 Paris
>
>* INSERM U106, Centre Medico-Chirurgical Foch
>   42 rue Desbassayns de Richemond, 92150 Suresnes, France
>
>** Institut Curie-Biologie, 91405 Orsay Cedex, France

Rous sarcoma virus (RSV) is a RNA virus which induces tumors in chick and other avian species. Its biological properties and genetics are well documented (Hanafusa, 1977). Briefly, the genome of the RSV is made up of four genes, of which three (gag, pol, env) are necessary for viral replication and one, src, is responsible for transformation. A sequence related to the viral src gene is present in the cells of all the vertebrate species that have been studied (Stehelin et al., 1976; Spector et al., 1978). The product of both the viral and the cellular src gene is a 60 Kd phosphoprotein (pp^{60} src), with an associated kinase activity (Collett and Erikson, 1978; Collett et al., 1978). The role, if any, of this protein in normal cells, is, at present, totally unknown.

We have previously reported that neuroretina (NR) cells taken from 7 day chick embryos differentiate in monolayer cultures: neurones spread on a "carpet" of Müller cells and mature as shown by the appearance of synapses, in particular ribbon synapses (Crisanti-Combes et al., 1977) and by the development of the enzymatic activities which have been measured, choline acetyl transferase (CAT) and glutamic acid decarboxylase (GAD) (Crisanti-Combes et al., 1978; Guérinot and Pessac, 1979). Neuroretina cultures from 7 day quail embryos follow

1. To whom reprint requests should be addressed.

Figure 1. Visualization of tetanus toxin binding cells by the double immunofluorescence technique. Many round cells are stained on their surface. (x 375)

a pattern of differentiation similar to that of chick cells (unpublished data), with differences due to the more rapid development of the quail NR (Moscona and Degenstein, 1981).

Infection of 7 days chick NR cells in monolayer cultures with RSV results in morphological transformation and induction of sustained proliferation (Pessac and Calothy, 1974; Calothy et al., 1980); similar results are obtained with quail NR cells (Pessac et al., manuscript in preparation). It was therefore decided to investigate the effects of RSV on the program of differentiation of chick and quail embryo neuroretina cells in monolayer cultures.

The markers which have been studied were the same as those selected for the differentiation of normal NR cells: i.e. appearance of synapses, and the specific activity of CAT and GAD. In addition, we have taken the binding of tetanus toxin to the plasma membrane as a specific marker for neurones (Mirsky et al., 1978; Pettmann et al., 1979). The percentage of cells labeled by the tetanus toxin and the specific activity of the enzymes can be quantified. As the "carpet" cells of normal, uninfected NR monolayer cultures are probably Müller cells (Crisanti-Combes et al., 1977) we have also looked for the presence of two glial markers: glial fibrillary acidic protein (GFAP) (Bignami et al., 1980) and glutamine synthetase (Linser and Moscona, 1979). Preliminary experiments have shown that these two molecules could not be detected in the monolayers of control and RSV infected

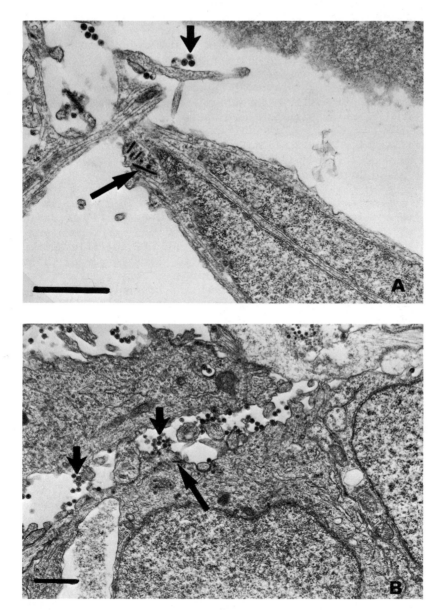

Figure 2 A and B. Electron micrographs of 7 day quail embryo neuro-retina cells infected with Rous sarcoma virus, after 14 days in culture.
Note the virus particles (➤) and the ribbon synapses (➤).
Each bar represents 1 μm.

NR cells by the double immunofluorescence technique. These data do not necessarily imply that the "carpet" cells are not astroglial, but rather show that the anti-human GFAP serum we have used did not react with the avian GFA, and confirm that glutamine synthetase cannot be shown in monolayer cultures (Linser and Moscona, 1979).

In most experiments, the RSV used to infect NR cells was ts NY-68, a thermosensitive transformation defective mutant, which replicates in NR cells at both the permissive (36°C) and non-permissive (41°C) temperatures, but transforms them only at 36°C. In some experiments, the wild type Schmidt-Ruppin RSV-A (SR-RSV-A) and the leukosis virus Rous associated virus-1 (RAV-1), a transformation defective mutant which lacks the src gene, have been used.

The effects of ts NY-68 on cells kept at 36°C, i.e. in cultures where morphological transformation takes place, were similar on both chick and quail primary cultures. The neuronal markers always followed two different patterns: synapses and the specific activity of CAT appeared in the transformed cells in the same way as in its control cultures, and subsequently disappeared. By contrast, tetanus toxin binding cells (which are present in controls) were seen in all transformed cultures, as well as in subcultures of quail cells (Fig. 1); and GAD specific activity, which was markedly stimulated in transformed NR cultures as compared to uninfected controls, persisted at a high level in quail NR cultures after several doublings of the cell population (Crisanti-Combes et al., manuscript submitted).

The ultrastructural studies of control and ts NY-68 infected chick and quail NR cultures were done after 7, 14, and 21 days in vitro. In infected culutres, synapses were present after 7 days (chick) or 3 days (quail) and were present until the 21st day in vitro. In the virus infected cultures, synapses (conventional and ribbon) were present on days 7 and 14. Virus particles were numerous, but no budding viral particles were detected in the vicinity of synapses (Fig. 2). No synapses were seen on the 21st day of culture in ts NY-68 infected NR cells. These data suggest that the synapting cells were infected by the virus, and that the synapses disappeared as a consequence either of cell rounding, or of more specific effects on the synaptic structures.

The kinetics of CAT specific activity in ts NY-68 infected chick and quail NR cultures was similar to that of their controls: the activity increased rapidly to its maximum on about day 7 of culture, then decreased, and was not detectable after 3 or 4 weeks. These data raise the interesting possibility, which cannot be answered at the moment, that the cholinergic cells were not infected by the virus. Alternatively, the virus may have no effect on the CAT activity.

The most striking result of the infection of NR cells by ts NY-68 was on the specific activity of GAD. In chick primary cultures in-

Table 1. Glutamic acid decarboxylase specific activity of control and ts NY-68 infected chick and quail NR primary cell cultures maintained at 36°C.

	Control		ts NY-68 infected cells	
Days in culture	Protein/flask (µg)	Specific activity	Protein/flask (µg)	Specific activity
CHICK				
7	794	763 ± 285	786	787 ± 59
14	897	572 ± 171	1170	3210 ± 410
21	1154	629 ± 203	1230	3297 ± 744
28	1100	400 ± 214	1300	2485 ± 512
QUAIL				
7	494	1004 ± 327	405	1129 ± 168
14	579	541 ± 222	885	1001 ± 94
21	220	331 ± 13	1234	4063 ± 643
28	295	517 ± 79	1233	4078 ± 552

Neuroretina cells from 7 day chick and quail embryos either uninfected (controls) or infected with ts NY-68 were plated in 25 sq cm Corning flasks with 5 ml of basal medium of Eagle plus 5% fetal bovine serum. All flasks were seeded with 1×10^7 cells dissociated from chick or quail neuroretinas. The initial difference in the protein content per flask between chick and quail cultures is due to the fact that the protein content of chick cells is higher than that of quail cells (Moscona and Degenstein, 1981). After 7, 14, 21 or 28 days in culture, the amount of protein per flask was measured by the Lowry method and the GAD specific activity determined as described in Guerinot and Pessac (1979). Each measurement was made in duplicate. The data represent the mean of four independent experiments.

fected with ts NY-68 and maintained at 36°C, the specific activity of GAD was maximal after 14 days, and 6 times that of the control (Table 1). On that day, the increase in cell protein was 50% over that measured on day 7, when the GAD specific activity was still similar to that of controls. In quail primary cultures, the GAD specific activity increased after day 14 and was maximal on day 21, when the values of GAD specific activity were 12 times, and the amount of cell protein, was 3 times that of controls (Table 1). This increase of GAD specific activity was correlated with the presence of neurones: in chick NR cultures, passaged 2 or more times, and where no tetanus toxin binding cells were seen, the GAD specific activity was no longer measurable, while in quail NR cultures, which can be passaged many times, numerous tetanus toxin binding cells were seen, and GAD activity was twice as high as in ts NY-68 infected primary cultures. Furthermore, no GAD activity was found in transformed fibroblasts. These data indicate that the GAD-ergic cells were infected by the virus, and that transformed, multiplying quail NR cells showed a GAD specific activity which was markedly increased as compared to controls. The next step was to determine which viral gene was responsible for the stimulation of the GAD specific activity. The following experiments showed that this effect was under the control of the src gene of RSV. The GAD activity of NR cells, infected with either wild type SR-RSV-A or with ts NY-68 which both contain the src gene, was markedly stimulated; while the GAD activity of NR cells infected with RAV-1, a virus lacking the src gene, was identical to that of uninfected controls.

To investigate if the continuous expression of the src gene - and in particular the protein kinase activity of the pp^{60} src - was required for the maintenance of the GAD activity, quail NR cells infected with ts NY-68 and kept at 36°C for 21 days, were shifted to 41°C for 72 hours. No significant differences were observed between these cultures, and their duplicates kept at 36°C, although the cells maintained at 41°C reverted to a "normal" morphology. These data indicate that the mechanism by which the src gene and its product (pp60 src) generate the transformed phenotype, and stimulate GAD activity, are probably different. Two different mechanisms may account for the stimulation of GAD activity by the src gene of RSV: the amount of enzyme synthesized by the neurones could be increased following an increased rate of transcription and/or translation of the GAD gene; alternatively, the proportion of GAD-ergic cells in these cultures might be increased as the result of a specific interaction between these cells and RSV. At present we have no data which could support either hypothesis. The limited increase in cell number in chick cultures does however show that a considerable stimulation of GAD activity can occur independently of cell multiplication. In this context, it has been reported that the globin gene is activated in chick embryo fibroblasts transformed by RSV, and that the activation requires the presence of the src gene (Groudine and Weintraub, 1980).

In summary, our results show that the src gene of RSV stimulates the activity of an enzyme (GAD), which is normally present in a population of neurones of the avian NR cells. Since the products of the viral and cellular src gene are related (Collett et al., 1978), our data raise the possibility that the product of the c-src genes may play a role in the regulation of a gene(s) coding for specific proteins at certain stages of differentiation. Recent experiments of Scolnick et al. (1981) which show that a mouse hemopoietic cell line expresses a high level of the sarc gene product, might support this hypothesis.

ACKNOWLEDGEMENTS

This work was supported by the Centre National de la Recherche Scientifique and the Institut National de la Santé et de la Recherche Médicale (ATP No 81-79-113).

REFERENCES

Bignami, A., Dahl, D., and Rueger, D. C., 1980, Glial fibrillary acidic protein (GFA) in normal neural cells and in pathological conditions, Adv. Cell. Neurobiol., 1:285.
Calothy, G., Poirier, F., Dambrine, G., Mignatti, P., Combes, P., and Pessac, B., 1980, Expression of viral oncogenes in differentiating chick embryo neuroretina cells infected with avian tumor viruses, Cold Spring Harb. Symp. quant. Biol., 44:983.
Collett, M. S., Brugge, J. S., and Erikson, R., 1978, Characterization of a normal cell protein related to the avian sarcoma virus transforming gene product, Cell, 15:1363.
Collett, M. S., and Erikson, R. L., 1978, Protein kinase activity associated with the avian sarcoma virus src gene product, Proc. Natl. Acad. Sci. U.S.A., 75:2021.
Crisanti-Combes, P., Pessac, B., and Calothy, G., 1978, Choline acetyl transferase activity in chick embryo neuroretinas during development "in ovo" and in monolayer cultures, Develop. Biol., 65:228.
Crisanti-Combes, P., Privat, A., Pessac, B., and Calothy, G., 1977, Differentiation of chick embryo neuroretina cells in monolayer cultures. An ultrastructural study. I. - Seven-day retina, Cell & Tiss. Res., 185:159.
Dowling, J. E., 1975, The vertebrate retina, in:"The Nervous System," D. B. Tower, ed., Raven Press, New York, Vol. 1, p.91.
Groudine, M., and Weintraub, H., 1980, Activation of cellular genes by avian RNA tumor viruses, Proc. Natl. Acad. Sci. U.S.A., 77:5351.
Guérinot, F., and Pessac, B., 1979, Uptake of gamma-aminobutyric acid and glutamic acid decarboxylase activity in chick embryo neuroretinas in monolayer cultures, Brain Res., 162:179.
Hanafusa, H., 1977, Cell transformation by RNA tumor viruses, in: "Comprehensive Virology," H. Fraenkel-Conrat and R. R. Wagner,

eds., Plenum Press, New York, p.401.

Linser, P., and Moscona, A. A., 1979, Induction of glutamine synthetase in embryonic neural retina: localization in Müller fibers and dependence on cell interactions, Proc. Natl. Acad. Sci. U.S.A., 76:6476.

Mirsky, R., Wendon, L. M. B., Blacks, P., Stolkin, C., and Bray, D., 1978, Tetanus toxin: a cell surface marker for neurons in culture, Brain Res., 148:251.

Moscona, A. A., and Degenstein, L., 1981, Normal development and precocious induction of glutamine synthetase in the neural retina of the quail embryo, Develop. Neurosci., 4:211.

Pessac, B., and Calothy, G., 1974, Transformation of chick embryo neuroretinal cells by Rous sarcoma virus "in vitro": induction of cell proliferation, Science, 185:709.

Pettmann, B., Louis, J. C., and Sensenbrenner, M., 1979, Morphological and biochemical maturation of neurones cultured in the absence of glial cells, Nature, Lond., 281:378.

Scolnick, E. M., Weeks, M. O., Shih, T. Y., Ruscetti, S. K., and Dexter, T. M., 1981, Markedly elevated levels of an endogenous sarc protein in a hemopoietic precursor cell line, Mol. Cell. Biol., 1:66.

Spector, D. H., Varmus, H. E., and Bishop, J. M., 1978, Nucleotide sequences related to the transforming gene of avian sarcoma virus are present in DNA of uninfected vertebrates, Proc. Natl. Acad. Sci. U.S.A., 75:4102.

Stehelin, D., Varmus, H. E., Bishop, J. M., and Vogt, P. K., 1976, DNA related to the transforming gene(s) of avian sarcoma virus is present in normal avian DNA, Nature, Lond., 260:170.

MOLECULES THAT DEFINE A DORSAL-VENTRAL AXIS OF RETINA

CAN BE USED TO IDENTIFY CELL POSITION

G. David Trisler, Michael D. Schneider, Joseph R Moskal
and Marshall Nirenberg

National Heart, Lung, and Blood Institute
National Institutes of Health
Bethesda, Maryland 20205

A hybridoma antibody was obtained that binds to cell membrane molecules distributed in a gradient in chick retina. Thirty-five fold more antigen was detected in dorsoposterior than ventroanterior retina. The concentration of antigen detected (F_x) is a function of the square of circumferential distance (D_x) from the ventroanterior margin towards the dorsoposterior margin of the retina; thus, the antigen can be used as a marker of cell position along the ventroanterior-dorsoposterior axis of the retina, i.e.,

$$D_x = D_{max} (F_x/F_{max})\ 0.5$$

where D_{max} and F_{max} are values for the dorsoposterior margin. Immunofluorescence and autoradiography revealed the antigen on the surface of most, or all, cell types in dorsoposterior retina with most of the antigen found in the synaptic layers. The antigen was detected in the optic cup of 48-hr chick embryos, and evidence for a gradient was found with retina from 4-day embryos, the earliest time examined, through the adult. The antigen was found in chick retina > cerebrum > thalamus >> cerebellum > optic tectum and retina pigment epithelium. The antigen was not detected in heart, liver, kidney, or cells from blood. Antigen gradients were found in chicken, quail, duck and turkey retina, but little or no antigen was detected in goldfish, toad, frog, or rat retina.

The antigen was inactivated at 100°C and was converted from a membrane bound to a soluble form by trypsin. No antigen was detected on trypsinized retina cells. However, the antigen was found on retina cells after 2 to 4 hours in culture. No antigen synthesis by cultured cells was detected in the presence of 7 μM cycloheximide or 0.8 μM

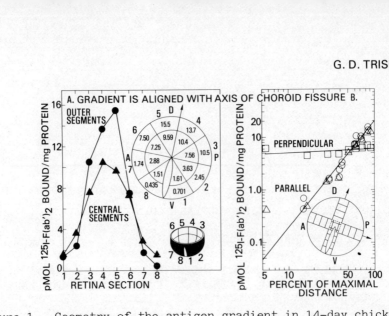

Figure 1. Geometry of the antigen gradient in 14-day chicken embryo retina. Specifically bound ^{125}I-$F(ab')_2$ (pmol/mg of protein) is shown on the ordinate in A and B and within the appropriate segment of retina tested in A. (A) Each retina (left eye) was cut into eight 45° sections (7.25 mm in length from the center to periphery of retina) and each was divided into central (4.9 mm) and outer (2.35 mm) segments. (B) Demonstration that antigen concentration detected depends on the square of distance from the ventroanterior margin of the retina. Percentage of maximal circumferential distance is shown on the abscissa; 100 % corresponds to 14.5 mm. △ , Strips of retina from ventroanterior (0 %) to dorsoposterior (100 %) retina margins, 14.5 x 2.5 mm, were removed from eight retinas (left eyes), and each was cut into nine segments (1.6 x 2.5 mm) as shown; each segment was assayed for antigen. □ , Strips of retina from dorsoanterior (0 %) to posterior (100 %) margins of the retina perpendicular to the choroid fissure were prepared and assayed as above. ○ , Data from A. The length of the arc from the ventral pole of the gradient to the center of each segment was calculated by assuming the retina to be a hemisphere and using equations for spherical triangles. (Reprinted from Trisler, G. D., M. D. Schneider and M. Nirenberg, 1981, Proc. Natl. Acad. Sci. U.S.A., 78:2145.)

actinomycin D. Cells that were dissociated with trypsin from dorsal, middle, or ventral retina and were cultured separately for 10 days contained the levels of antigen expected for cells from the corresponding regions of retina in ovo. Cells from dorsoposterior and ventroanterior retina dissociated with trypsin were mixed in different proportions and cocultured for 6 days; the levels of antigen were almost additive, without evidence of induction or suppression of antigen synthesis. These results show that dissociated cells in culture synthesize the amount of antigen expected for their position of origin in the retina despite the absence of other embryonic tissues.

^{35}S-Protein from retina was solubilized with 0.2% SDS and 1% Triton X-1 and fractionated by protein A-Sepharose or hybridoma antibody-Sepharose column chromatography, and by SDS-polyacrylamide gel electrophoresis. Forty percent of the radioactivity recovered was in a single band approximately 55 000 M . Further work is needed to determine whether the antigen plays a role in the coding of positional information in the retina and the specification of synaptic connections.

REFERENCES

Barbera, A. J., Marchase, R. B., and Roth, S., 1973, Adhesive recognition and retinotectal specificity, Proc. Natl. Acad. Sci. U.S.A., 70:2482.
Barnstable, C. J., 1980, Monoclonal antibodies which recognize different cell types in the rat retina, Nature, Lond., 286:231.
Cafferata, R., Panosian, J., and Bordley, G., 1979, Developmental and biochemical studies of adhesive specificity among embryonic retinal cells, Develop. Biol., 69:108.
DeLong, G. R., and Coulombre, A. J., 1965, Development of retinotectal topographic projection in the chick embryo, Expl. Neurol., 13:351.
Eisenbarth, G. S., Walsh, F. S., and Nirenberg, M., 1979, Monoclonal antibody to a plasma membrane antigen of neurons, Proc. Natl. Acad. Sci. U.S.A., 76:4913.
Galfre, G., Howe, S. C., Milstein, C., Butcher, G. W., and Howard, 1977, Antibodies to major histocompatibility antigens produced by hybrid cell lines, Nature, Lond., 266:550.
Galifret, Y., 1968, Les diverses aires fonctionelles de la retina du pigeon, Zellforsch., 86:535.
Goldschneider, I., and Moscona, A. A., 1972, Tissue-specific cell-surface antigens in embryonic cells, J. Cell Biol., 53:435.
Gottlieb, D. I., Rock, K., and Glaser, L., 1976, A gradient of adhesive specificity in developing avian retina, Proc. Natl. Acad. Sci. U.S.A., 73:410.
Hausman, R. E., and Moscona, A. A., 1979, Immunologic detection of retina cognin on the surface of embryonic cells, Expl. Cell Res., 119:191.
Hirose, G., and Jacobson, M., 1979, Clonal organization of the central nervous system of the frog. I. Clones stemming from individual blastomeres of the 16-cell and earlier stages, Develop. Biol., 71:191.
Köhler, G., and Milstein, C., 1975, Continuous cultures of fused cells secreting antibody of predefined specificity, Nature,

Lond., 256:495.
Marchase, R. B., 1977, Biochemical investigations of retinotectal adhesive specificity, J. Cell Biol., 75:237.
Mintz, B., and Sanyal, S., 1970, Clonal origin of the mouse visual retina mapped from genetically mosaic eyes, Genetics, 64:43.
Rutz, R., and Lilien, J., 1979, Functional characterization of an adhesive component from the embryonic chick neural retina, J. Cell Sci., 36:323.
Sanyal, S., and Zeilmaker, G. M., 1977, Cell lineage in retinal development of mice studied in experimental chimaeras, Nature, Lond., 265:731.
Schneider, M. D., and Eisenbarth, G. S., 1979, Transfer plate radioassay using cell monolayers to detect anti-cell surface antibodies synthesized by lymphocyte hybridomas, J. Immunol. Methods, 29:331.
Silver, J., and Sidman, R. L., 1980, A mechanism for the guidance of nerve fiber patterns and connections, J. comp. Neurol., 189:101.
Sperry, R. W., 1963, Chemoaffinity in the orderly growth of nerve fiber patterns and connections, Proc. Natl. Acad. Sci. U.S.A., 50:703.
Subtelny, S., and Konigsberg, I. R., eds., 1979, "Determinants of spatial organization," 37th Symposium of the Soc. Developmental Biol., Academic Press, New York.
Theiry, J. P., Brackenbury, R., Rutishauser, U., and Edelman, G. M., 1977, Adhesion among neural cells of the chick embryo. II. Purification and characterization of a cell adhesion molecule from neural retina, J. biol. Chem., 252:6841.
Trisler, G. D., Donlon, M. A., Shain, W. G., and Coon, H. G., 1979, Recognition of antigenic difference among neurons using antiserums to clonal neural retina hybrid cells, Fedn. Proc., 38:2368.
Trisler, G. D., Schneider, M. D., and Nirenberg, M., 1981, A topographic gradient of molecules in retina can be used to identify neuron position, Proc. Natl. Acad. Sci. U.S.A., 78:2145.
Yazulla, S., 1974, Inraretinal differentiation in the synaptic organization of the inner plexiform layer of the pigeon retina, J. comp. Neurol., 153:309.

ANALYSIS OF MYOGENESIS WITH RECOMBINANT DNA TECHNIQUES

D. Yaffe, U. Nudel, H. Czosnek, R. Zakut, Y. Carmon, and M. Shani

Department of Cell Biology
The Weizmann Institute of Science
Rehovot 76100
Israel

SUMMARY

Recombinant phages containing the rat skeletal muscle α-actin gene and the cytoplasmic β-actin gene were isolated and the structure of these genes was determined. Both genes contain a large intron in the 5' untranslated region and smaller introns at codons 41, 267 and 327. In addition, the α-actin contains introns at codons 150 and 204 not present in the β-actin gene, whereas the β-actin gene contains an intron at codon 121. The evolutionary aspects of these findings are discussed.

Active genes are organized in chromatin in a conformation which renders them preferentially sensitive to digestion with nucleolytic enzymes. The DNAase I sensitivity of genes programmed to be expressed during myogenesis was tested in a cloned cell population of a myogenic cell line. It was found that these genes are not preferentially sensitive to DNAase I in the chromatin of proliferating mononucleated cells. They become DNAase I sensitive during terminal differentiation.

INTRODUCTION

Recent progress in recombinant DNA techniques provides new tools to investigate questions concerning the control of specific gene expression during differentiation at the chromatin and DNA levels. In order to study the molecular aspects of the control of gene expression during myogenesis, we constructed recombinant plasmids containing cDNA sequences homologous to various parts of actin, myosin heavy chain, and myosin light chain 2 mRNAs (Katcoff et al., 1980;

Nudel et al., 1980; Shani et al., 1981a). These were used as probes to screen libraries of recombinant phages containing inserts of rat genomic DNA sequences for phages containing genes expressed during myogenesis.

The present communication describes two lines of investigation carried out in our laboratory: (a) A comparative study of the structure of the skeletal muscle α-actin gene and the nonmuscle β-actin gene, and (b) a study concerned with the conformational changes in muscle-specific genes associated with the terminal differentiation of cultured muscle cells.

THE STRUCTURE OF THE GENES CODING FOR THE RAT SKELETAL MUSCLE α-ACTIN AND THE CYTOPLASMIC β-ACTIN

The actins comprise a family of highly conserved proteins found in all eukaryotic cells. They play an important role in cell motility, cytoskeleton structure and muscle contraction. In higher vertebrates three major isoelectric forms of actin have been observed. The most acidic, α-actin, is the muscle-specific type, while β- and γ- are the nonmuscle "cytoplasmic" actins found in many cell types. Amino acid sequence analysis showed that β - and γ-actins differ from each other by 5 amino acids (out of 374) located at the N terminus of the protein. They differ from the skeletal muscle actin by 25 amino acids which are spread throughout the protein. Amino acid sequencing data showed also that the skeletal muscle actin differs from cardiac muscle actin by 4 residues and from smooth muscle by 6 residues (Vanderkerckhove and Weber, 1979). The great conservation of actins during evolution and their occurrence in remote taxonomic groups can provide an excellent model for the study of the evolution of genes. In addition, since different actin genes are expressed in different tissues, a comparison of the structure of actin genes and their organization may be helpful for the understanding of how their expression is controlled. The skeletal muscle actin gene is of particular interest, since it becomes expressed during terminal differentiation of muscle cells (Shani et al., 1981b). Thus, structural analysis of this gene and its flanking regions may provide information on the control of its expression during development.

We have isolated recombinant bacteriophages containing DNA sequences of at least 8 different actin genes. One of them (Act 15) contains the gene coding for the skeletal muscle actin and another phage (Act 1) was found to contain the gene for the cytoplasmic β-actin (Nudel et al., 1982). The actin gene in the phage Act 15 was completely sequenced and its structure is schematically described in Fig. 1a. As can be seen, this gene is split by 5 small introns in the coding region (at codons 41, 150, 204, 267, 327) and one large intron of 970 bp in the 5' untranslated region 12 bp upstream from the translation initiation codon (Table 1).

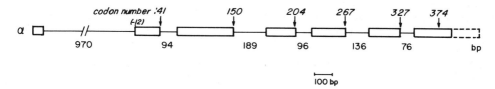

Figure 1. α – The structure of the skeletal muscle α-actin gene.
Open bars – exons. Solid line – introns. Dashed bar – a
portion of the 3' untranslated region of the gene which is
missing in the insert of Act 15. The structure of this
part of the gene was deduced from the nucleotide sequence
of p106 which contains the entire 3' untranslated region of
the gene (Shami et al., 1981a). The numbers below the
scheme indicate sizes of the introns. Bp = base pair. The
numbers above the scheme indicate the codon numbers. The
numbers in parentheses = the number of base pairs 5' to the
initiator codon.

β – The structure of the cytoplasmic β-actin gene. The
sequencing of the 3' untranslated region and the large
intron in the 5' untranslated region was not completed
(see also α above).

The coding region of the insert in the recombinant phage Act 1
is almost completely sequenced. The sequence codes for the cytoplasmic β-actin. The structure of this gene is shown in Fig. 1b. Similar
to the α-actin, the β-actin gene contains a large intron (over 1 000
bp) in the 5' untranslated region located 6 bp upstream to the initiator codon. The coding region is interrupted by introns at codons
41, 121, 267 and 327. Thus, α- and β-actin genes have in common a
large intron in the 5' untranslated region and at codons 41, 267 and
327. The introns at codons 150 and 204 are found only in the α-actin
gene, while the intron at codon 121 is found only in the β-actin gene.
The splice sites of the three homologous introns in the coding region
are exactly in the same position in relation to the reading frame.
There are no amino acid replacements at those exon/intron junctions
(Table 1). The difference of 6 nucleotides in the splice site of
the large intron in relation to the coding region may be attributed
to the known rapid divergence of non-coding regions (including insertions and deletions) observed in many genes. It is possible that the
intron in the 5' untranslated region (8 nucleotides upstream from
the initiator codon) found in one of the Drosophila actin genes
(Table 2) is homologous to this intron.

Table 1. DNA sequence in the exon-intron junctions of α and β rat actin genes.

Intron site (codon No.)*	Actin gene	Sequence at splicing site** 5' exon/ intron /exon 3'
5' UT (-12)	α	CAGCAG/GTAGGGT-----TTTGCAG/AAACTA
5' UT (-6)	β	N.D. -----TTGATAG/TTCGCC
41/42	α	*HisGln* *GlyVal* CACCAG/GTCAGGC-----TCTGCAG/GGTGTC
	β	CACCAG/GTACCAG-----TCCGCAG/GGTGTC
121/122	α	No intron
	β	*ThrGln* *IleMet* ACCCAA/GTTAGTA-----TCTACAG/ATCATG
150	α	*ThrThrG* *lyIle* ACCACCG/GTGAGTG-----TGCACAG/GCATC
	β	No intron
204	α	*ThrThrA* *laGlu* ACCACAG/GTGCGTG-----CCTGCAG/CTGAA
	β	No intron
267	α	*PheIleG* *lyMet* TTTATCG/GTGGCCT-----CCACCAG/GTATG
	β	*Leu* TTCCTGG/GTAAGTT-----CTTTCAG/GTATG
327/328	α	*IleLys* *IleIle* ATCAAG/GTGGATG-----CTTGCAG/ATCATC
	β	*Val* GTCAAG/GTAAGCA-----TCTTCAG/ATCATT

*The numbers indicate the codons at which the introns are located. Numbers in parentheses indicate the position of introns in the 5' untranslated region (the number of nucleotides upstream from the initiator codon).

**The amino acid sequence is marked (in italics) above the nucleotide sequence of the α-actin gene. In the β-actin DNA sequence, only the amino acid replacements are indicated.

N.D. = not determined.

(From Zakut et al., manuscript in preparation.)

Several studies have indicated that although the nucleotide sequence of introns diverged rapidly during evolution, the position of introns was highly conserved, e.g., all known active globin genes in vertebrates have 2 introns in very similar positions. Based on this high degree of conservation and the relationship of intron sites to the protein structure, it has been suggested that introns are located between DNA sequences coding for different protein domains (Gilbert, 1978; Blake, 1981). The actin gene family was considered to be an exception to this rule. In spite of the great conservation of amino acid sequence of actins along a very wide evolutionary scale, earlier studies showed a great variability in intron sites (Ng and Abelson, 1980; Gallwitz and Sures, 1980; Fyrberg et al., 1981; Schuler and Keller, 1981). No common intron sites were observed among the 6 Drosophila actin genes, as well as between Drosophila, yeast and sea urchin (Table 2). The present investigation and recent data on the chick skeletal muscle actin gene (Ordahl, personal communication) show a correlation between the evolutionary relatedness and intron sites. As can be seen from Table 2, all intron sites observed so far in the echinoderms are present in the vertebrates in α- or β-actin gene or in both genes. The very high resemblance of the actins of vertebrates and the other organisms for which data are available excludes the possibility of convergent evolution from different ancestral genes. It can therefore be concluded that the intron sites, although very conserved when examined on a small evolutionary scale (i.e. within the vertebrates), do change during evolution. Although it is not yet possible to determine whether the differences in intron pattern arose by addition or deletion of introns, or both, these differences may be used as parameters in the study of evolution of genes on a large chronological scale.

It seems that differences in intron sites between actin isogenes in the same organism can be taken as an indication of an ancient separation of genes. Thus, the great heterogeneity of intron sites among the Drosophila actin genes suggests that these genes have separated from each other very early in their evolution. On the other hand, the conservation of several intron sites in the deuterostome actin gene (sea urchin, rat and chick) suggests that the separation of actin genes in this line of evolution is a relatively recent event and that it occurred much later than the separation between the two evolutionary lines.

Table 2. Positions of introns in actin genes.

Source and type of actin gene		Intron location					
Dictyostelium (several genes)[1]		None					
Yeast[2]		4					
Drosophila[3]	DmA 6	–					
	DmA 4	13					
	DmA 2	5' UT(–8)	–				
Sea urchin gene J[4]		41	121	–	203	267	–
Chicken skeletal muscle[5]		41	–	150	203	267	327
Rat skeletal muscle[6]	5' UT(–12)	41	–	150	204	267	327
Rat β cytoplasmic[6]	5' UT(–6)	41	121	–	–	267	327

*The numbers indicate the codons at which introns are located. Numbers in parentheses indicate the position of introns in the 5' untranslated region (the number of nucleotides upstream from the initiator ATG). No introns were found in the sites indicated by dashes. UT = untranslated.

Data were taken from: (1) Firtel et al., 1979; (2) Gallwitz and Sures, 1980; Ng and Abelson, 1980; (3) Fyrberg et al., 1981; (4) Scheller et al., 1981; (5) C. Ordahl, personal communication; (6) Nudel et al., 1982 and unpublished results.

(From Zakut et al., manuscript in preparation.)

It is of interest that the amino acid sequences for all 6 Drosophila actins is similar to that of the cytoplasmic (β and γ) actins of vertebrates. Although all Drosophila actin genes have been isolated and partially sequenced, no gene coding for a protein similar to vertebrate muscle actin has been found (Fyrberg et al., 1981). Based on these data and the intron pattern, we suggest that in spite of the great similarity in function, the vertebrate muscle actin and the insect muscle actin genes evolved independently from non-muscle actin genes.

CHANGES IN CHROMATIN ASSOCIATED WITH TERMINAL DIFFERENTIATION

Investigations using a variety of cell types have shown that active genes have altered chromatin structures which render them preferentially sensitive to digestion with nucleolytic enzymes. Several studies suggested that the preferential sensitivity to nuclease digestion may reflect the potentiality of genes to be expressed in a cell rather than their actual transcriptional activity (Stalder et al., 1980; Groudine and Weintraub, 1981; reviewed by Mathis et al., 1980).

Studies on the DNAase I sensitivity of the avian β-globin gene showed that in primitive erythroblasts (present in the embryonic circulation between 2 and 5 days) producing only embryonic β-globin chains both embryonic and adult β-globin genes are sensitive to DNAase I. In erythroid progenitor cells from 20-23 h embryos, which did not synthesize embryonic nor adult globin mRNA, the whole β-globin gene complex was DNAase I insensitive. In this system it was difficult to prove unequivocally that the cells isolated from 20-23 h embryos in which the β-globin genes showed relative resistance to DNAase I were indeed the precursors of the globin-producing cells in which this gene was found to be preferentially DNAase I sensitive (Groudine and Weintraub, 1981).

We have used cloned DNA probes to investigate the question whether in proliferating muscle precursor cells genes which are programmed to be expressed later in development are distinguishable in their DNAase I sensitivity from genes which are never expressed in these cells. In the present study we examined the DNAase I sensitivity of genes expressed during myogenesis in cloned cell populations of the myogenic cell line L8 (Yaffe, 1969; Yaffe and Saxel, 1977). Mononucleated cells of this line can be kept in log phase for many generations in standard culture conditions. When the cultures are allowed to reach confluence, or by switching to nutritional medium which stimulates cell fusion, these cells cease dividing and fuse into multinucleated fibers. As in primary muscle cultures, the cell fusion is associated with synthesis of large amounts of a number of muscle-specific proteins (Coleman and Coleman, 1968: Yaffe and Dym, 1972; Patterson and Strohman, 1972; Yablonka and Yaffe, 1977; Carmon et al., 1978; Shainberg et al., 1971; Dym and Yaffe, 1979). Hybridization of cloned DNA probes specific for the myosin heavy chain, myosin light chain 2 and skeletal muscle actin with mRNA extracts from differentiating cultures showed that mRNA sequences coding for these proteins are undetectable in the proliferating mononucleated cells but accumulate rapidly during the process of cell fusion (Shani et al., 1981b). Thus, using the myogenic cell lines, it is possible to test the DNAase I sensitivity of genes programmed to be expressed later in development, in the same cell lineage, and to determine at which stage of cellular differentiation the change in the conformation of the genes in chromatin occurs (Carmon et al., in preparation).

Nuclei were isolated from cultured proliferating myoblasts. Parallel cultures from the same cell line were induced to differentiate by changing the nutritional medium and the nuclei were isolated 100 h later. At this stage, the great majority of the cells had fused into multinucleated fibers. Nuclei were also isolated from rat brain, to provide information on the DNAase I sensitivity of muscle-specific genes in a non-muscle tissue. The nuclei were incubated with increasing concentrations of DNAase I. The isolated DNA was digested with EcoRI. The fragments were separated by electrophoresis and transferred onto nitrocellulose sheets. The blotted DNA was hybridized with cloned DNA probes for the muscle-specific proteins myosin light chain 2 (plasmid p103) and α-actin (plasmid p749) and for the cytoplasmic β-actin (plasmid 72). A probe for an immunoglobulin constant region genes (plasmid B1) was used as an internal control to measure the DNAase I sensitivity of a gene not expressed in the muscle tissue. The autoradiograms obtained after hybridizations were scanned and the differentiation-related sensitivity of the muscle-specific genes quantitated as described in Fig. 2.

This study indicated that (i) the genes coding for the skeletal muscle α-actin and myosin light chain 2 are DNAase I sensitive in differentiated cultures but not in proliferating myoblasts; (ii) the gene coding for the cytoplasmic β-actin is DNAase I sensitive in proliferating myoblasts, in differentiated cultures, and in brain; and (iii) the immunoglobulin gene is DNAase I resistant in nuclei from all three sources. For some unknown reason in the brain cells the DNA fragment which contains the gene coding for myosin light chain 2 seemed to be mildly DNAase I sensitive, although myosin light chain 2 mRNA was undetectable in these cells (Katcoff et al., 1980).

Thus, in muscle precursor cells, the potentiality of tissue-specific genes to be expressed is not reflected in their DNAase I sensitivity. Changes in DNAase I sensitivity of these genes, associated with terminal differentiation, indicate a qualitative change in their transcriptional activity. This suggests that the accumulation of muscle-specific mRNA, which takes place during terminal dif-

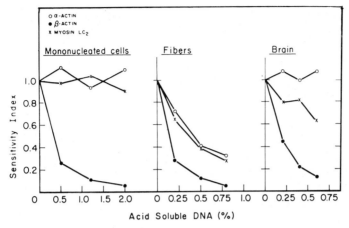

Figure 2. DNAase I sensitivity of the α-actin, β-actin and myosin light chain 2 genes in differentiating muscle cells. DNA extracted from DNAase I treated nuclei was digested with EcoRI and run on a 1% agarose gel, blotted on nitrocellulose paper, and hybridized with radiolabeled probes specific for α-actin gene, β-actin gene, myosin light chain 2 and the immunoglobulin gene. The blots were exposed to X-ray film and the autoradiograms were scanned in a Gilford spectrophotometer. The DNAase I sensitivity index was determined by dividing the intensity of the bands corresponding to fragments of the α-actin, β-actin and myosin light chain 2 genes in each track by the value of the immunoglobulin gene band in the same track. The values are normalized; the ratio obtained for DNA from untreated nuclei equals 1 (from Carmon et al., manuscript in preparation).

ferentiation, indicate a qualitative change in their transcriptional activity. This suggests that the accumulation of muscle-specific mRNA, which takes place during terminal differentiation, is primarily the result of a change in transcription rather than a change in the processing or stability of these mRNAs.

In proliferating myogenic cells, genes which are programmed to be expressed during terminal differentiation are indistinguishable in their DNAase I sensitivity from genes which are not expressed in this cell lineage. This indicates that the stable maintenance of the myogenic cell line latent program of gene expression is not associated with chromatin modifications reflected in preferential DNAase sensitivity of muscle-specific genes. However, this does not rule out the possibility that other differences between the genes, not manifested in the DNAase I sensitivity, do exist. It is, of course, also possible that other genes which control the activation of the set of genes involved in the phenotypic expression of myogenesis do become DNAase I sensitive much earlier in development, and are responsible for the maintenance of the determined state.

ACKNOWLEDGEMENTS

The authors thank Ms. Z. Levy, Ms. S Neuman and Ms. O. Saxel for technical assistance. Supported by grant NIH # R01 GM 22767 and by the Muscular Dystrophy Association, Inc., New York. U.N. holds an A. and E. Blum Career Development Chair for Cancer Research. H.C. holds a post-doctoral fellowship from the Muscular Dystrophy Association.

REFERENCES

Blake, C. C. F., 1981, Exons and the stucture, function and evolution of haemoglobin, Nature, 291:616.
Carmon, Y., Neuman, S., and Yaffe, D., 1978, Synthesis of tropomyosin in myogenic cultures and in RNA-directed cell-free systems: Qualitative changes in the polypeptides, Cell, 14:401.
Coleman, J. R., and Coleman, A. N., 1968, Muscle differentiation and macromolecular synthesis, J. Cell Physiol., 72 (Suppl. 1):19.
Dym, H., and Yaffe, D., 1979, Expression of creatine kinase isoenzymes in myogenic cell lines, Develop. Biol., 68:592.
Gallwitz, D., and Sures, I., 1980, Structure of a split yeast gene: Complete nucleotide sequence of the actin gene in Saccharomyces cerevisiae, Proc. Natl. Acad. Sci. U.S.A., 77:2546.
Gilbert, W., 1978, Why genes in pieces? Nature, 271:501.
Firtel, R. A., Timm, R., Kimmel, A. R., and McKeown, M., 1979, Unusual nucleotide sequences at the 5' end of actin genes in Dictyostelium discoideum, Proc. Natl. Acad. Sci. U.S.A., 76:6206.
Fyrberg, E. A., Bond, B. J., Hershel, D. N., Mixter, K. S., and Davidson, N., 1981, The actin genes of Drosophila: Protein coding regions are highly conserved but intron positions are not, Cell, 24:107.
Groudine, M., and Weintraub, H., 1981, Activation of globin genes during chicken development, Cell, 24:393.
Katcoff, D., Nudel, U., Zevin-Sonkin, D., Carmon, Y., Shani, M., Lehrach, H., Frischauf, A. M., and Yaffe, D., 1980, Construction of recombinant plasmids containing rat muscle actin and myosin light chain DNA sequences, Proc. Natl. Acad. Sci. U.S.A., 77:960.
Mathis, D., Oudet, P., and Chambon, P., 1980, Stucture of transcribing chromatin, Proc. Nucleic Acids Res. Mol. Biol., 24:1.
Ng, R., and Abelson, J., 1980, Isolation and sequence of the gene for actin in Saccharomyces cervisiae, Proc. Natl. Acad. Sci. U.S.A., 77:3912.
Nudel, Y., Katcoff, D., Carmon, Y., Zevin-Sonkin, D., Levi, Z., Shaul, Y., Shani, M., and Yaffe, D., 1980, Identification of

recombinant phages containing sequences from different rat myosin heavy chain genes, Nucleic Acids Res., 8:2133.

Nudel, U., Katcoff, D., Zakut, R., Shani, M., Carmon, Y., Finer, M., Czosnek, H., Ginsburg, I., and Yaffe, D., 1982, Isolation and characterization of rat skeletal muscle and cytoplasmic actin genes, Proc. Natl. Acad. Sci. U.S.A., in press.

Patterson, B. M., and Strohman, R. C., 1972, Myosin synthesis in culture of differentiating chicken embryo skeletal muscle, Develop. Biol., 29:113.

Scheller, R. H., McAllister, L. B., Crain, W. R., Durica, D. S., Posakony, J. W., Thomas, T. L., Britten, R. J., and Davidson, D. H., 1981, Organization and expression of multiple actin genes in the sea urchin, Mol. Cell. Biol., 1:609.

Schuler, A. M., and Keller, E. B., 1981, The chromosomal arrangement of two linked actin genes in the sea urchin S. purpuratus, Nucleic Acids Res., 9:591.

Shainberg, A., Yagil, G., and Yaffe, D., 1971, Alterations of enzymatic activities during muscle differentiation in vitro, Develop. Biol., 25:1.

Shani, M., Nudel, Y., Zevin-Sonkin, D., Zakut, R., Givol, D., Katcoff, D., Carmon, Y., Reiter, J., Frischauf, A. M., and Yaffe, D., 1981a, Skeletal muscle actin mRNA, Characterization of the 3' untranslated region, Nucleic Acids Res., 9:579.

Shani, M., Zevin-Sonkin, D., Saxel, O., Carmon, Y., Katcoff, D., Nudel, Y., and Yaffe, D., 1981b, The correlation between the synthesis of skeletal muscle actin, myosin heavy chain, myosin light chain and the accumulation of corresponding mRNA sequences during myogenesis, Develop. Biol., 86:483.

Stalder, J., Groudine, M., Dodgson, J.B., Engle, J. D., and Weintraub, H., 1980, Hemoglobin switching in chickens, Cell, 19:973.

Vanderkerckhove, J., and Weber, K., 1979, The complete amino acid sequence of actins from bovine aorta, bovine heart, bovine fast skeletal muscle and rabbit slow skeletal muscle, Differentiation, 14:123.

Yablonka, Z., and Yaffe, D., 1977, Synthesis of myosin light chains and accumulation of translatable mRNA coding for light chain-like polypeptides in differentiating muscle cultures, Differentiation, 8:133.

Yaffe, D., 1969, Cellular aspects of muscle differentiation in vitro, Curr. Top. Dev. Biol., 4:37.

Yaffe, D., and Dym, H., 1972, Gene expression during the differentiation of contractile muscle fibers, Cold Spr. Harb. Symp. quant. Biol., 37:543.

Yaffe, D., and Saxel, O., 1977, A myogenic cell line with altered serum requirements for differentiation, Differentiation, 7:159.

SUMMARY OF DISCUSSION ON THE MOLECULAR BASIS

OF DIFFERENTIATION AND COMPETENCE

In reply to a question from Grosveld about whether any of his transformed cells lacked the capacity for induction of globin synthesis, Paul replied that not all had been tested, but some did not appear to be inducible, though the reasons could be technical or some feature of the cells themselves. The copy numbers of the recombinants were of the order of 10 to 20 and varied from one transformant to another. It would be difficult to establish whether the transformants were integrated. There may be some integration but the data was not clear either way. Birnie also commented on the systems of Grosveld and of Paul in which cloned genes are introduced into mammalian cells as naked DNA. To what extent did this naked DNA become incorporated into the cell's nucleus, integrated into the chromosomes, and associated with chromosomal proteins? If the genes remained as naked DNA in the recipient cells it was questionable to what extent the effect detected in vitro corresponded to behaviour in vivo. Paul acknowledged the validity of the scepticism. The response obtained from the transformed cells was similar to the response of the genomic globin gene. A specific question that could be asked was whether the factors that induce an increase in the accumulation of mouse globin mRNA also induce an increase in human globin mRNA. Grosveld stated that there was evidence that the location at which a gene was integrated into the chromosome could determine its level of expression. Buckingham asked what were the possibilties of accumulating populations of cells in the human red blood cell lineage for transformation experiments. Had experiments concerned with the switching of the β-globin genes, rather than with quantitative changes in the expression of a β-gene been attempted yet in the mouse or human cell systems? Paul emphasised the difficulties of obtaining quantities of pure early human erythroid precursors. The human cell line K562, which behaves like a Friend cell, is available, and is derived from a granulocytic leukaemia, so it is very atypical. In using mouse cells containing the human globin gene it is possible to distinguish the gene products and there is good reason, from studies of interspecific cell hybrids, to believe that the controls of gene regulation are not species-specific. Schibler, in a question to Grosveld, pointed out that when a non-functional mRNA was produced, as by a frame-shift mutation, there would be no feed-back to the synthesising system from the accumulated product. Did the non-functional mRNA accumulate to about the same level as normal RNA? Grosveld replied that in his HeLa system it did, the only noticable difference being an accumulation of more precursor before transport out of the nucleus

Following Radbruch's paper on class-switching in cells producing immunoglobulins, there was an extensive discussion of the extent to which the findings from the immune system might be extrapolated to other cells, or whether the extent of class switching and gene deletion in this system was a unique property associated specifically with immunoglobulin production. Clayton suggested that the large number of genomic components which were rather similar to each other permitted chromosomal rearrangements which otherwise might not be possible. Radbruch suggested that selection of gene expression by switching might subsequently be made permanent by the gene deletion step. The building up of rather similar transcriptional units of the genome by gene duplication may make deletion more likely. It might be that gene rearrangement had not yet been found in other systems through the lack of intensive investigation. However, the point was made by Paul that the globin gene family had been very thoroughly investigated and showed no sign of deletions during differentiation.

Sueoka questioned Harrison about the significance of his term 'undetectable'. Harrison replied that this meant less than 10^{-4}% of the poly(A)$^+$RNA. It would not exclude one or two molecules per cell, so that such genes could be expressed at what Sueoka had called the 'rare class level'.

Sueoka asked Mirsky whether any of the surface antigens on the cells that she had studied had been found on both neuronal and glial cells. Mirsky replied that she had not studied any with this distribution but that Cohen and Bender had found that the A4 antibody of the central nervous system was found on neurones and on the surface of some astrocytes. In cultures of optic nerves there were astrocytes which react with tetanus toxin and for glial fibrilary acidic protein which is an astrocyte marker, and which also possess the A4 antigen McDevitt noted that Mirsky's 38D7 antibody seemed to react predominantly with cells of neural crest origin, including adrenal medulla. Did this have any special significance for differentiation? Mirsky believed that the nodose ganglion was of neural crest origin, but was not willing to generalise from her data. Balazs asked whether the distribution of antigens described by Mirsky correlated with other characteristics indicating a specific subpopulation of nerve cells. Mirsky replied that though they wished to study neurotransmitters as well as antigens, this had not yet been done. There is a subpopulation of the ganglia cells that had been studied.

Paul asked Trisler about the duration of his cultures of retinal cells and the fluctuations of antigens during that time. Trisler replied that the cells would live for 4 weeks in culture and that there

SUMMARY OF DISCUSSION

was a disturbance of the antigens during the first two days after which they returned to normal. He believed that virtually every cell in the retina was carrying a slightly different quantitative signal. Mirsky asked Trisler if there was any evidence of gradients in the nerves of the optic tectum that might be related to the antigenic gradients in the retina cell, but was told that such experiments had not yet been done. Clayton asked whether there was any reason to believe in a probability gradient of forming connections with a given region of tectum which corresponds to the gradient across the retina. If one particular retina cell had a particular quantity of the glycoprotein, did this in fact relate to its probability of joining up with a particular tectum cell. There might well be a stochastic element in the establishment of such connections. Trisler replied that he had no evidence of his own but believed that it was likely that the gradient might be responsible for ensuring that connecting cells were in the right area, but that there may be some other form of fine tuning.

Owen commented that there were many instances of glycoprotein molecules, detected by monoclonal antibodies, which were to be found with spatial gradients. Examples were to be found among the cells of the lymphoid system. In many instances the functions of the molecules were unknown, but frequently gradients could arise because of a gradient of maturation of the cells with a consequent gradient in cell surface markers. Was it not possible that in the retina, the gradient was also due to varying degrees of maturation? Paul pointed out that the retina cells retained their gradient in culture even where there was proliferation. Clayton doubted whether the question of maturity could arise in the retina cells. The antigenic concentration was stable: even if the cells were trypsinised they returned to their characteristic levels of the antigen. Harrison commented on the high degree of genetic programming which appeared to exist in the retina cells. Iscove also questioned the biological significance of the gradient of antigens shown in the retina by Trisler. What functions of the retina were correlated with a dorso-ventral gradient? Trisler replied that there were differences in the visual field, in the colour of oil droplets and in the lineages of the cells. They had chosen to study retinas at 14 days of development because this was around the time of synapse formation which begins at around day 13 and reaches the maximum number by day 18. If there are molecules that are used for coding or directing appropriate synapse formation this would be the time at which they would be present. Chader commented that the gradient may not only be significant for normal development but also for pathological development. Most retinal degenerative diseases, including Retinitis Pigmentosa show specific or preferential regional patterns of degeneration that must in some way reflect the underlying biochemical lesion.

Balazs asked Pessac whether they had followed the effect of the viral transformation on other neuronal properties besides GAD activity. Since the indication is that GABA-ergic specification of certain cells has been promoted it may be worth studying using autoradiography to find whether or not parallel changes occur in GABA uptake. This is a fairly specific indication of GABA-ergic neurones. Pessac replied that virus infection did not appear to change these properties. There was a possibility that the virus had not entered the cell, or that it had no effect. Mirsky asked Pessac whether GAD activity might be found in glial cells as well as neurones. Pessac replied that, at least as far as birds were concerned GAD activity was confined to the neurones and absent from Müller cells. This was confirmed by Moscona. Balazs commented that a special feature of Pessac's results was that they showed that a specific sub-population of nerve cells could be triggered into differentiation.

With regard to muscle cell differentiation, Rifkind asked Yaffe what significance he attributed to the single cell division which follows the change of medium which induces fusion of myoblasts. Was it possible that, in a fashion analogous to the situation with murine erythroleukemia cells, there might be events, related to the cell cycle, which must occur in the presence of 'inductive' medium if differentiation is to occur. Yaffe replied that theoretically the division may not be vital. There was evidence that cells were most sensitive to signals from the medium during the later part of the S phase of the cycle. Even cells which have been in medium which normally stimulates cell fusion can be manipulated so that they will go back to a proliferative phase, so that it appears that if the stimulation is associated with replication it is not an irreversible process. Recent evidence suggests that the position of the gene in the chromatin is important in determining whether or not it is active.

INTRODUCTORY REVIEW:

TUMOURS, TRANSPOSITION AND TRANSDIFFERENTIATION

A. H. Wyllie, R. M. Clayton*, and D. E. S. Truman*

Department of Pathology
University of Edinburgh
Edinburgh   EH8 9AG

*Department of Genetics
University of Edinburgh
Edinburgh   EH9 3JN

Is the transition from normality to malignant neoplasia a process analogous to normal differentiation? Are tumour cells subject to restrictions on gene expression similar to those pertaining in the tissue from which they arose? The contributions in this section do not provide solutions to these questions, but they may provoke some ideas on the relationship between neoplasia and differentiation, at a time when our understanding of both these processes is increasing rapidly. This brief review outlines some of the deficiencies inherent in conventional views of the nature of tumours, summarises recent evidence that neoplasia represents altered expression of normal genes, gives an account of observations suggesting that the patterns of gene expression accessible to human tumours in different sites are both restricted and predictable, and finally compares the adoption of the neoplastic phenotype with the phenomenon of transdifferentiation, which is discussed in the subsequent articles.

THE NATURE OF TUMOURS: SOME CONVENTIONAL VIEWS AND THEIR INHERENT PROBLEMS

Malignant tumours are recognised by their abnormal behaviour. Their growth is uncoordinated with that of the surrounding tissue in which they arose, and - with few exceptions - they are capable of shedding cells through the blood stream and lymphatics, or across tissue spaces, to form secondary tumour masses (metastases). Frequently this abnormal behaviour is accompanied by altered cytological

characteristics, loss of the normal orientation of cell to cell, loss
of features - both morphological and biochemical - considered typical
of normal differentiation in the tissue of origin, and the acquisition
of types of differentiation inappropriate to that tissue. It must
be emphasised that these new, acquired features are commonly observed.
For example squamous differentiation is a feature of about half of
all bronchial carcinomas of man (the commonest cancer in the male)
although the normal bronchial epithelium is columnar and mucous-
secreting. Similarly a high proportion of tumours of the human ovary
consist of mucin-secreting cells, a type not normally observed there.
Sometimes the inappropriate differentiation represents expression of
genes normally active at an earlier phase of the development of the
tissue in which the tumour arises, such as the secretion of alpha
fetoprotein by certain hepatic carcinomas. Other examples of inappropriate differentiation are more bizarre, such as the secretion of
peptide hormones by small-cell carcinoma of the human bronchus.

In an early attempt to rationalize the abnormal differentiation
seen in tumours, Cohnheim (1889) suggested that tumours arose from
residual islands of embryonic cells persisting amongst the committed
and differentiated cells of the normal tissue. There is no evidence
that such "embryonic rests" exist, but two more modern concepts retain
the view that tumours may arise from a distinct subset of cells within
the tissue. Firstly, the observation that most cells in normal tissues are committed to terminal differentiation and death, has led to
the widely held view that tumours arise from the small subpopulation
of stem cells (Lajtha, 1981). Secondly, the linked expression of certain "inappropriate" differentiated features in some tumours led to
the theory that tumours of this particular class arise from cells of
a common lineage, different from that of the majority in the tissue.
The lineage in question is that of the widely dispersed APUD cells
which are recognised histochemically inter alia by their capacity for
amine precursor uptake and decarboxylation. These cells are apparently involved in neuro-endocrine activity and thought to be of neural
crest derivation (Pearse, 1969). Both concepts require that pre-
existing normal cells, of restricted developmental potential, acquire
neoplastic behaviour whilst retaining some of their developmental restriction. The question thus arises as to the nature of carcinogenesis - the change permitting acquisition of this new behaviour.

A formidable array of circumstantial evidence suggests that carcinogenesis requires genetic mutation. It is pointed out, for example, that - like mutations - carcinogenesis involves an alteration
in cell characteristics which is maintained through cell replication
and is usually irreversible. The active forms of many potent chemical
carcinogens react with and alter DNA, either directly, as a result
of base mispairing, or secondarily through error-prone repair (Lawley,
1980). Few chemical carcinogens are not mutagens, so that it is prac-
ticable to screen unknown compounds for carcinogenicity on the basis
of their mutagenicity (McCann et al., 1975). Human patients with xer-

oderma pigmetosum, a disease in which deficiency in excision repair leads to persistence of DNA base modifications, is associated with a high incidence of carcinoma of the skin, where such modifications are produced through exposure to the ultraviolet radiation in sunlight.

Nonetheless the relationship between DNA modification, mutation, and exposure to chemical carcinogens is an association only: agents capable of modifying DNA to cause mutations may also cause other classes of change, the latter alone being responsible for carcinogenesis. Cairns (1981) has argued that patients with xeroderma pigmentosum might be expected to have an increased incidence of carcinoma of the respiratory and gastrointestinal tracts, through exposure to environmental carcinogens, since the DNA repair defect is present in all their cells. However, it appears that the incidence of such tumours is no different from that in normal persons. Further, formal comparison of the effectiveness of one carcinogen in evoking either transformation (an *in vitro* model of carcinogensis) or mutation of a single copy gene in the same cell line, has revealed that the size of the DNA target affected in transformation is many fold larger than that involved in the mutation (Huberman, Mager and Sachs, 1976). Finally, there are two features of gene expression in tumours which a simple mutational model of carcinogenesis does not address. First, as discussed below, tumour phenotype is subject to limitations determined by the tissue of origin. Secondly, tumour phenotypes tend to undergo sequential changes, so that established tumours eventually accumulate wide deviations from normality in fundamental properties such as karyotype. This phenomenon, known as tumour progression, is secondary to the initiation of neoplasia (Foulds, 1954; Farber and Cameron, 1980). With few exceptions, the abnormal karyotypes associated with tumours of the same site and histological type are heterogeneous, and they may change with the passage of time (Rowley, 1978).

Neoplasia and the Expression of Normal Genes

Studies with oncogenic viruses have demonstrated that *in vitro* transformation can be attributed to the activity of specific genes. Sometimes transformation requires the activity of a viral gene such as the A gene of SV40 (a DNA virus), and the *src* gene of the DNA copy of the RNA-containing retroviruses (see Sharp, 1980 for review). In other examples the critical agent appears to be a sequence repeated at the ends of the integrated retroviral DNA, which initiates an open reading frame extending into adjacent host sequences (Temin, 1982). In these examples the transforming DNA is foreign to the host cell genome, but there is good evidence that transformation arises through expression of host cell functions. Thus, maintenance of the transformed state induced by SV40 is associated with synthesis of coded protein of molecular weight 55Kd. A similar if not identical protein appears in cells in which transformation arose spontaneously or was induced by other DNA and RNA tumour viruses, or chemical carcinogens. Further evidence for activation of pre-existing host genes in carcinogenesis derives from studies on retroviruses recovered from tumours which were produced by infecting animals with mutant viruses deficient in the src gene. The defective viruses acquired a functional src gene from endogenous sequences in the host (Karess and Hanafusa, 1981). Hybridization studies show that genes homologous to the trans forming genes of retroviruses are widespread in normal cells. Great interest has centred recently on the observation that the molecular architecture of retroviruses is closely similar to that of transposable elements ("transposons"): a single copy sequence placed between two long repeated sequences, flanked by short repeated sequences (Finnegan, 1981). It now appears likely that integrated retrovirus and endogenous retrovirus-like genomes are similar to transposons in function also.

The implications of this to our understanding of neoplasia are profound. Previously unexpressed DNA, including endogenous retroviral genomes, on transposition to a new site might initiate new patterns of gene expression some of them including neoplastic transformation. There is now direct evidence in support of these suggestions in human neoplasia. Endogenous retroviral genomes are expressed apparently selectively, in human tumours (Eva et al., 1982). Human genomes include small circular DNA molecules whose sequences are repeated in sites which vary from tissue to tissue, suggesting that they are indeed mobile, transposable elements (Calabretta et al., 1982). One of the characteristic nucleotide sequences in these supposedly mobile elements is a repeated sequence recognised by the restriction enzyme Alu I. Fragments rich in Alu sequences from human tumour DNA cause transformation of cells from other species in vitro (Murray et al., 1981). We do not yet know whether the circular DNA molecules also include endogenous retroviral sequences.

A further intriguing suggestion derives from studies of replication-defective retroviruses. In these, there is a single copy genetic element, homologous to host DNA, inserted 3' to a truncated viral gag gene. These viruses have transforming potential with distinct tissue specificities. For example, avian myeloblastosis virus applied to chick marrow cultures leads to proliferation of myeloblasts, whereas another defective virus, that of avian myelocytomatosis, applied to the same cultures leads to proliferation of macrophage-like cells. Both viruses infect precursor cells and their expression in transformation appears to be linked to the particular state of differentiation of the host cell (Boettinger and Durban, 1980).

Transformation which occurs only in cells at certain stages in their differentiation is also observed in mouse embryonal carcinoma cells infected with SV40. These cells are the stem cells of murine teratoma and bear strong similarities to certain cells of the post-implantation embryo, including pluripotency in differentiation. They

can be induced to differentiate by various stimuli. Integration of
SV40 DNA into embryonal carcinoma cells is not followed by expression
of T antigen, but the antigen is expressed in differentiated cells
(Linnenbach, Huebner and Croce, 1980). In this instance, the regulation in gene expression does not appear to be a simple on-off control
of transcription, in that T antigen message appears in both cell
types.

The mouse teratoma system presents a further demonstration that
expression of a malignant phenotype may be directed by normal genes.
The pluripotent embryonal carcinoma cells, if inserted into a blastocyst, can differentiate to the complete range of normal tissues (Mintz
and Illmensee, 1975). Conversely cells from the normal inner cell
mass, when injected into an inappropriate environment such as the
renal capsule, give rise to teratomas which grow progressively and
kill their host. It is conventional to explain these findings by postulating that embryonal carcinoma cells and their normal homologues
are programmed to divide until receiving a differentiation stimulus
from the environment of the developing embryo; an abnormal environment
does not provide this stimulus and so permits the cells to proliferate
continuously (Martin, 1980). This explanation is, of course, quite
compatible with the hypothesis that the programme for proliferation
includes an endogenous oncogene or transposon whose expression is disallowed by the genomic changes involved in differentiation.

Transposons and mutagenesis may complement each other in carcinogenesis: the existence of a mutation may promote the expression of
movement of pre-existing transposons. Unlike simple mutational models
of carcinogenesis, however, transposons allow explanation of both progression and restricted gene expression in tumours. Thus, tumour cell
populations are envisaged as being selected for the presence of a
mobile transposon. Precedent from other eukaryotic systems suggests
that mobile elements are liable to continue to insert new copies of
themselves at various points in the genome (Roeder et al., 1980),
hence providing a molecular basis for tumour progression. Further,
as discussed above, the transposon is considered to initiate transcription of linked groups of normal host genes, but may be active
only in certain states of differentiation of the cell. This provides
a basis for the restricted patterns of gene expression to be discussed
in more detail below.

Gene Expression in Tumours is Predictably Restricted

The phenotype of tumour cells from patients with lymphocytic leukaemia and lymphoma can be delineated with precision by a battery of
methods including probes for surface receptors for complement and immunoglobulin, histochemical identification of marker enzymes such as
terminal deoxynucleotide transferase, detection of the synthesis and
secretion of immunoglobulins of different types, and characterization

by monoclonal antibodies specific for various phases in lymphocyte
maturation. Examination of such tumours, arising "spontaneously" in
a large number of patients, has revealed striking limitation in the
patterns of phenotypic expression (Habeshaw et al., 1979; Reinherz
and Schlossman, 1981). The constellations of features observed are
those which occur together at some stage in the development and maturation of normal lymphocytes; mixtures of features from different
developmental stages, or from both of the principal (B and T) lineages
are not found. Thus in this tumour family, in which more detailed
information on phenotype is available than in any other, the patterns
of gene expression are shown to be restricted to entirely predictable
combinations.

In most other observations on human tumours, differentiation is
recognised by morphological features alone, but similar conclusions
pertain. The only tumours in which one is liable to find features
characteristic of almost all tissues of the body are the teratomas,
which arise from germ cells. Even here 'hybrid' cells expressing
features of more than one normal cell type simultaneously do not
appear to exist, although it is possible that morphological methods
are not adequate to detect them. Tumours of embryonic tissue, detected early in infancy, sometimes include cell types inappropriate
to the organ of origin. Thus nephroblastomas (tumours of the developing kidney) frequently contain striated muscle, cartilage, bone and
adipose tissue, in addition to epithelial elements forming recognisable tubular and glomeruloid structures, but it is argued that other
types of differentiation (such as neural tissue or respiratory and
keratinising epithelium) are never observed (Willis, 1967). Similarly
hepatoblastomas, tumours of the developing liver, commonly include
mesodermal tissues such as bone and cartilage, but ectodermal elements
are not found.

Epithelial tumours of the ovary may adopt a range of cytological
characteristics, which appear to mimic the cell types found in other
Mullerian derivatives. Thus the predominant cell type in some tumours
is mucin-secreting, similar to that of the endocervix, whilst in
others is reveals characteristics of Fallopian tube, and endometrial
epithelium (Scully, 1977). Other types of epithelial differentiation
in tumours of this class are not observed.

Finally, there is the example of peptide hormone secretion by
various tumours, particularly of the bronchus, which is discussed by
Ratcliffe in these proceedings. Although these peptides are not found
in abundance at any stage in the development of the normal tissue
from which the tumours arise, defined sets of peptides are found in
certain classes of tumours and not in others. Thus some tissue specific phenotypic restriction is present here also.

As discussed earlier, tumour progression often leads to gross
karyotypic abnormalities. These restrictions in tumour phenotype -

many apparently related to the state of differentiation of the cells from which the tumour arose - are thus the more striking and deserving of explanation.

THE NEOPLASTIC PHENOTYPE AND TRANSDIFFERENTIATION

"Transdifferentiation" is the term given to conversions from one distinct cell type to another developmentally unrelated type. This was first observed to take place in vivo in Wolffian regeneration of the lens from the dorsal iris (Colluci, 1891; Wolff, 1895), and there have been very numerous elegant experimental approaches to the process, but the problem was opened up when Eguchi and Okada (1973) first demonstrated the phenomenon under defined cell culture conditions, when they obtained transdifferentiation of the pigment epithelium of retina into lens. Transdifferentiation of neural and pigmented retina and iris to lens are not the only types of change possible, but remain the best studied examples, and in all these cases transdifferentiation occurs when certain apparently essential conditions are met. First, the gene products which at high levels characterise the final cell type, whether produced normally, or by transdifferentiation, are also found at low levels in the tissues which are capable of undergoing transdifferentiation (reviewed Clayton, 1982). Secondly, cell contacts must be disrupted. Following this, certain other conditions affect the selection of the subsequent pathway of transdifferentiation taken. In the case of transdifferentiation to lens, new extensive cell-cell contacts are amongst the requirements.

In theory the appearance of cells synthesising gene products inappropriate to the tissue of origin at high levels might be due to the expansion of clones of cells of a different lineage from the majority in the tissue, to differentiation of stem cells following an intrinsic but unexpressed programme, or to reprogramming of gene expression in cells of the original lineage. There is good evidence that transdifferentiation in the lens system is not due to expansion of clones of cells of a different origin: there is no migration of lens cells at any stage of development to retina, or to adenohypophysis (which can also transdifferentiate to lens as discussed in these proceedings by several authors), nor to limb buds, some cells of which can be induced to synthesise crystallins (Kodama and Eguchi, 1982). Direct evidence for the reprogramming of individual differentiated cells was first shown by Eguchi and Okada (1973) and further evidence is reviewed by Okada and by Eguchi in these proceedings. However, it is also clear from clonal culture that retinal cells are heterogeneous in transdifferentiation potential since the percentage of clones which transdifferentiate is governed by species, age, tissue, and the conditions of cell culture. Although this cellular heterogeneity in the starting population suggests that cell selection may be important, the results of various manipulations of culture conditions suggest that the number of cells which could potentially transdifferentiate may be larger than the number which do so in a given situation. This unexpressed potential may depend on the molecular constitution of a cell, its opportunities for receiving necessary signals, or both together. It appears that new and increased cell-cell contacts are required for transdifferentiation to lens, and this implies that the likelihood of making such contacts may be limited by the number of cells in the population with transdifferentiation potential (presumably those already expressing certain levels of crystallin mRNA). This possibility is supported by Eguchi's finding that those single clones from iris cells which transdifferentiate do so simultaneously, and far more rapidly than they do in mass culture: conditions which facilitate transdifferentiation are described by several authors in this publication.

Several aspects of transdifferentiation are relevant to discussion of the nature of gene expression in tumours. First both transdifferentiation and neoplasia involve the expression of normal endogenous genes. A corollary is that transdifferentiation might also be responsible for reversion from neoplastic to normal phenotype, and this has been observed: retinoblastoma cells can be induced to transdifferentiate to normal lens (Okada, 1978).

Secondly, in both transdifferentiation and neoplasia the new combinations of expressed genes are not random, but represent linked activities associated with particular phenotypes. It is implausible that conventional single base mutations, even affecting regulatory elements, could be responsible for the highly organised and restricted patterns of gene expression observed in transdifferentiation, a point already discussed for neoplasia. Similarly the random insertion of mobile sequences is unlikely to be the basis of such organised patterns of gene expression (and even less likely to explain the relationship between the age of the embryo and the proportions of crystallins in the transdifferentiated lens cells). Transdifferentiation affords a means whereby new but restricted gene combinations may occur in tissues (including tumours) without the need to postulate the presence of stem cells of a minority lineage. We have already discussed theories of tumour origin which require such postulates, notably the theory linking peptide hormone secreting tumours with the APUD cell lineage (Pearse, 1969). It is clear that histochemistry alone does not denote the embryological origin of a cell, since lens cells of normal and transdifferentiation origin cannot be distinguished on the basis of their crystallin content or ultrastructure. Hence the similar histochemistry and ultrastructure of cells of APUD type does not necessarily imply a shared lineage. Indeed, recent experiments with quail-chick chimaeras have shown that at least some of the sites in which APUD cells are found do not receive migrating neural crest cells, and there is evidence that in some tissues APUD cells originate in situ (reviewed by Baylin and Mendelsohn, 1980).

A third point of similarity between some instances of neoplasms synthesising ectopic products, and some types of transdifferentiation is that the modified cells with superabundant products arise from precursor cell populations which normally express a low abundance of mRNA for the product which later comes to predominate. Examples of this include normal adult testis and gut containing low levels of HCG, which is produced at high levels by some tumours arising at these sites. Similarly, the synthesis of α-fetoprotein in normal and neoplastic liver and carcino-embryonic antigen in normal and neoplastic intestine shows the same pattern. The increase in abundance of these compounds may also be observed in other non-neoplastic states of increased cell renewal in these tissues, including regeneration and inflamation (reviewed by Baylin and Mendelsohn, 1980).

Finally, there is increasing evidence that certain carcinogens can cause transdifferentiation and other discrete shifts in phenotypic expression. Thus carcinogens have been found to induce supernumerary limb formation (Breedis, 1952; Tsonis and Eguchi, 1981), and lens formation from ventral iris (Eguchi and Watanabe, 1973). Effects *in vitro* include the differentiation of neurites from neuroblastoma cells (Yoda and Fumimura, 1979), alteration in gastric epithelial cell types (Fukamuchi and Takoyama, 1979) and several transdifferentiation-like events and specific effects on transdifferentiation pathways, in lens and neural retinal cells (Clayton et al., 1980; Clayton and Patek, these proceedings).

In conclusion, transdifferentiation and the anomalous gene expression of neoplastic tissues appear to share certain important features. Although we do not yet know how transdifferentiation is initiated, factors which influence its probability and outcome are becoming defined, such as the regulation of the abundance of particular mRNA species, the significance of cell-cell relationships, the rate of cell division and various metabolic conditions. It is a hopeful prospect that studies such as those in the following contributions will throw some further insights on the problems of neoplasia as described by Ratcliffe in this section and by Woodruff elsewhere in these proceedings.

REFERENCES

Baylin, S. B., and Mendelsohn, G., 1980, Ectopic (inappropriate) hormone production by tumours: mechanisms involved and the biological and clinical implications, Endocr. Rev., 1:45.

Boettinger, D., and Durban, E. M., 1980, Progenitor-cell populations can be infected by RNA tumor viruses, but transformation is dependent on the expression of specific differentiated function, Cold Spring Harb. Symp. quant. Biol., 44:1249.

Breedis, C., 1952, Induction of accessory limbs and of sarcoma in the newt (*Triturus viridescens*), with carcinogenic substances, Cancer Res., 12:861.

Cairns, J., 1981, The origin of human cancers, Nature, Lond., 289:353.
Calabretta, B., Robberson, D. L., Barrera-Saldana, H. A., Lambrou, T. P., and Saunders, G. F., 1982, Genome instability in a region of human DNA enriched in Alu repeat sequences, Nature, Lond., 296:219.
Clayton, R. M., 1982, Cellular and molecular aspects of differentiation and transdifferentiation of ocular tissues in vitro, in: "Differentiation in Vitro," M. Yeoman and D. E. S. Truman, eds., Cambridge University Press, p.83.
Clayton, R. M., Bower, D. J., Clayton, P. R., Patek, C. E., Randall, F. E., Sime, C., Wainwright, N. R., and Zehir, A., 1980, Cell culture in the investigation of normal and abnormal differentiation of eye tissues, in:"Tissue Culture Methods in Medicine," R. Richards and K. Rajan, eds., Pergamon Press, Oxford, p.185.
Clayton, R. M., and Patek, C. E., 1982, Apparent redifferentiation of chicken lens epithelium by N-methyl-N'-nitro-N-nitrosoguanidine in vitro, these proceedings.
Cohnheim, J., 1889, Lectures on General Pathology (translated by A. B. McKee), New Sydenham Society, London, p.764.
Colluci, V., 1891, Sulla rigenerazione parziale dell'acchio nei Tritoni. Istogenesi e sviluppo, Mem. roy. Accad. Sci. Ist. Bologna, Ser. 5, T. 1:593.
Eguchi, G., and Okada, T. S., 1973, Differentiation of lens tissue from the progeny of chick retinal pigment cells cultured in vitro: a demonstration of a switch of cell types in clonal cell culture, Proc. Natl. Acad. Sci. U.S.A., 70:1495.
Eguchi, G., and Watanabe, K., 1973, Elicitation of lens formation from the 'ventral iris' epithelium of the newt by a carcinogen, N-methyl-N'-nitro-N-nitrosoguanidine, J. Embryol. exp. Morph., 30:63.
Eva, A., Robbins, K. C., Andersen, P. R., Srinivasan, A., Tronick, S. R., Reddy, E. P., Ellmore, N. W., Galen, A., Lautenberger, J. A., Papas, T. S., Westin, E. H., Wong-Staal, F., Gallo, R. C., and Aaronson, S. A., 1982, Cellular genes analogous to retroviral onc genes are transcribed in human tumour cells, Nature, Lond., 295:116.
Farber, E., and Cameron, R., 1980, The sequential analysis of cancer development, Adv. Canc. Res., 31:125.
Finnegan, D. J., 1981, Transposable elements and proviruses, Nature, Lond., 292:800.
Fould, L., 1954, The experimental study of tumor progression: a review, Cancer Res., 14:327.
Fukamuchi, H., and Takoyama, S., 1979, Promotion of epithelial keratinization by N-methyl-N'-nitro-N-nitrosoguanidine in rat forestomach in organ culture, Experientia, 35:666.
Habeshaw, J. A., Catley, P. F., Stansfeld, A. G., and Brearley, R. L., 1979, Surface phenotyping, histology and the nature of non-Hodgkin lymphoma in 157 patients, Br. J. Cancer, 40:11.

Huberman, E., Mager, R., and Sachs, L., 1979, Mutagenesis and transformation of normal cells by chemical carcinogens, Nature, Lond., 264:360.

Karess, R. E., and Hanafusa, H., 1981, Viral and cellular Src genes contribute to the structure of recovered avian sarcoma virus transforming protein, Cell, 24:155.

Kodami, R., and Eguchi, G., 1982, Dissociated limb bud cells of chick embryos can express lens specificity when reaggregated or cultured in vitro, Develop. Biol., 90 (in press).

Lajtha, L. G., 1981, Which are the leukaemic cells? Blood Cells, 7:45.

Lawley, P. D., 1980, DNA as a target of alkylating agents, Br. Med. Bull., 36:19.

Linnenbach, A., Huebner, K., and Croce, C. M., 1980, DNA-transformed murine teratocarcinoma cells: regulation of expression of simian virus 40 tumor antigen in stem versus differentiated cells, Proc. Natl. Acad. Sci. U.S.A., 77:4875.

McCann, J., Choi, E., Yamasaki, E., and Ames, B. N., 1975, Detection of carcinogens as mutagens in the Salmonella microsome test: assay of 300 chemicals, Proc. Natl. Acad. Sci. U.S.A., 72:5135.

Martin, G. R., 1980, Teratocarcinogenesis and mammalian embryogenesis, Science, 209:768.

Mintz, B., and Illmensee, K., 1975, Normal gentically mosaic mice produced from malignant teratocarcinoma cells, Proc. Natl. Acad. Sci. U.S.A., 72:3585.

Murray, M. J., Shild, B.-Z., Shih, C., Cowing, D., Hsu, H. W., and Weinberg, R. A., 1981, Three different human tumor cell lines contain different oncogenes, Cell, 25:355.

Okada, T. S., 1978, Redifferentiation and transdifferentiation of retinoblastoma cells, in:"Proceedings of the VIth International Congress on Eye Research," Osaka, Japan.

Pearse, A. G. E., 1969, The cytochemistry and ultrastructure of polypeptide hormone-producing cells of the APUD series and the embryologic, physiologic and pathologic implications of the concept, J. Histochem. Cytochem., 17:303.

Reinherz, E. L., and Schlossman, S. F., 1981, Derivation of human T-cell leukemias, Cancer Res., 41:4767.

Roeder, G. S., Farabaugh, P. J., Chaleff, D. T., and Fink, G. R., 1980, The origin of gene instability in yeast, Science, 209:1375.

Rowley, J. D., 1978, Chromosomes in leukemia and lymphoma, Sem. Hematol., 15:301.

Scully, R. E., 1977, Ovarian tumours. A review. Amer. J. Pathol., 87:686.

Sharp, P. A., 1980, Molecular biology of viral oncogenesis, Cold Spring Harb. Symp. quant. Biol., 44:1305.

Temin, H. M., 1982, Function of the retrovirus long terminal repeat, Cell, 28:3.

Tsonis, D. A., and Eguchi, G., 1981, Effects of carcinogens on limb regeneration, Differentiation, 20:52.

Willis, R. A., 1967, Pathology of Tumours, 4th ed., Butterworths,

London, p. 944.
Wolff, G., 1895, Entwicklungsphysiologische Studien I. Die Regeneration der Urodelenlinse, Arch. EntwMech. Org., 1:380.
Yoda, K., and Fujimura, S., 1979, Induction of differentiation in cultured mouse meuroblastoma cells by N-methyl-N'-nitro-N-nitrosoguanidine, Biochem. Biophys. Res. Comm., 87:128.

ECTOPIC HORMONES

J. G. Ratcliffe

Department of Biochemistry
Royal Infirmary
Glasgow G4 OSF

INTRODUCTION

The idea that tumours arising in tissues conventionally regarded as 'non-endocrine' (e.g. lung) can synthesise and secrete hormones has only been clearly recognised by clinicians in the last 20 years. When first described, ectopic hormone production was considered to be a bizarre and rare phenomenon, but increased clinical awareness and the application of sophisticated methods of hormone assay have fundamentally changed this perspective (Liddle et al., 1969). Indeed, there is now evidence that ectopic hormone production is a common and even a general feature of neoplasia (Odell et al., 1977). From the clinical point of view ectopic hormones may give rise to severe metabolic derangements (hypokalaemia, hyponatraemia, hyper-calcaemia, hyper- and hypo-glycaemia) which may be more malignant than the tumour itself, and adversely affect prognosis. There is also considerable interest in the clinical application of assays for ectopic hormones as markers of malignancy in diagnosis and monitoring and in the functional classification of tumours.

The underlying pathogenesis of ectopic hormone production is poorly understood. As will be discussed later, this is mainly due to lack of knowledge of normal mechanisms of differentiation, but progress has also been hindered by deficient and inaccurate observational data. The types of validation criteria used for ectopic hormone production are shown in Table 1. The most commonly advanced, but weakest, evidence rests on the association of a tumour with an endocrine syndrome and/or elevated or inappropriate plasma hormone levels. More stringent criteria (Table 1, items 2, 3, 4, 5) are often not feasible clinically. Very rarely has hormone synthesis been demonstrated _in_ _vitro_ or the appropriate mRNA been extracted

Table 1. Criteria for ectopic hormone production.

1. Association of tumour with
   a) endocrine syndrome
   b) elevated or inappropriate plasma hormone levels.

2. Reversal of endocrine abnormality after removal of tumour.

3. Persisting endocrine abnormality after removal of the normal gland of origin of the hormone in question.

4. Arterio-venous hormone gradient across the tumour bed.

5. Demonstration of hormone in tumour tissue in amounts greater than adjacent tissue.

6. Demonstration of in vitro hormone synthesis by tumour cells and/or extraction of appropriate mRNA from tumour.

from the tumour (Table 1, item 6). Failure to satisfy the more stringent criteria has given rise to a literature of uneven quality, uncritical acceptance of which provides a poor basis for theoretical speculation on pathogenesis.

Recent work has also highlighted semantic difficulties with the term ectopic. Hormones previously considered ectopic (e.g. human chorionic gonadotrophin, HCG) may be produced in small amounts outside the normal gland of origin (i.e. placenta) in adult tissues or non-neoplastic proliferative states. Hence it can be questioned whether any hormone is truly ectopic according to the classical definition. It may be more correct to describe ectopic hormone production as 'tumour associated hormone production' and specify the organ of origin, or redefine 'ectopic' to mean a hormone synthesised by a tumour arising from a tissue other than the major gland of origin of that hormone. I shall use the term 'ectopic' in the latter sense.

Useful general reviews on ectopic hormone production are those by Rees and Ratcliffe (1974), Sherwood and Gould (1979), Baylin and Mendelsohn (1980) and Imura (1980).

PREVALENCE AND TUMOUR ASSOCIATIONS

Applying the more stringent criteria listed in Table 1 there is good evidence for ectopic production of most of the recognised peptide and protein hormones, but not catecholamines, steroid or thyroid hormones. It has been suggested that ectopic production of the latter hormones is unlikely as it would require simultaneous expression of several enzymes of specialised biosynthetic pathways. In the few cases where ectopic steroid production has been suggested, it was attributable to conversion of a circulating precursor by a single tumour enzyme (e.g. aromatase or sulphatase).

Accurate figures for the prevalence of ectopic hormone production are not available. An estimate of the relative frequency and tumour associations is given in Table 2. Clinical syndromes attributable to ectopic hormone secretion occur in about 10% of unselected patients with lung cancer, the malignancy for which data is most complete (Azzopardi et al., 1970). However, florid clinical features of endocrine syndromes are often lacking so that clinical recognition requires a high index of suspicion with confirmatory biochemical investigation. Systematic investigation of patients with neoplasms suggests that ectopic hormone production is considerably more common than indicated by clinical evidence alone. For example, impaired suppression of corticosteroids occurs in about half the patients with small cell carcinoma of lung in the absence of clinically apparent ectopic ACTH syndrome. Positive ACTH immunoactivity in tumour extracts and by immunoperoxidase staining is also detected in the majority of small cell and carcinoid tumours of lung (Bloomfield et al., 1977). There are many reasons for the relative insensitivity of clinical methods of recognising ectopic hormone production. Many hormones lack obvious clinical effect in the short term, or lack significant biological activity (e.g. precursor forms of hormones such as pro-ACTH, calcitonin, HCG and subunits). Circulating levels of ectopic hormones may also be inadequate to produce clinical effects, and the effects of one hormone (e.g. ACTH) may mask the clinical effects of another (e.g. antidiuretic hormone, ADH). Conversely, abnormal biochemical data (e.g. serum electrolyte or hormone levels) may be incorrectly attributed to ectopic hormone secretion.

CLASSIFICATION OF ECTOPIC HORMONES

With these pitfalls in interpreting clinical and biochemical data in mind, an attempt can be made to classify ectopic hormones on the basis of their associations and tumour origins (Levine and Metz, 1974).

Group 1

ACTH and related peptides, vasopressin, calcitonin, somatostatin and biogenic amines, (e.g. 5-hydroxytryptamine, histamine) are associated typically with tumour cells having endocrine features (secretory granules, and Amine Precursor Uptake and De-carboxylation (APUD) characteristics). These include small cell carcinoma of lung and other tumours of foregut origin (medullary carcinoma of thyroid, pancreatic islet cell tumours, phaeochromocytoma, ganglioneuroma,

Table 2.

Ectopic hormone	Clinical syndrome	Associated tumours	Relative prevalence	
			Clinical syndrome	Biochemical evidence
?Osteolysins/PTH	Hypercalcaemia without metastases	Lung (epidermoid) Breast, Renal	+++	?++
ACTH/LPH/CRF	Cushing's syndrome	Lung (small cell)	++	+++
AVP	SIADH	Lung (small cell)	++	++
HCG and subunits	Gynaecomastia, Precocious puberty	Lung, Gut, Hepatoblastoma	+	+++
Prolactin	Galactorrhoea	Kidney	rare	?+
NSILA(S)	Hypoglycaemia	Mesenchymal	rare	?
Enteroglucagon	Malabsorption	Renal	rare	?
VIP	Water diarrhoea	Lung	rare	?
Erythropoietin	Erythrocytosis	Uterus	rare	?

ECTOPIC HORMONES

	Acromegaly			
GH/GHRF		Lung	rare	?+
GHRIH	–	Thymus	NA	?
GnRH	–	Thymus	NA	?
Calcitonin	–	Lung, Breast	NA	+++
HPL	–	Lung, Breast	NA	+++

Abbreviations:

ACTH – adenocorticotrophic hormone
AVP – argenine vasopressin
CRF – corticotrophin releasing factor
GH – growth hormone
GHRF – growth hormone releasing factor
GHRIH – growth hormone release inhibiting hormone (somatostatin)
GnRH – gonadotrophin
HCG – human chorionic gonadotrophin
HPL – human placental lactogen
LPH – lipotrophic hormone
NSILA(s) – non-suppressible insulin-like activity soluble in acid ethanol
PTH – parathyroid hormone
SIADH – syndrome of inappropriate antidiuretic hormone
VIP – vasoactive intestinal peptide

NA – not applicable as no clinical syndrome recognised for these hormones

carcinoids of lung and thymus). Some of these hormones are often
produced together by the same tumour though it is not clear whether
each hormone is produced by a separate cell line. Recent evidence
suggests that some tumour cells may elaborate more than one hormone;
thus calcitonin and ACTH may be produced by the same medullary carcinoma cell (Goltzman et al., 1979).

Group 2

The placental proteins (HCG and its subunits, HPL) occur in a
wide variety of tumours particularly of gonadal and gastrointestinal
origin (especially gastric, pancreatic and hepatic) but also lung
(usually non-small cell) and breast. The typical histological pattern
of HCG secreting tumours includes syncytial giant cells or frankly
choriocarcinomatous elements. Although the origin of these cells is
unknown the association of HCG with placental enzymes and foetal
proteins (e.g. CEA and AFP) suggests that these tumours may be derived
from cells at a more primitive stage of differentiation than tumours
of endocrine cell origin and may be related to endoderm.

Group 3

The other ectopic hormones are more difficult to classify.
Ectopic PTH is especially associated with epidermoid carcinoma of
lung, and ectopic erythropoietin and other growth factors (e.g. non-
suppressible insulin-like activity, NSILA) are associated with tumours
of mesodermal origin including retroperitoneal connective tissue
tumours, and tumours of adrenal cortex, kidney and gonads.

Secretory granules and APUD features which are characteristic
of Group 1 tumours are not prominent in tumours of Groups 2 and 3,
though there may be heterogeneous cellular inclusions (liposomes,
lipofuscin granules, filaments and fibrils) suggesting secretory as
well as phagocytic activity. Very few hormone-secreting tumours of
these groups have been well studied ultra-structurally.

THEORETICAL BASIS OF ECTOPIC HORMONE PRODUCTION

Although the pathogenesis of ectopic hormone production remains
speculative, there are several key experimental observations which
must be accomodated by any hypothesis:

1) ectopic hormone production is not random either _in vivo_ or _in
vitro_ but tends to be associated with specific types of tumour
(Azzopardi and Williams, 1968, Rosen et al., 1980). Striking clinical
examples mentioned above are the associations of ACTH and related
peptides and ADH with small cell lung cancer, and foregut tumours.

2) the primary structures of ectopic hormones are closely similar to their normal counterparts. Apparent structural differences suggested by abnormal ratios of immuno- to bio-activity or immunoactivity detected with different region-specific antisera can be accounted for by different patterns of tumour hormone metabolism compared to the normal gland of origin i.e. altered post-translational processing rather than abnormalities in coding for the primary gene product. Thus, whereas the authentic hormone is the main eutopic storage form, precursor, subunit and fragment forms often predominate in an ectopic hormone producing tumour and its secreted products (Ratcliffe, 1980).

3) characteristic groups of ectopic hormones are produced together e.g. ACTH and calcitonin.

4) hormone production outside the major gland of origin may not be unique to neoplasia e.g. HCG appears to be synthesised in normal adult gastrointestinal tissues and some proliferating non-malignant tissues (Yoshimoto et al., 1977).

The following hypotheses will be considered in relation to these observations:

a) Abnormal genome

Since all the genetic information is present in every diploid cell it is possible that extensive base substitutions in nuclear DNA might fortuitously give rise to the synthesis of peptides which possess amino acid sequences required for hormone activity. However, this fails to explain the structural similarity of ectopic hormones with their normal counterparts, the particular tumour associations of ectopic hormones, the production and association of multiple ectopic hormones or the high prevalence of certain ectopic hormones. It can therefore be discounted as the usual cause of ectopic hormone production.

b) Derepression

This takes as its starting point the assumption that most nuclear DNA is repressed in adult cells. Derepression of previously inactive genes could give rise to gene products identical to their normal counterparts and could explain production of multiple hormones due to simultaneous derepression of hormone coding genes which are located close together, and the high prevalence of ectopic hormone production. However, it does not readily predict specific tumour associations of ectopic hormones. Further, there is no experimental evidence for a general increase in gene transcription in transformed cells (Shields, 1977), and the presence of low hormone concentrations in normal 'non-endocrine' adult tissues suggests that the responsible genes are not fully repressed.

c) Abnormal differentiation

The difficulties in accounting for the observed tumour specificities of ectopic hormones and the probable production of hormones by normal 'non-endocrine' adult tissues has led to re-examination of hypotheses based on altered differentiation in neoplasia towards cell types normally occurring during development and/or non-malignant proliferation. It has also been prompted by recognition that many tumour products are also found in foetal life or are of foetal type, and that cells with endocrine characteristics are normally present in many tissues.

i) Endocrine cell and/or APUD hypothesis. This suggests a switch in gene function in a cell which already possesses many features of a peptide secreting cell (e.g. secretory granules) and which may normally synthesise a different hormone. Hormone coding DNA may be more readily derepressed in such cells (Williams, 1969). Endocrine-type cells are widely distributed in normal tissues and many show APUD properties (Gould, 1978). The APUD hypothesis suggests that only tumours derived from tissues containing APUD cells produce hormones (Pearse and Welbourn, 1973). An APUD origin of ectopic hormones of Group 1 (vide supra) certainly seems plausible for small cell carcinomas, foregut carcinoids, islet cell tumours, medullary carcinoma of thyroid, thymoma, adrenal medulla, ganglioneuroma and carcinoma of prostate which may arise from APUD cells. However, this hypothesis does not readily account for the production of Group 2 or Group 3 hormones by tumours which do not generally show APUD features and which are normally produced by non APUD cells. The APUD concept has perhaps been mainly of use in drawing attention to the hormone producing capability of cells with APUD characteristics. Certainly the original claim that all APUD cells are derived from the neural crest is untenable and extensions of the concept to 'neuroendocrine programmed' cells cannot be readily tested.

The cell hybridisation modification of the APUD hypothesis suggests that neoplastic cells derived from non-APUD cells hybridise with adjacent non-neoplastic APUD cells, and that the resulting hybrid cells retain the ability to produce hormones as they proliferate (Warner, 1974). While this could account for ectopic hormone production by apparent non-APUD tumours, there is little evidence either for or against it.

ii) Dysdifferentiation (Baylin and Mendelsohn, 1980). The similarity in regulation of gene expression between tumour cells and immature developing cells has led to the suggestion that the neoplastic event(s) may affect proliferative cells with hormone producing capabilities which are present in small number in normal mature epithial tissue. The subsequent tumour then contains increased numbers of early cell forms capable of producing hormones, only a proportion of which continue to differentiate. The development of APUD characteristics may be one of several pathways of differentiation available to primitive uncommitted epithelial cells. In the context of ectopic hormone production in lung cancer, proliferation of cells which have progressed partially in an endocrine direction might give rise to small cell cancer producing ACTH (APUD positive) whereas proliferation of more primitive foetal type cells could give rise to non-small cell cancer producing for example placental and foetal peptides (e.g. HCG and CEA) and lacking APUD features. The processes involved in ectopic hormone production may thus involve variable degrees of proliferation and maturation arrest along one or more lines of differentiation. Abnormal ratios of proliferation to maturation in developing tumours may reflect differences between tumour differentiation and that in the normal tissue of origin. These abnormal ratios amplify differentiation features which are usually only a minor aspect of the mature phenotype of normal epithelium. Clearly the resolution of these ideas will depend on establishing the mechanisms and pathways of normal and neoplastic cellular differentiation in specific tissues.

ACKNOWLEDGEMENTS

I am grateful to the Scottish Hospitals Endowment Research Trust for financial support and Mrs A. McKinnon for secretarial help.

REFERENCES

Azzopardi, J. G., and Williams, E. D., 1968, Pathology of non-endocrine tumours associated with Cushing's syndrome, Cancer, 22:274.
Azzopardi, J. G., Freeman, D., and Poole, G., 1970, Endocrine and metabolic disorders in bronchial carcinoma, Br. Med. J., 4:528.
Baylin, S.B., and Mendelsohn, G., 1980, Ectopic (inappropriate) hormone production by tumours: mechanisms involved and the biological and clinical implications, Endocr. Rev., 1:45.
Bloomfield, G. A., Holdaway, I. M., Corrin, B., Ratcliffe, J. G., Rees, G. M., Ellison, M. L., and Rees, L. H., 1977, Lung tumours and ACTH production, Clin. Endocr., 6:95.
Goltzman, D., Huang, S. N., Browne, C., and Solomon, S., 1979, ACTH and calcitonin in medullary thyroid carcinoma: frequency of occurrence and localisation in the same cell type by immunocytochemistry, J. Clin. Endocr. Metab., 49:364.
Gould, R. P., 1978, The APUD cell system, in: "Recent Advances in Histopathology," P. P. Anthony and N. Woolf, eds., Churchill Livingstone, Edinburgh.

Imura, H., 1980, Ectopic hormone production viewed as an abnormality in regulation of gene expression, Adv. Canc. Res., 33:39.

Levine, R. J., and Metz, S. A., 1974, A classification of ectopic hormone producing tumours, Ann. N.Y. Acad. Sci., 230:533.

Liddle, G. W., Nicholson, W. E., Island, D. P., Orth, D. N., Abe K., and Lowder, S. C., 1969, Clinical and laboratory studies of ectopic humoral syndromes, Rec. Prog. Horm. Res., 25:283.

Odell, W. D., Wolfsen, A. R., Yoshimoto, Y., Weitzman, R., Fisher, D. A., and Hirose, F., 1977, Ectopic peptide synthesis: a universal concomitant of neoplasia, Trans. Ass. Am. Phys., 90:204.

Pearse, A. G. E., and Welbourn, R., 1973, The Apudomas, Br. J. Hosp. Med., 10:617.

Ratcliffe, J. G., 1980, Ectopic hormones: biochemical aspects. Scott. Med. J., 25:146.

Rees, L. H., and Ratcliffe, J. G., 1974, Ectopic hormone production by non-endocrine tumours, Clin. Endocr., 3:263.

Rosen, S. W., Weintraub, B. D., and Aaronson, S. A., 1980, Non-random ectopic protein production by malignant cells: direct evidence in vitro, J. Clin. Endocrinol. Metab., 50:834.

Sherwood, L. M., and Gould, V. E., 1979, Ectopic hormone syndromes and multiple endocrine neoplasia, in: "Endocrinology," L. J. DeGroot, ed., Grune and Stratton, New York, Vol. 3:1733.

Shields, R., 1977, Gene derepression in tumours, Nature, Lond., 269:752.

Warner, T. F. C. S., 1974, Cell hybridisation in the genesis of ectopic hormone secreting tumours, Lancet, i:1259.

Williams, E. D., 1969, Tumours, hormones and cellular differentiation, Lancet, ii:1108.

Yoshimoto, Y., Wolfsen, A. R., and Odell, W. D., 1977, Human chorionic gonadotrophin-like substance in non-endocrine tissues of normal subjects, Science, 197:575.

NEURONAL-GLIAL DIFFERENTIATION OF A STEM CELL LINE

FROM A RAT NEUROTUMOR RT4 - BRANCH DETERMINATION

Noboru Sueoka, Masaru Imada*, Yasuko Tomozawa, Kurt Droms,
Theresa Chow and Thomas Leighton

Department of Molecular, Cellular,
   and Developmental Biology
University of Colorado
Boulder CO 80309

*Department of Pathology
University of Colorado
Denver CO 80262

Our understanding of the regulability of differentiation will be greatly enhanced by the detailed description of specific biological systems which represent basic patterns of differentiation. As we understand them at present, the basic patterns common to eukaryote differentiation and development are: a) branching of developmental paths leading to different cell types (branch determination); b) linear progression of different cell types with several states of differentiation (transdifferentiation); and c) maturation processes which lead to a full expression of potentially activated genes by the former two differentiation processes.

In 1974, M. Imada and N. Sueoka made an observation in a rat peripheral neurotumor (RT4) induced by ethylnitrosourea that one of the cloned cell types from the tumor repeatedly gave rise to three other cell types in culture. This observation led to our long-term study of the branching mechanism in neuronal-glial cell differentiation in culture and in vivo.

A NEUROTUMOR, RT4

RT4 is a peripheral neurotumor induced by injecting ethylnitrosourea into a newborn BDIX rat according to the method developed by Druckrey and his collaborators (1966) (Imada and Sueoka, 1978). The

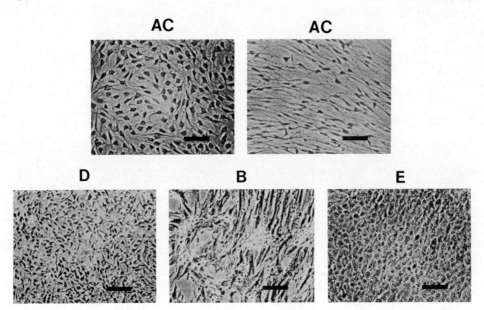

Figure 1. Morphologies under a phase-contrast microscope of the four cell types from RT4 (from Imada and Sueoka, 1978). The RT4-B culture often has lumps of refractile cells as shown here. The bars represent 100 μm.

tumor can readily be transplanted by subcutaneous injection of tumor cells (10^6-10^7 cells) into BDIX rats and primary cell cultures can easily be established from the tumor. We have made an extensive cloning of the primary culture of the tumor, RT4, and established four homogeneous cell lines of four distinct cell types (RT4-AC, RT4-B, RT4-D, and RT4-E). Morphological differences of these cell types are shown in Fig. 1. RT4-AC, RT4-D and RT4-E have more or less similar generation times (19-24 hours), and RT4-B cells seem to have a longer generation time (36 hours) (Imada and Sueoka, 1978).

CELL TYPE CONVERSION

Whereas RT4-D and RT4-E are stable in culture and always give rise to a homogeneous population, RT4-B and RT4-AC generate patches of different cell types with low frequencies (approximately one in 10^6 cells). This change in cell type in culture is termed "cell type conversion" (Imada and Sueoka, 1978). RT4-B shows the conversion to a new type, RT4-F, with a larger cell size, whose characteristics have not been studied in detail. In contrast, RT4-AC, gives rise to three other cell types, RT4-B, RT4-D, and RT4-E. We therefore define RT4-A as a multipotential stem cell type. The rates of conversions, though their accurate estimations are difficult, are of the order of one in 10^5-10^6 cells in all three directions. One group of experiments

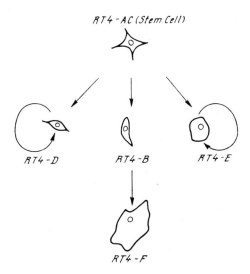

Figure 2. A schematic presentation of cell type conversion.

showed that the rate of conversions to RT4-E may be an order of magnitude higher than those to RT4-D and RT4-B (Table 1; Imada and Sueoka, 1978). This means that for each single cell clonal culture of RT4-AC in a culture dish of 10 cm diameter, 1-3 patches of the new cell types - RT4-B, RT4-D or RT4-E - appear. We can, therefore, establish a number of cell lines of these derivative cell types which are independently derived by cell type conversions from the stem cell type. The cell type conversions of RT4-AC are schematically presented in Fig. 2.

CHARACTERIZATION OF THE RT4 CELL TYPES

The RT4 cell types have been examined for a number of properties.

1. Cell Morphology

Morphological criteria are initially used in establishing cell lines of each cell type. Differences in morphology of these cell types are so distinct that these derivative cell types make clear boundaries with RT4-AC and with each other. Some conspicuous features of RT4 cell types in the standard medium described by Imada and Sueoka (1978) will be presented. RT4-AC cells take two forms, presumably in reponse to serum concentrations or for unknown physiological reasons: A-type has a flat cell body and multipolar short processes, and C-type a spindle-shaped cell body and mostly bipolar, long processes. These two cell types do not breed true indicating that they are

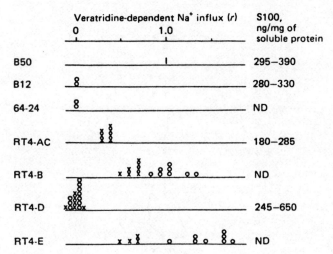

Figure 3. Summary of Na⁺-influx and S100 protein content of standard cell lines and clonal cell lines of the RT4 family (from Tomozawa and Sueoka, 1978). The standardized veratridine response ratio (r) of Na⁺-influx is a relative measure of the extent of Na⁺-influx. The clonal cell lines, B50 (neuronal) and B12 (glial), originally isolated from the rat central nervous system by Schubert et al. (1974) were used as neural cell standards. A rat mammary gland tumor cell line, 64-24 (Kano-Sueoka and Hsieh, 1973), was used as a non-neural cell control. Na⁺-influx values of B50 were used as the reference; its value is designated as 1.0. o, stem-cell derived cell lines; x, tumor-derived cell lines. For S100 protein, only the ranges are shown. ND, not detectable. For details, see Imada and Sueoka (1978) and Tomozawa and Sueoka (1978).

interchangeable. RT4-B cells are larger and have fibroblastic morphology. Exponentially growing RT4-D cells have a similar morphology to that of growing RT4-AC cells, but towards confluence they become small spindle-shaped cells with two short processes or spherical cells. RT4-D cells tend to migrate in the plate and grow in an overlapping manner. RT4-E cells grow in tight colonies and form a contacted monolayer pattern of confluency.

2. Neural Properties

An interesting feature of the RT4 system in culture is that neuronal and glial properties are both expressed in the stem cell type (RT4-AC) and they are segregated among the derivative cell lines in a consistent pattern (Imada and Sueoka, 1978; Tomozawa and Sueoka,

1978). Thus, cell types RT4-B and RT4-E show high levels of voltage-dependent Na^+-influx (a neuronal property), while RT4-D does not. In contrast, RT4-D produces a large amount of S100 protein (a glial protein), whereas the protein is not detectable in RT4-B and RT4-E (Fig. 3). The stem cell type RT4-AC is positive for both Na^+-influx and S100 protein, but to a lesser extent than RT4-B or RT4-E for Na^+-influx and than RT4-D for S100 content. We have preliminary evidence from immunofluorescent studies that RT4-AC and RT4-D also contain glial fibrino acidic protein (GFA), but GFA is not detectable above background in RT4-B or RT4-E.

3. Karyotype

Chromosome numbers of RT4-AC, -B, -D and -E had a modal value of 42, which is the normal diploid number of rat (Imada and Sueoka, 1978). More detailed analyses of Dr. Shirley Soukup and Mary Haag (personal communication) using the technique of G-banding have shown that all of these four cell types can have normal banding patterns, except that in all cell types one of the fourth chromosomes has an extra band at the terminus of its long arm. The karyotype of the cells from syngeneic rat BDIX does not exhibit this extra band on the fourth chromosome. These results show that no systematic changes in the karyotype accompany cell type conversions in RT4 system.

4. Tumorigenicity

We have examined the tumorigenicity of RT4 cell types using three criteria: a) Tumorigenesis *in vivo*; b) Presence of 250 K cell surface protein; and c) Plasminogen activator activity (Imada et al., 1978). Tumorigenicity was examined by injecting subcutaneously 10^6 - 2.5×10^7 cultured cells of each cell type into syngeneic BDIX rats. In all cases RT4-AC and RT4-D gave rise to tumors, whereas RT4-B and RT4-E did not form tumors within 30-40 days. In two out of eleven cases RT4-B cells formed tumors after 220 days. Proteins with apparent molecular weight close to 250 000 were examined by labeling cell surface proteins with ^{125}I followed by electrophoresing lysed cells in polyacrylamide gels. The results show that the high molecular weight surface proteins are virtually absent in RT4-AC and RT4-D and exist in large amounts in RT4-B and RT4-E. This is consistent with the report by Hynes (1977), which correlates the paucity of these proteins with tumorigenicity of cell lines. Intracellular activity of plasminogen activator of RT4 cell types was examined by Dr. D. B. Rifkin, showing that lysates of RT4-AC and RT4-D cells are much more active than those of RT4-B and RT4-E (Imada et al., 1978). These results indicate that RT4-AC and RT4-D are tumorigenic and RT4-B and RT4-E are non-tumorigenic.

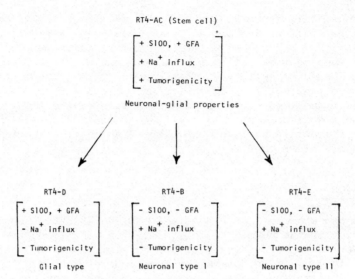

Fig. 4. Conversion coupling in the RT4 family. In cell type conversion, tumorigenicity is coupled with the expression of S100 protein, whereas cells with positive Na-influx are not tumorigenic. The expression of these and morphological characteristics are "coupled" in that whenever one characteristic (e.g. the presence of S100 protein) is observed, tumorigenicity is observed. In all cases so far examined, when cells converted to neuronal (i.e. absence of S100 and positive Na influx), tumorigenicity is negative. The coupling feature is consistent in all cell lines so far examined. Diagram based on present and previous results (Imada and Sueoka, 1978; Imada, et al., 1978; Tomozawa and Sueoka, 1978).

CONVERSION COUPLING

Characteristic changes of cell properties which accompany cell type conversions of RT4-AC are summarized in Fig. 4. A salient feature of the RT4 system is the fact that upon cell type conversion the expression of a number of genes is consistently altered in a coupled fashion. For example, whenever, the RT4-AC stem cell type converts to the RT4-D cell type, the RT4-D cell type invariably retains both tumorigenicity and the production of the glial protein S100, whereas cells converted to either the RT4-B, or the RT4-E phenotype lose both properties. This consistency is also observed for voltage dependent Na^+-influx where conversion to RT4-B or RT4-E is correlated with enhanced Na^+-influx while RT4-D does not exhibit this property. These co-ordinate changes have not been violated in any case so far examined of derivative cell types independently generated by cell type conversion from RT4-AC.

Table 1. Appearance of RT4-B, -D, and -E cells from RT4-AC clonal lines by Day 40.

Cell lines[a]	Differentiated morphologies[b]		
	RT4-B	RT4-D	RT4-E
RT4-51-AC14-1			+
RT4-51-AC14-2	+		
RT4-51-AC14-3	+	+	
RT4-51-AC20-1		+	
RT4-51-AC23-1			+
RT4-51-AC23-3			+
RT4-51-AC23-4			+
RT4-51-AC24-2	(+)[c]	+	+
RT4-51-AC24-3			+
RT4-51-AC24-4			+

[a] Single RT4-AC cells were isolated from RT4-51-2, a subclone of RT4-51, by two sequential single cell isolation procedures with glass cloning cylinders.

[b] Single RT4-AC cells were propagated for 40 days. In 10 of the 12 RT4-AC cultures, areas containing cells of RT4-B, -D, and -E morphologies were observed. The presence of such differentiated cells on Day 40 is indicated by a +. The cells initially manifested homogeneous RT4-AC morphologies and gave rise to small areas of other types, except for RT4-51-AC23-3 and RT4-51-24-3, which converted to -E cells almost completely by Day 40. The RT4-B, -D, and -E cells in this table are the source of the "stem cell-derived" clonal cell lines.

[c] As described in the text, RT4-51-AC24-2 gave rise to RT4-D and RT4-E by Day 40 and to RT4-B by Day 55.

DISCUSSION

The significance of the RT4 system in studying the mechanism of cell differentiation lies in the possibility that cell type conversion and its coupling to the expression of certain genes may represent a branching point in the developmental process. Specifically it may represent a branch point from neural stem cells into glial and neuronal cells. The consistency of coupled conversion with regard to morphology, tumorigenicity, and glial-neuronal characteristics hitherto examined suggests that coordinate expression and repression of a number of genes is likely to be involved in cell type conversion in this system.

A unique feature of the RT4 system is the fact that the stem cell type (RT4-AC) itself expresses both a glial property (synthesis of S100 protein) and a neuronal property (voltage dependent Na^+-influx). Branching, therefore, is accompanied by repression of gene functions: in differentiation to the glial type (RT4-D) the Na^+-influx is repressed, and in differentiation to the neuronal types (RT4-B and RT4-E) the S100 protein production is repressed (Fig. 4). In this sense, the regulation is negative. It is interesting to note that the majority of cell lines from ethyl-nitrosourea induced rat brain tumors studied by Schubert et al. (1974) show both neuronal and glial properties, as is seen in RT4-AC. This indicates that cells such as RT4-AC which exhibit dual characteristics are not a rare phenomenon found only in rat peripheral and central neurotumors. Our results using immunofluorescent staining of RT4-AC cells show that practically all cells of RT4-AC culture have S100 protein. Three basic types of mechanisms are possible to explain these results: a) production of repressor molecules, for some glial genes in RT4-B and RT4-E and for Na^+-influx genes in RT4-D, b) changes in DNA, such as sequence rearrangement or methylation, or c) selective processing of two classes of mRNAs, one for glial proteins and the other neuronal proteins.

Branching determination of the progeny of individual stem cells could take any one of the three possible patterns shown in Fig. 5. To examine this point, ten single cell clones of RT4-AC were made and cell type conversions were examined (Imada and Sueoka, 1978). The results (Table 1) show that after 40 days in culture eight clones gave rise to a patch of one cell type (one RT4-B, one RT4-D, and six RT4-Es), one clone gave two patches (RT4-B and RT4-D), and one clone gave three patches (RT4-D, RT4-E, and RT4-B). The fact that some clones can generate more than one cell type demonstrates the multipotential nature of individual stem cells. In addition, these result eliminate the branching pattern in which a stem cell divides into glial and neuronal types in a single critical cell division (pattern in Fig. 5). Distinction between patterns II and III in Fig. 5 is not possible from these experiments. This implies that branching into glial and neuronal cell types does not occur at the individual stem cell level but in a stem cell population.

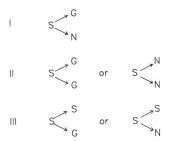

Figure 5. Three patterns of cell type conversion of a stem cell at the critical cell division. 1) one daughter cell (G) of a stem cell (S) becomes a cell committed to be glial and the other (N) to be neuronal. 2) both daughter cells are committed to be glial or neuronal. 3) only one of the daughter cells is committed to be glial or neuronal and the other remains as a stem cell.

Branching determination during development can be contrasted to maturation and possibly transdifferentiation processes whereby external or internal agents induce cell properties to be expressed in the majority of cells which are already determined for a developmental pathway. Branch determination, at least in a number of cases in vertebrate development, seems to occur by probabilistic change of multipotential stem cells. This distinction between branch determination and maturation can best be seen in a rat mammary carcinoma cell line (Rama25) reported by Bennett, et al. (1978). In this case all stem cells (cuboidal) undergo maturation in response to dimethylsulfoxide to give rise to non-dividing milk producing cells (droplet cells). In contrast, without dimethylsulfoxide, stem cells proliferate and undergo cell type conversion at a low frequency to generate a different cell type (fusiform), which is an in vitro equivalent of the myoepithelial cell. Branch determination also is in contrast to linear differentiation as exemplified in lens regeneration where differentiation proceeds in a cascading fashion from one cell type to another (see articles in this volume by R. M. Clayton, G. Eguchi, A. A. Mosocona, and T. S. Okada). Studies on branch determination, linear differentiation, and maturation processes at the levels of DNA and chromatin structures, transcription, RNA processing, and translation are obviously a direction of research whose results will greatly advance our understanding of the mechanism of differentiation and development. The use of a transformed stem cell line and its derivative cell lines has an advantage for molecular biological studies, since the indefinite growth of the cells provides quantities of homogeneous cells of each cell type.

The rate of cell type conversion in the RT4 system (of the order of 10^{-5} to 10^{-6}) may seem to be too low to account for the normal differentiation of glias and neurons. Evaluation of this point is currently difficult. It is possible that the cell type conversion occurs only after the culture reaches confluence. Also one must consider that transformation of the RT4-AC cell line may have affected the rate of conversion. Moreover, in vivo there could be a factor or factors which influence the overall frequency of cell type conversion. To investigate this possibility we have tested a number of compounds and tissue extracts on the RT4-AC culture, so far without evidence of enhancement of the cell type conversion rate. This point, however, needs further investigation.

We have been examining development of normal rat peripheral ganglia in comparison with RT4 in vitro differentiation. The critical point of investigation is to seek cells in ganglia which express both neuronal and glial properties during normal development and show that those cells segregate later into glial and neuronal cells. Our study on the rat superior cervical ganglion so far has not shown such cells. This is because S100 protein is not detectable in the newborn rat sympathetic ganglion and when it is detected five days after birth, cells are already segregated in the expression of the glial property (S100 protein) and the neuronal property tested (catecholamine synthesis detected by the formaldehyde-induced fluorescence method). Recently, however, DeVitry et al. (1980) showed that during hypothalamus development in mice the ependymal cells individually possess both a glial protein (S100 protein) and a neuronal protein (14-3-2), and later these properties segregated into different cells. These results support the notion that the main features of the RT4 system are also found in the normal development of the nervous system. Our study on the sympathetic ganglia indicates that the branch determination cannot necessarily be observed as a segregation of different gene products already expressed, but rather the potential of gene expression may undergo branching. In other words, actual expression or repression of genes may not be concomitant with opening or closing of the genes at the level of determination. Many cases of branching are found where derivative cell types from a common stem cell type show their unique phenotypes later through a series of maturation processes. In the case of the Rama 25 system, Thy-1 antigen is found in the cytoplasm but not on the cell surface of the cuboidal stem cells, whereas the Thy-1 antigen is found on the cell surface of the myoepithelial-like cells (fusiform) derived from the stem cell type (Lennon et al., 1978). Thus, at the determination level, the type of regulation seen in RT4, namely negative regulation at a branch point, may be more general.

Among the RT4 cell types RT4-AC and RT4-D are highly tumorigenic. The primary culture of the original tumor, however, showed all of the four cell types, indicating a heterogeneity of cell types in the tumor. The primary culture of a tumor obtained by injecting clonal RT4-AC cells also showed other RT4 cell types. In contrast the tumor from RT4-D gave a homogeneous primary culture consisting of RT4-D cells (unpublished results). These results and those of karyotype analyses strongly support the notion that the cell type heterogeneity of the RT4 tumor originated with transformation of a single stem cell (RT4-AC) and that other cell types were generated by cell type conversions of the transformed stem cell population as seen in the RT4-AC culture. It is also interesting to note that the synthesis of S100 protein and tumorigenicity are coupled in the RT4 system, and that in general the majority of neurotumors are gliomas.

ACKNOWLEDGEMENTS

This work has been supported by NIH CA16856, NS15304, and ACS CD-1. We thank S. Beckmann, M. Brilliant and J. Errick for improving the manuscript.

REFERENCES

Bennett, D. C., Peachey, L. A., Durbin, H., and Rudland, P. S., 1978, A possible mammary stem cell line, Cell, 15:283.
DeVitry, F., Picart, R., Jacque, C., Legault, L., Duponey, P., and Tixier-Vidal, A., 1980, Presumptive common precursor for neuronal and glial cell lineages in mouse hypothalamus, Proc. Natl. Acad. Sci. U.S.A., 77:4165.
Druckrey, H., Ivankovic, S., and Preussman, R., 1965, Selektive Erzeugung Maligner Tumoren im Gehirn und Ruckenmark von Ratten durch N-Methyl-N-Nitorsoharnstoff, Z. Krebsfforsch, 66:389.
Hynes, R. O., 1978, Alteration of cell-surface proteins by viral transformation and by proteolysis, Proc. Natl. Acad. Sci. U.S.A., 70:3170.
Imada, M., and Sueoka, N., 1978, Clonal sublines of rat neurotumor RT4 and cell differentiation. I. Isolation and characterization of cell lines and cell type conversion, Develop. Biol., 66:97.
Imada, M., Sueoka, N., and Rifkin, D. B., 1978, Clonal sublines of rat neurotumor RT4 and cell differentiation. II. A. Conversion coupling of tumorigenicity and a glial property, Develop. Biol., 66:109
Kano-Sueoka, T., and Hsieh, P., 1973, Study on a rat mammary carcinoma in vitro and in vivo. 1. Establishment of clonal lines of the tumor, Proc. Natl. Acad. Sci. U.S.A., 70:1922.
Lennon, V. S., Unger, M., and Dulbecco, R., 1978, Thy-1: A differentiation marker of potential mammary myoepithelial cells in vitro, Proc. Natl. Acad. Sci. U.S.A., 75:6093.
Schubert, D., Heinemann, S., Carlisle, W., Tarikas, H., Kimes, B., Pattrick, J., Steinbach, J. H., Culp, W., and Brandt, B. L., 1974, Clonal cell lines from the rat central nervous system, Nature, Lond., 249:224.
Tomozawa, Y., and Sueoka, N., 1978, In vitro segregation of different cell lines with neuronal and glial properties from a stem cell

line of rat neurotumor RT4, Proc. Natl. Acad. Sci. U.S.A., 75:6305.

THE CRYSTALLINS OF NORMAL AND REGENERATED NEWT EYE LENSES

David S. McDevitt

Department of Animal Biology
University of Pennsylvania
School of Veterinary Medicine

INTRODUCTION

The Spotted Newt of the eastern United States, Notophthalmus viridescens, was chosen as a subject for investigation of the characteristic structural proteins of the eye lens, the crystallins, because of its developmental applications and its unique ability to regenerate a new lens from the dorsal pupillary margin after lentectomy. The source of this regenerated lens is now known to be the dorsal iris epithelial cells, which have withdrawn from the cell cycle (for reviews see Reyer, 1977; Yamada, 1982).

The crystallins of the normal adult N. viridescens lens have now been characterized in our laboratories as a basis for study of the crystallins produced by the regenerating/regenerated lens, whose morphogenesis is at least superficially similar to that of the embryonic lens rudiment. The iris epithelial cells responsible for lens regeneration would never have produced crystallins in the normal course of events. In our attempt to determine the fidelity imposed by the differentiation process in two developmentally different systems, each resulting in a lens, we have determined the ontogeny and localization of all the crystallin classes in development and regeneration, by means of the immunofluorescence technique. We have also begun to characterize the crystallins of the fully-regenerated lens.

METHODS

Normal adult and 5-6 month regenerated N. viridescens lenses were pooled, homogenized, and the 6 600 g total soluble protein supernatant fractionated by Sephadex G-200 (Superfine) and DEAE-Sephacel (Pharmacia) chromatography, according to McDevitt (1967) and McDevitt

et al. (1969). S.D.S. (0.1%)-polyacrylamide gel (13%) electrophoresis
of the crystallins was carried out using a slab gel modification of
the Laemmli method (1970). Antibodies were produced to the crystallin
classes, and the indirect immunofluorescence technique carried out
on 3 μm median sections through the developing and regenerating
lenses, according to McDevitt and Brahma (1981a, b). Peptide mapping
and amino acid analyses were performed using the method of McDevitt
and Croft (1977). Animals were collected in the field (Mountain Lake
Biological Station of the University of Virginia, Giles Co., Va.) or
obtained commercially (Bill Lee Newt Farm, Oak Ridge, Tenn.); and
lentectomized (McDevitt and Brahma, 1981b); adult newts were kept in
aquaria up to six months before use.

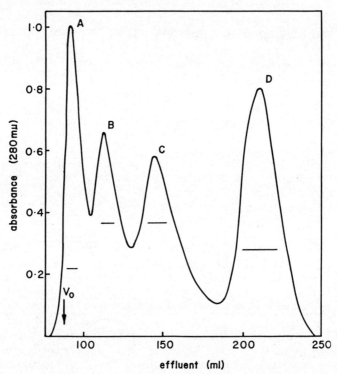

Figure 1. Representative gel-filtration chromatography of adult N. viridescens total soluble lens proteins on Sephadex G-200, Superfine. 74 mg of protein was added to a 2.6 cm x 48.5 cm column, and eluted with 0.05 M tris-Cl buffer, pH 7.2, with 0.1 M KCl, 5 mM β mercaptoethanol and 1 mM EDTA. Fractions were collected as 1.9 ml aliquots. Vo = Void Volume, determined using Blue Dextran 2 000; bovine thyroglobulin (M.W. 6.7×10^5) also eluted before peak A. The bars indicate the fractions pooled for further analysis (Figure 3).

RESULTS AND DISCUSSION

The results obtained from gel filtration chromatography of adult normal N. viridescens crystallins can be seen in Figure 1. The classic profile of 4 major fractions - α, β High molecular weight, β Low molecular weight, and γ crystallins (for review see McDevitt and Brahma, 1982) as seen in the bovine lens - was obtained. This method yields β_H, β_L, and γ crystallin fractions suitable for use without additional purification schemes, according to immunoelectophoresis (McDevitt et al., 1969) and S.D.S. polyacrylamide gel electrophoresis criteria. The first peak to be eluted, after the void volume, unexpectedly, contained β as well as α crystallins; it was necessary to employ ion-exchange chromatography (Figure 2) to obtain an α crystallin fraction free of β crystallins. Calf cortex preparations fractionated on Sephadex G-200 yield a first fraction, of high molecular weight, of solely α crystallins; bovine lens nucleus extracts also exhibit an α/β fraction eluting after the void volume (Manski et al.,

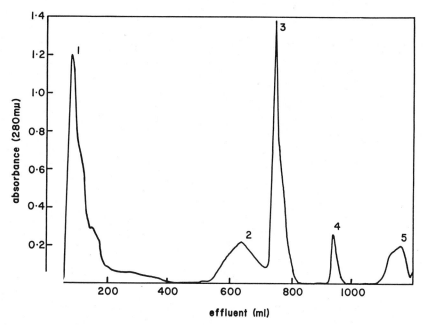

Figure 2. Representative ion-exchange (DEAE-Sephacel) chromatography of adult N. viridescens total soluble lens proteins. 275 mg of protein was added to a 2.6 cm x 25 cm column, and eluted with the following buffers: I, 100 ml 0.005 M phosphate, pH 7.0; II, 200 ml 0.0075 M phosphate, pH 6.5; III, 150 ml 0.01 M phosphate, pH 6.0; IV, 200 ml 0.02 M phosphate pH 5.7; V, 200 ml 0.02 M phosphate pH 5.7, + 0.1 M NaCl; VI, 150 ml 0.1 M phosphate, pH 5.7, + 0.3 M NaCl. Fractions were collected as 10 ml aliquots.

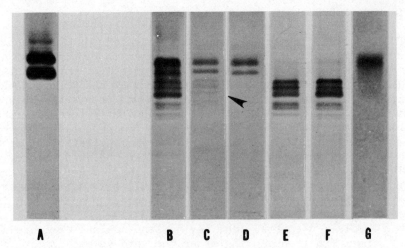

Figure 3. S.D.S. (0.1%) - polyacrylamide gel (13%) electrophoresis of normal adult N. viridescens crystallins (B-G) demonstrating the effectiveness of the gel filtration and ion-exchange fractionations. A) adult bovine cortex purified α crystallins (as a reference), obtained from successive G-200 Superfine and DEAE-Sephacel chromatographies (sampl overloaded); B) total soluble lens proteins; C) α crystallins, G-200 fraction (ARROW indicates β crystallin polypeptide presence/contamination); D) α crystallins derived from DEAE-Sephacel chromatography (Figure 2, No. 4), free of β crystallin polypeptides; E) G-200 $_H$ fraction; F) G-200 β_L fraction; G) G-200 γ crystallin fraction. Polypeptide M.W.'s range from 17.5 K (top) to 36 K (bottom).

1979; Malinowski and Manski, 1980a, b) as do embryonic calf lens crystallins (Bloemendal, 1981; personal communication). Thus, the High Molecular Weight crystallin aggregates associated with ageing and senile cataractogenesis (Spector et al., 1971; Jedziniak et al., 1975) are not present in the mature N. viridescens lens.

S.D.S. data (Figure 3) thus reveal 9 crystallin polypeptides present in the adult newt lens, ranging in M.W. from 17.5 K to 36 K. Purified α crystallin (as above) exhibits 2 bands at 17.5 Kd and 19. Kd; γ crystallin usually appears as a single band comigrating with the 17.5 Kd A chain of α crystallin; and β_H and β_L crystallins demon strate 6 and 7 polypeptides respectively, 22.5 K to 36 K (with the 36 K polypeptide only apparent in β_L). The similarity of β_H and β_L in this respect was unexpected, yet bovine cortex total lens protein separated on the same Sephadex G-200 column under the same condition yielded the classic S.D.S. profile (c.f. Harding and Dilley, 1976) of 6 β_H and 4 β_L (data not shown).

STAGE	Anti-α E	Anti-α F	Anti-β E	Anti-β F	Anti-γ E	Anti-γ F
V	−	−	−	+	−	−
VI	−	−	−	+	−	+
VII	−	+	−	+	−	+
VIII	−	+	−	+	−	+
IX	−	+	+	+	−	+
X	−	+	+	+	−	+

E = Epithelium (includes external layer of early lens rudiment)
F = Fibers (includes internal layer of early lens rudiment)
+ = Positive immunofluorescence for designated crystallin (No distinction made for intensity of immunofluorescence or % of total area positive)
− = Negative; no immunofluorescence observed

Figure 4. Detection of crystallins in the embryonic lens of N. viridescens by immunofluorescence. From McDevitt and Brahma, 1981a; with the permission of Academic Press.

The amino acid analyses obtained for the purified α, $β_H$, $β_L$ and γ crystallins obtained are in broad agreement with published reports for those of total cystallin fractions, mostly mammalian, as were the peptide maps of the crystallin classes (reviewed in Harding and Dilley, 1976). Significantly, the γ crystallins peptide map revealed the peptide cluster characteristic for all previously-known γ crystallins (Croft, 1973), fastest-moving in the electrophoresis direction.

Antibodies to the specific crystallin classes used in the immunofluorescence technique revealed differences in their ontogeny and localization in the developing and regenerating lens (McDevitt and Brahma, 1981a, b; summarized in Figures 4 and 5). β crystallins are the first of the crystallin classes to appear in lens development, at the lens vesicle stage, with γ crystallins first evident at a slightly later stage and α crystallins the last to be detectable, only in the lens fibers. No immunofluorescence reaction in the epi-

STAGE		ANTI-α E F	ANTI-β E F	ANTI-γ E F
V		− −	− +	− +
VI		− −	− +	− +
VII		− +	− +	− +
VIII		+ +	+ +	+ +
IX		+ +	+ +	+ +
X		+ +	+ +	− +
XI		+ +	+ +	− +

E = Epithelium (includes external layer of early lens regenerate; stalk ALWAYS negative)
F = Fibers (includes internal layer of early lens regenerate)
+ = Positive immunofluorescence for designated crystallin (No distinction made for intensity of immunofluorescence or % of total area positive)
− = Negative; no immunofluorescence observed

Figure 5. Detection of crystallins in the regenerating lens of N. viridescens by immunofluorescence. From McDevitt and Brahma, 1981b; with the permission of Academic Press.

thelium for α crystallins is evident in the lens of this organism through the beginning of metamorphosis. In lens regeneration, β-crystallins were first detectable in the thickening internal layer of the lens vesicle; γ also first appeared, albeit erratically, at this time in the same location. α crystallins did not appear until later, in a few cells of the developing lens fiber region, and later yet in the external layer/lens epithelium.

Sephadex G-200 Superfine chromatography of crystallins derived from 5-6 month regenerated lenses (not until this time does the size approximate that of the normal lens) can be seen in Figure 6. It agrees with results obtained for the adult normal lens in that the first crystallin peak elutes after the void volume, indicating a lack of High Molecular Weight α or other crystallins in the regenerated

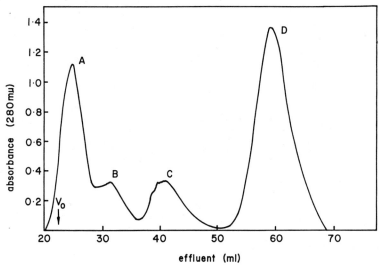

Figure 6. Sephadex G-200 Superfine chromatography of total soluble lens proteins derived from six-month regenerated N. viridescens lenses. 23 mg of protein was added to a 1.6 cm x 35 cm column, and eluted with 0.05 M tris-Cl buffer, pH 7.2, with 0.1 M KCl, 5 mM β mercaptoethanol and 1 mM EDTA. Fractions were collected as 0.5 ml aliquots. Vo = Void Volume, determined using Blue Dextran 2 000 and thyroglobulin.

lens; but differs from the normal lens in that the β crystallins elute as markedly more heterogeneous in size and polydisperse than the $β_H$ and $β_L$ crystallins of the normal adult lens. Amino acid analyses have thus far revealed the larger, putatively $β_L$ fraction from regenerated lenses to have significantly lower values for tyrosine and histidine (2.2 and 1.3 residues/100 residues, respectively) than the comparable normal $β_H$ and $β_L$ crystallins; further analysis of this and other fractions derived from the regenerated lens are ongoing. Peptide mapping of the γ crystallins derived from normal and regenerated lenses (Figure 7), which are similar in chromatographic behaviour reveals, however, distinct and intriguing differences. Each has the characteristic γ crystallin N-terminal Gly-Lys peptide seen in purified mammalian γ crystallins (Croft, 1973), but the regenerated crystallins sample exhibits "embyonic/fetal" γ crystallin subfractions (Croft, 1972; Slingsby and Croft, 1973) not seen in the γ crystallins of the adult lens, i.e. γIII and γIV (bovine). This cannot be directly confirmed, since not enough material is available for fractionation of the total γ crystallin sample. Thus, 20 peptides were common to both regenerated and normal total γ crystallins; 3 peptides were found only in the regenerated γ's and 4 peptides only in those γ crystallins derived from adult normal lenses.

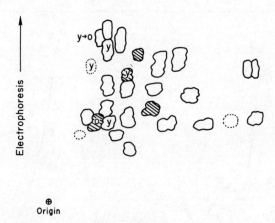

Figure 7. Peptide maps of tryptic digests of N. viridescens adult normal and regenerated γ crystallins (Figures 1 and 6,D), oxidized with performic acid. Low-voltage electrophoresis was performed with pH 3.6 pyridine-acetate buffer; chromatography was performed at 90° to the electrophoresis direction in the solvent BAWP (butan-1-ol-acetic acid water-pyridine 15:3:12:10, v/v). Cross-hatched circles, peptides present only in normal γ crystallins; broken circles, peptides present only in regenerated γ crystallins; solid circles, peptides common to both. (From McDevitt and Brahma, 1982; with the permission of Academic Press.)

Planimetry of the areas under the G-200 regenerated and normal crystallin peaks (Figures 1 and 6) also reveals that, although the proportion of α crystallins (Peak A) to the total remains near 25% in each, the proportion of γ crystallins to the total increases markedly from 375 in the normal lens to 52% in the regenerated lens.

The above results suggest that the regenerated lens exhibits features considered to be "embryonic" or immature, i.e. increased percentages of γ crystallins; the presence of embryonic/fetal crystallin subfractions; and decreased amounts of β_H (formation of β_H (mammalian) appears to be related to aging of the lens cell, Asselbergs et al., 1979). It is also apparent that, at least with regard to crystallin presence per se, the effect of alteration in genome expression, - reversion in vivo of the iris epithelial cells to an amitotic state in the adult - does not affect re-utilization of that genome. It remains for new approaches in molecular biology to elucidate the mechanism(s) by which new information is made available to the iris epithelial cell, enabling its transformation into a lens cell after lentectomy.

ACKNOWLEDGEMENTS

Samir K. Brahma, Laurie Croft, Sharon Di Riehzo, and Tuneo Yamada provided invaluable collaboration on much of the work discussed. This research was supported by the National Eye Institute, N.I.H. (EY-02534).

REFERENCES

Asselbergs, F. A. M., Koopmans, M., van Venrooij, W. J., and Bloemendal, H., 1979, β crystallin synthesis in Xenopus oocytes, Expl. Eye Res., 28:475.
Croft, L. R., 1972, The amino acid sequence of γ crystallin (fraction II from calf lens), Biochem. J., 128:961.
Croft, L. R., 1973, Low molecular weight proteins of the lens, in: "The Human Lens in Relation to Cataract," CIBA Foundation Symposium, 19:207, Amsterdam.
Harding, J. J., and Dilley, K. J., 1976, Structural proteins of the mammalian lens: A review with emphasis on changes in development, aging and cataract, Expl. Eye Res., 22:1.
Jedziniak, J. A., Kinoshita, J. H., Yates, E. M., and Benedek, G. B., 1975, The concentration and localization of heavy molecular weight aggregates in aging normal and cataractous lenses, Expl. Eye Res., 20:367.
Laemmli, U. K., 1970, Cleavage of structural proteins during the assembly of the head of bacteriophage T4, Nature, Lond., 227:680.
McDevitt, D. S., 1967, Separation and characterization of the lens proteins of the amphibian, Rana pipiens, J. exp. Zool., 164:21.
McDevitt, D. S., and Brahma, S. K., 1981a, Ontogeny and localization of the α, β and γ crystallins in newt eye lens development, Develop. Biol., 84:449.
McDevitt, D. S., and Brahma, S. K., 1981b, α, β and γ crystallins in the regenerating lens of Notophthalmus viridescens, Expl. Eye Res., in press.
McDevitt, D. S., and Brahma, S. K., 1982, Ontogeny and localization of the crystallins in eye lens development and regeneration, in:"Cell Biology of the Eye," D. S. McDevitt, ed., Academic Press, New York.
McDevitt, D. S., and Croft, L. R., 1977, On the existence of γ crystallin in the bird lens, Expl. Eye Res., 25:473.
McDevitt, D. S., Meza, I., and Yamada, T., 1969, Immunofluorescence localization of the crystallins in amphibian lens development, with special reference to the γ crystallins, Develop. Biol., 19:581.
Malinowski, K.,and Manski, W., 1980a, An immunochemical study of the different proteins in the beta and gamma crystallin families, Expl. Eye Res., 30:519.
Malinowski, K., and Manski, W., 1980b, Immunochemical studies on lens protein-protein complexes. III. Age-related changes in the

stability of lens alpha crystallin-containing complexes, Expl. Eye Res., 30:537.

Manski, W., Malinowski, K., and Bonitsis, G., 1979, Immunochemical studies in lens protein-protein complexes I. The heterogeneity and structure of complexed α crystallin, Expl. Eye Res., 29:625.

Reyer, R. W., 1977, The amphibian eye, development and regeneration, in:"Handbook of Sensory Physiology, VII/5; Visual System in Evolution," Springer, Berlin.

Slingsby, C., and Croft, L. R., 1973, Developmental changes in the low molecular weight proteins of the bovine lens, Expl. Eye Res., 17:369.

Spector, A., Freund, T., Li, L. K., and Augusteyn, R. D., 1971, Age dependent changes in the structure of alpha crystallin, Invest. Ophthalmol., 10:677.

Yamada, T., 1982, Transdifferentiation of lens cells and its regulation, in:"Cell Biology of the Eye," D. S. McDevitt, ed., Academic Press, New York.

FORMATION OF LENTOIDS FROM NEURAL RETINA CELLS:

GLIAL ORIGIN OF THE TRANSFORMED CELLS

A. A. Moscona and Linda Degenstein

Cummings Life Science Center
Chicago
Illinois 60637

INTRODUCTION

The question before this conference is whether differentiated cells retain a latent capacity for phenotype alteration, i.e., trans-differentiation. Under normal conditions, specialized cells remain "committed" to their restricted characteristics and can undergo only limited modulations. However, the genome of, at least, some kinds of differentiated cells retains much of its original information (Gurdon, 1977); hence, it is conceivable that such cells could be converted into another phenotype by altering mechanisms that control phenotype stability. This problem is of fundamental biological and biomedical importance and deserves exploration in vertebrate systems.

Among mechanisms that regulate cell differentiation, contact-mediated cell interactions play an especially significant role (Moscona, 1980). I will review experiments that implicate contact interactions in maintaining the phenotypic properties of retinal glia cells. These experiments were carried out with chick embryo cells. They demonstrated that by altering their contact-relationships glia cells become predisposed to rapid transformation into lens-type cells.

STATEMENT OF THE PROBLEM

It should be recalled that in the course of normal embryogenesis, the lens develops from the head ectoderm, while the neural retina arises as an extension from the brain; thus, the cells that give rise to these two tissues have different origins and destinies. However, it has long been known (Moscona, 1957, 1960) that, under certain conditions, dissociated neural retina cells of the chick embryo can form in vitro lens-like bodies or "lentoids". The demonstration by Okada et al. (1975) that such lentoids contained lens-specific antigens convincingly confirmed the identification of these structures and renewed our interest in the manner of their derivation from neural retina cells, and in the type of cells involved in this transformation.

In the work of Okada's group, cells dissociated from neural retinas of early chick embryos (3 to 8 days) were grown in monolayer cultures; after several weeks, some of the cells spontaneously formed lentoid clusters (Nomura and Okada, 1979). However, their frequency of formation was low and random. Because of the long cultivation in monolayer the identity of cells that gave rise to lentoids could not be determined. Furthermore, since the cells were obtained from early embryonic retinas that contain undifferentiated neuroepithelial blast cells, it could not be stated whether lentoids arose from blast cells, from definitive neurons, or glia cells.

Our work (Moscona and Degenstein, 1981) addressed the following questions: 1) Is it possible to obtain lentoids from differentiated retina cells of late embryonic and fetal stages? 2) Can the transformation be expedited and the time required reduced from several weeks to days? 3) Which of the several cell types in the retina gives rise to the lentoids?

The avian retina contains only a single kind of glia cells, known as Müller cells. Earlier work demonstrated that various phenotypic characteristics of these glia cells depend on their contact interactions with the neurons (Kaplowitz and Moscona, 1976; Linser and Moscona, 1979). This suggested that modification of normal cell interactions may predispose retina cells to lentogenic transformation. Since cell interactions can be effectively studied in vitro by rotation-mediated cell aggregation (Moscona, 1974), we turned to this procedure, as in the original work on lentoids (Moscona, 1960). Using this approach, we were able to obtain lentoids rapidly, consistently, and in quantity from retina cells in advanced stages of differentiation (Moscona and Degenstein, 1981).

Our experiments were guided by the following working hypothesis (Moscona, 1960; Moscona and Degenstein, 1981), borne out by the present results: 1) Dissociation and separation of retina cells from their original contact relationships "destabilizes" phenotypic controls in glia cells and predisposes them for the transformation. 2) Aggregation of the modified cells by rotation expedites formation of contact between them, and this enhances expression of their new phenotype. 3) The cells that undergo lentoid transformation are derived from definitive Müller gliocytes; the transformation can be elicited in cells from late embryonic and fetal retinas.

EXPERIMENTAL RESULTS

Neural retinas from 13- or 16-day chick embryos were dissociated by mild trypsinization into suspensions of single cells; the cells were plated in culture flasks in Medium 199 with 5% fetal bovine serum, under conditions that minimized cell clustering. After 7 days, the monolayers were dissociated and the cells were aggregated by rotation. Two days later, the aggregates were sectioned and examined histologically; other sections were immunostained with antiserum to lens δ-crystallins. Detection of antibody binding to lens-specific proteins was by indirect immunofluorescence. (For details of methods, see Moscona and Degenstein, 1981).

Neural retinas from 13- and 16-day embryos were used because by then the retina had completed its growth phase (Moscona and Moscona, 1979). The cells no longer divide; they are diversified into definitive phenotypes and have entered (or are entering) the final stages of functional differentiation.

When cells dissociated from 13- or 16-day neural retina are plated in monolayer, glia cells re-initiate cell divisions (Kaplowitz and Moscona, 1976). After 7 days, the cultures contain, in addition to residual non-dividing neurons, an abundant population of flattened, epithelioid glia-derived cells (gliocytes). Cultures of 16-day cells contain much fewer neurons than those of 13-day cells, because most of the neurons readily detach and are lost during medium changes.

The glia derivation of the proliferating cells was confirmed by showing that they contained carbonic anhydrase C (CA-C). This enzyme is a marker of definitive Müller glia cells and is present in these cells in 13- and 16-day retinas (Linser and Moscona, 1981a). CA-C was detected by immmunostaining of tissue sections, or of cells in monolayer cultures with antiserum to purified avian CA-C (Linser and Moscona, 1981a). The retention of CA-C by the proliferating epithelioid cells in monolayer cultures identified them as being derived from retinal glia.

Monolayer cultures were dissociated (after 7 days) and the cells were aggregated by rotation. For brevity, cells from monolayers derived from 13-day retina will be referred to as "younger" cells, and those from 16-day retina as "older" cells. After 2 days, the aggregates were examined histologically. Virtually each of the numerous aggregates formed by the younger cells contained an inner round cell core with typical lentoid morphology, and an outer zone of small, loosely adherent cells (Figs. 1, 2). As will be explained below, the lentoidal cores were formed by the glia-derived cells, and the outer zone cells were predominantly - but not exclusively - of neuronal origin. Recently we determined that in early phases of aggregated formation the two kinds of cells are at first intermingled, and that they rapidly sort out, the modified gliocytic cells segregating internally to form the lentoidal cores.

Aggregates of the older cells consisted only of core-like cell clusters (Fig. 3); only few outer cells were present because of the paucity of neurons in monolayer cultures of the older cells.

The inner-core cells with lentoid morphology immunostained with anti-lens antiserum and with antiserum to δ-crystallin; the outer cells showed only background level of immunofluorescence (Figs. 3, 4). Also aggregates of the older cells immunostained with these antisera (Figs. 5, 6); often they showed more intense immunofluorescence than the lentoidal cores in aggregates of the younger cells.

These results led us to conclude that, under the experimental conditions employed, glia-derived cells from 13- and 16-day neural retinas rapidly become competent for transformation into a lens phenotype; when reaggregated, they preferentially associated with one another into lentoidogenic structures.

That the lentoid-froming cells were derived from gliocytes was further supported by the finding that the inner core cells contained

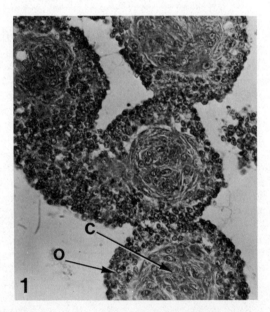

Figure 1. Histological section of 2-day aggregates of cells derived from neural retina of 13-day chick embryos. Before aggregation the cells were precultured for 7 days in monlayer. C - inner core consisting of lentoidal cells; O - outer zone of non-lentoidal cells. x400 (Modified from Moscona and Degenstein, 1981.)

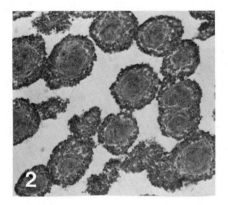

Figure 2. Culture similar to that described in Fig. 1 showing aggregates at low magnification. Note the presence of an inner lentoidal core in virtually all the aggregates. x160 (Modified from Moscona and Degenstein, 1981.)

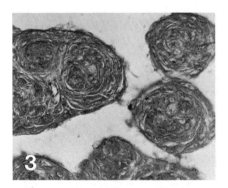

Figure 3. 2-day aggregates of cells derived from neural retina of 16-day chick embryos. Before aggregation, the cells were precultured in monolayer for 7 days by which time the culture consisted predominantly of glia-derived epithelioid cells and very few neuronal cells. Note the virtual absence of outer zone around these aggregates. x400 (Modified from Moscona and Degenstein, 1981.)

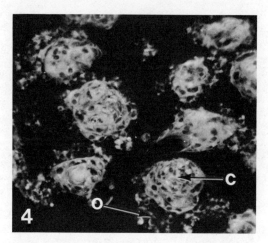

Figure 4. Section of 2-day aggregates of cells derived from neural retina of 13-day embryos, as in Fig. 1. Immunostained with anti-lens antiserum and FITC-GAR. Note immunofluorescence of core cells (C), compared with outer zone cells (O). (See control in Fig. 5). x270 (Modified from Moscona and Degenstein, 1981.)

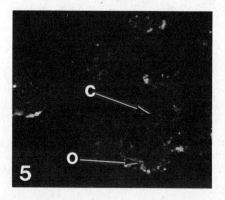

Figure 5. Similar to Fig. 4, but treated with control (non-immune) serum and RITC-GAR. Note absence of immunofluorescence in core of aggregates. x270 (Modified from Moscona and Degenstein, 1981.)

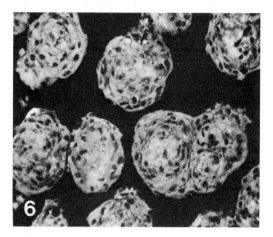

Figure 6. Section of 2-day aggregates of cells derived from neural retina of 16-day chick embryos, as in Fig. 3. Immunostained with anti-lens antiserum and RITC-GAR. x270 (Modified from Moscona and Degenstein, 1981.)

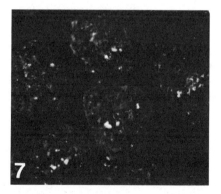

Figure 7. Control for Fig. 6, treated with non-immune serum and FITC-GAR. x270 (Modified from Moscona and Degenstein, 1981.)

CA-C; as explained above, this enzyme is a Müller glia cell marker, and it is retained in glia-derived cells in monolayer cultures. It is of interest that CA-C is present also in normal early developing lens of the chick embryo (Linser and Moscona, 1981a); as the lens fibers elongate and their content of lens antigens increases, the amount of CA-C declines. Also in the lentoids, the expression of CA-C showed an inverse relationship to that of lens antigens and declined with time. Aggregates derived from the older cells immunostained intensely with antiserum to adult lens proteins and showed a lesser reaction for CA-C than the cores in aggregates of younger cells.

The gliocytic origin of the lentoid-forming cells was corroborated by experiments with the gliatoxic agent α-aminoadipic acid. In the retina of chick embryos, this agent is toxic only to definitive Müller glia cells (Linser and Moscona, 1981b). Exposure of retina tissue from 13-day embryos to this agent results in destruction of the majority of Müller cells. Such glia-depleted retinas were dissociated and the cells were cultured in monolayer. The resulting monolayers contained during the first two days only sparse gliocytes. Accordingly, aggregates obtained from such cultures contained only few small lentoids.

Taken as a whole, these results provide convincing evidence that the cells which transform into lens phenotype and give rise to lentoids are derived from retinal glia.

CONCLUSIONS AND COMMENTS

Based on the above, we suggest that the phenotype of glia cells in the retina is stabilized by histotypic contact-interactions with neurons. Disruption of these contacts and cell dispersion in vitro trigger in definitive gliocytes (from retinas of 13- and 16-day chick embryos) changes conducive to transformation into a lens phenotype (Moscona, 1960). The rate at which these changes occur and their expression depend on culture conditions, and on the embryonic age of the cells; under the experimental conditions used by us, they appear faster in fetal than in early embryonic cells. The new phenotype is expressed when the modified cells contact one another and associate in 3-dimensional aggregates (Moscona and Degenstein, 1981). There is no evidence that the transformant cells originate from a few undifferentiated blast cells; our results show that definitive gliocytes give rise to the cells that become transformation competent.

The above general concept agrees - in principle - with related views of Okada et al. (1975), Nomura and Okada (1979), and Clayton et al. (1979). However, these workers tentatively assumed that neurons were the transforming cell type; our evidence conclusively points to gliocytes. Another difference is that, in Okada's experiments several weeks were required before lentoids spontaneously appeared in the monolayer cultures, while we could detect transformation-competent cells already after a few days. This difference is readily traceable to two conditions. 1) We used cells from older retinas which contain gliocytes that, apparently, transform faster than at an earlier stage. 2) Rather than wait for spontaneous, random formation of cell aggregates in monolayer cultures, we aggregated monolayer-precultured cells by rotation; since this causes rapid assembly of the modified cells into lentoidal structures, the presence of abundant transformation-competent cells in the monolayers could be detected already after 5 to 7 days.

The detailed mechanisms that cause retinal gliocytes to rapidly switch into lens phenotype under these experimental conditions remain to be examined. However, what little is known about this system points to the cell surface as a likely initiation site of this change. Trypsin-dissociation of 13-day and older retina tissue modifies various surface properties of gliocytes. In contrast to younger cells, cells dissociated from older retinas no longer can effectively regenerate surface molecules that, earlier in development, mediate recognition and histotypic contacts between gliocytes and neurons (Ben-Shaul et al., 1980); following dissociation, older gliocytes acquire greater contact-affinity for each other than for neurons, especially after some time in monolayer culture. Thus, when the monolayer-precultured cells are aggregated, the modified gliocytes segregate from neurons and adhere together to form lentoid cores, as described above.

Cell separation and the resulting surface changes probably initiate mitogenesis in postmitotic retina gliocytes (Kaplowitz and Moscona, 1976); it is not known if the gliocytes must undergo cell divisions in order to become transformed into lens phenotype. Surface changes resulting from cell separation may also be implicated in the loss by gliocytes in monolayer cultures of susceptibility to α-aminoadipic acid (Linser and Moscona, 1981b); in the decline of corticosteroid receptors in dispersed retina cells (Saad et al., 1981); and in the rapid loss in dispersed retinal glia cells of inducibility for glutamine synthetase (Linser and Moscona, 1981a).

Why such altered gliocytes become predisposed to accumulation of lens proteins is not clear; perhaps the genes for these proteins are especially "leaky" in these cells and their expression takes over when the original phenotype is destabilized. Other possibilities can be envisaged. The experimental approaches described here should facilitate analysis of this important problem.

ACKNOWLEDGEMENT

Research supported by grant HD01253 from the National Institute of Child Health and Human Development and grant 1-733 from the March of Dimes-Birth Defects Foundation.

REFERENCES

Ben-Shaul, Y., Hausman, R. E., and Moscona, A. A., 1980, Age-dependent differences in cognin regeneration on embryonic retina cells: immunolabeling and SEM studies, Develop. Neurosci., 3:66.

Clayton, R. M., Thomson, I., and De Pomerai, D. I., 1979, Relationship between crystallin mRNA expression in retina cells and their capacity to re-differentiate into lens cells, Nature, Lond., 282:628.

Gurdon, J. B., 1977, Egg cytoplasm and gene control in development, Proc. R. Soc. B., 198:211.

Kaplowitz, P. B., and Moscona, A. A., 1976, Lectin-mediated stimulation of DNA synthesis in cultures of embryonic neural retina cells, Expl. Cell Res., 100:177.

Linser, P., and Moscona, A. A., 1979, Induction of glutamine synthetase in embryonic neural retina: localization in Müller fibers and dependence on cell interactions, Proc. Natl. Acad. Sci. U.S.A., 76:6476.

Linser, P., and Moscona, A. A., 1981a, Carbonic anhydrase-C in the neural retina: transition from generalized to glia-specific cell localization during embryonic development, Proc. Natl. Acad. Sci. U.S.A., in press.

Linser, P., and Moscona, A. A., 1981b, Induction of glutamine synthetase in embryonic neural retina: its suppression by the gliatoxic agent α-aminoadipic acid, Develop. Brain Res., 1:103.

Moscona, A. A., 1957, Formation of lentoids by dissociated retinal cells of the chick embryo, Science, 125:598.

Moscona, A. A., 1960, Patterns and mechanisms of tissue reconstruction from dissociated cells, in:"Developing Cell Systems and their Control," D. Rudnick, ed., Ronald Press, New York, p.45.

Moscona, A. A., 1974, Surface specification of embryonic cells: lectin receptors, cell recognition and specific cell ligands, in:"The Cell Surface in Development," A. A. Moscona, ed., John Wiley & Sons, New York, p.67.

Moscona, A. A., 1980, Embryonic cell recognition: cellular and molecular aspects, in:"Membranes, Receptors and the Immune Response," E. P. Cohen and H. Kohler, eds., Alan R. Liss, New York, Vol. 42, p.171.

Moscona, A. A., and Degenstein, L., 1981, Lentoids in aggregates of embryonic neural retina cells, Cell Differentiation, 10:39.

Moscona, M., and Moscona, A. A., 1979, The development of inducibility for glutamine synthetase in embryonic neural retina: inhibition by BrdU, Differentiation, 13:165.

Nomura, K., and Okada, T. S., 1979, Age-dependent change in the transdifferentiation ability of chick neural retina in cell culture, Develop. Growth and Differ., 21:161.

Okada, T. S., Itoh, Y., Watanabe, K., and Eguchi, G., 1975, Differentiation of lens in cultures of neural retina cells of chick embryos, Develop. Biol., 45:318.

Saad, A. D., Soh, B. M., and Moscona, A. A., 1981, Modulation of cortisol receptors in embryonic retina cells by changes in cell-cell contacts: correlations with induction of glutamine synthetase, Biochem. Biophys. Res. Commun., 98:701.

SERUM FACTORS AFFECTING TRANSDIFFERENTIATION

IN CHICK EMBRYO NEURO-RETINAL CULTURES

D. I. de Pomerai and M. A. H. Gali

Department of Zoology
University of Nottingham
Nottingham NG7 2RD

SUMMARY

Adult sera (horse, chicken and newborn bovine serum) do not support extensive transdifferentiation of lens cells in cultures of 9-day chick embryo neural retina. Conversely, both chick embryo extract and foetal calf serum promote the accumulation of δ-crystallin (a marker for lens cells) in such cultures. Dialysed foetal calf serum does not allow transdifferentiation into lens, whereas the dialysis medium is able to do so in the absence of macromolecular serum components, suggesting one or more active factors of low molecular weight. Choline acetyltransferase activity (cholinergic neuronal marker) is generally maintained for longer under conditions which do not permit extensive transdifferentiation. Glutamine synthetase activity (a glial marker) is inducible by hydrocortisone in dense neuro-retinal cultures, and this hormone also reduces the extent of later lens development.

INTRODUCTION

Monolayer cultures of chick embryo neuro-retinal (NR) cells transdifferentiate extensively into foreign tissue types (lens and pigment cells; reviewed by Okada, 1980) when maintained for several weeks in media containing foetal calf serum (FCS) but not horse serum (HS; de Pomerai and Gali, 1981a). The extent of transdifferentiation in FCS-containing media can also be modulated by altering the levels of various non-serum medium components (Itoh, 1976; Araki and Okada, 1978), including glucose (de Pomerai and Gali, 1981a, b). In this report we further investigate the role of serum factors in promoting or inhibiting normal and foreign differentiation pathways in cultures of 9-day chick embryo NR. Lens cell development was monitored as accumulation of δ-crystallin (a lens fibre marker in chick embryos), and normal retinal differentiation patterns were assessed in terms of glutamine synthetase activity (GSase; a glial marker; Linser and Moscona, 1979) inducible by hydrocortisone, and of choline acetyltransferase activity (CAT; a cholinergic neuronal marker; Crisanti-Combes et al., 1978).

MATERIALS AND METHODS

Fertile chick eggs were from Ross Poultry Ltd (Bilsthorpe, Notts). All culture procedures for 9-day chick embryo NR cells were as detailed previously (de Pomerai and Gali, 1981a). FCS, HS, chicken serum (CS), chick ebmryo extract (CEE), newborn bovine serum (NBS) and all other medium components were obtained from GIBCO-Europe Ltd (Paisley, Scotland). The techniques for detecting δ-crystallin by means of haemagglutination inhibition assays with monospecific anti-δ-crystallin antiserum (a generous gift from Mrs R. M. Clayton, Institute of Animal Genetics, Edinburgh), and for determination of choline acetyltransferase activity (CAT; Crisanti-Combes et al., 1978) were exactly as described previously (de Pomerai and Gali, 1981a, b) except that a lower concentration of [^{14}C]-Acetyl CoA was used in the CAT assays (0.01 mM final; specific activity 58 Ci/μmole). Assays for glutamine synthetase followed standard procedures (Moscona and Hubby, 1963; Iqbal and Ottaway, 1980).

Dialysis of FCS or HS was performed at 4°C in 50 ml batches against four twelve-hourly changes of 500 ml complete medium lacking only the normal serum supplement (i.e. Eagle's MEM with Earle's salts, 2 mM L-Glutamine, 26 mM NaHCO, 100 I.U./ml penicillin and 100 μg/ml streptomycin). The first batch of dialysis medium was designated F_{DM} or H_{DM}, according to whether FCS or HS had been dialysed against it. After the fourth dialysis, the material which remained in the dialysis bag, containing macromolecular serum components (about 90% of the acid-precipitable protein originally present, as shown by Lowry assays), was similarly designated F_{MAC} or H_{MAC}. MAC material was used in the same way as normal serum (10% to supplement medium as above), while DM medium was used either alone or supplemented with one of the MAC fractions.

All cultures involving chicken serum (CS) were grown on collagen-coated plates (de Pomerai and Gali, 1981c).

RESULTS AND DISCUSSION

As shown in Fig. 1, 10% newborn bovine serum (NBS) does not support transdifferentiation into lens, and 15% chicken serum (CS) allows only low levels of δ-crystallin to accumulate (de Pomerai and

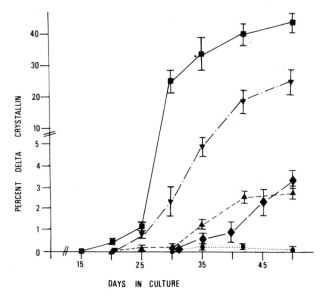

Figure 1. Effect of serum type on transdifferentiation into lens. 9-day chick embryo NR cultures were set up at initial densities of 3×10^6 cells per ml and maintained in MEM supplemented with various types of serum. At the indicated times, cultures were homogenised in saline and the δ-crystallin content determined by haemagglutination inhibition assays (see methods). δ-crystallin levels have been calculated as percentages of total soluble protein, and each point gives the mean ± standard error (vertical bar) from at least 6 such assays on two or more cultures. All data in this figure derive from a single large batch of cultures.

■——■, 10% CS + 5% CEE; ▼—·—▼, 10% FCS; ▲-----▲, 15% CS; ●······●, 10% NBS; ◆——◆, 10% FCS + 5 μM hydrocortisone.

Gali, 1981c). By contrast, a mixture containing 10% CS plus 5% chick embryo extract (CEE) promotes δ-crystallin accumulation to an even greater extent than 10% FCS, and lentoids are abundant in such cultures from 30 days onwards. The effects of hydrocortisone on transdifferentiation in cultures with 10% FCS are discussed below. To date, no adult serum (NBS, CS; also HS, Fig. 2, de Pomerai and Gali, 1981a) has proved able to support extensive transdifferentiation into lens, in contrast to embryonic sera or extracts (FCS, CEE). This is not directly related to overall culture growth or cell density, since CS at least is able to support growth rates comparable with those in FCS media (de Pomerai and Gali, 1981c).

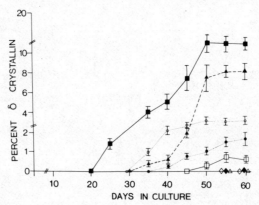

Figure 2. Effect of dialysed serum fractions on transdifferentiation. Dialysed serum fractions (F_{MAC} and H_{MAC}) and initial dialysis media (F_{DM} and H_{DM}) were prepared from FCS and HS as described in methods. Cultures were set up at 3×10^6 cells per ml and maintained in various combinations of the MAC and DM fractions. δ-crystallin levels were determined as for Fig. 1, each point giving the mean and standard error (vertical bar) from at least 6 assays on two or more cultures. Again all data in this figure derive from a single large batch of cultures.

■——■, F_{MAC} + F_{DM}; □——□, H_{MAC} + H_{DM}; ▲---▲, F_{DM}; △---△, H_{DM}; ◆, F_{MAC}; ◇, H_{MAC}; ●·····●, H_{MAC} + F_{DM}; ○·····○, F_{MAC} + H_{DM}.

Over the past two years we have used FCS from at least six different batches, HS and CS from four different batches each, and NBS from two (all supplied by GIBCO or Flow). Although the time of onset of crystallin accumulation varied considerably (between 18 and 33 days of culture), all FCS batches yet tested in this laboratory have supported extensive lens cell formation, the content of δ-crystallin ranging from 20 to 40% of total soluble protein in 50-day cultures. By contrast, two different batches of HS allowed no significant δ-crystallin accumulation (<0.1%), while the other two gave 0.7 and 0.4% δ respectively after 45-50 days (de Pomerai and Gali, 1981a). CS generally allowed limited transdifferentiation into lens, the content ranging from 3.5 to 8.5% by 50 days (de Pomerai and Gali, 1981c). With both batches of NBS, less than 0.2% δ-crystallin was detectable between 35 and 50 days of culture. Chick embryo extract, whether commercial (GIBCO) or prepared in our laboratory, promoted extensive transdifferentiation into lens when used as a supplement together with DS, giving over 40% δ-crystallin by 50 days in vitro. The above data includes cultures set up on many occasions, at widely differing initial cell densities, using several unrelated commercial chick strains, and embryonic ages ranging from 8 to 10 days of development. Since each of these variables can also influence the extent of transdifferentiation (reviewed in Okada, 1980), the maintenance of significant differences between serum types (as documented above) suggests a key role for serum factors in transdifferentiation.

In Fig. 2, the highest levels of δ-crystallin accumulation are found with reconstituted F medium (i.e. F_{DM} (Foetal calf dialysate) + 10% F_{MAC} (Foetal calf macromolecular fraction)), which is almost as effective as standard F medium (10% undialysed FCS; compare Fig. 1). Unsupplemented F_{DM} also permits extensive transdifferentiation, although the onset of δ-crystallin production is delayed by some 10 days, and the final level of δ-crystallin accumulated in late (50-60 day) F_{DM} cultures is only about 60% of that in comparable F_{DM} + F_{MAC} cultures. This may be attributable in part to slower growth in the absence of F_{MAC}. Very little δ-crystallin accumulates in cultures with reconstituted H medium (H_{MAC} (horse serum macromolecular fraction) + H_{DM} (horse serum dialysate)), and none with H_{DM} alone. Similarly, media containing only dialysed serum (H_{MAC}, F_{MAC} do not support transdifferentiation into lens (as observed previously; Itoh, 1976), and no net culture growth is observed. The heterologous combinations of F_{MAC} + H_{DM} and H_{MAC} + F_{DM} both permit low levels of δ-crystallin accumulation. H_{MAC} + F_{DM} gives less than 20% of the final δ level attained in F_{DM} alone, which might suggest that the H_{MAC} fraction contains factor(s) inhibiting transdifferentiation. The converse combination (F_{MAC} + H_{DM}) also allows significant levels of δ-crystallin accumulation, although neither F_{MAC} nor H_{DM} are able to do so alone. With all these medium combinations, raising the glucose level (to 18 mM instead of 6 mM final) abolishes all detectable δ-crystallin accumulation, and no lentoids are produced even after 60 days <u>in vitro</u> (data not shown; de Pomerai and Gali, 1981a, b).

The effects of these serum fractions on a marker of normal retinal differentiation, choline acetyltransferase (CAT; Crisanti-Combes et al., 1978), have also been studied, as shown in Fig. 3. CAT activity is very rapidly lost in F_{DM} alone, but reaches higher levels and is maintained for longer in F_{MAC} medium (3A). For H_{DM} and H_{MAC}, this contrast is much less marked, with significant CAT activity remaining in the H_{DM} cultures (though less than in H_{MAC}; 3B). Reconstituted F medium (F_{DM} + F_{MAC}) permits very little CAT activity, whereas in reconstituted H medium (H_{DM} + H_{MAC}), significant CAT levels are maintained until 15 days (3D). With F_{DM} + H_{MAC}, CAT activity is barely detectable after 5 days, but F_{MAC} + H_{DM} permits higher CAT levels until 15 days (3C). A similar pattern is obtained when supplementary glucose is present (results not shown). These data suggest that CAT-containing cholinergic neuronal cells are either lost or dedifferentiate rapidly when cultured in the presence of F_{DM}, but are maintained in the presence of F_{MAC} or H serum fractions. However, the results of combining MAC and DM fractions imply that these effects are not simply additive.

It remains to be established whether cholinergic neuronal cells ever transdifferentiate directly into lens cells in 9-day embryonic NR cultures (discussed in Okada, 1980). However, the retinal epithelial cells (putative Müller glia) from such cultures can do so extens-

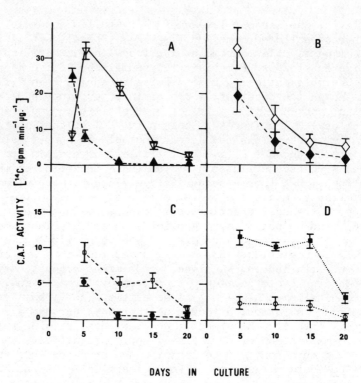

Figure 3. Effect of dialysed serum fractions on CAT activity. These data were derived from the same batch of cultures as that used for Fig. 2. CAT activity was assayed as described previously (see methods), and each point gives the mean and standard error (vertical bar) for at least 4 assays on two or more cultures. All results are given as d.p.m. of [^{14}C]-acetylcholine synthesised per min. per µg protein during a 10 minute assay.

A. ▲---▲, F_{DM}; ▽——▽, F_{MAC}.

B. ◆---◆, H_{DM}; ◇——◇, H_{MAC}.

C. ●----●, $F_{DM} + H_{MAC}$; □----□, $H_{DM} + F_{MAC}$.

D. ■······■, $H_{DM} + H_{MAC}$; ○······○, $F_{DM} + F_{MAC}$.

SERUM FACTORS AFFECTING TRANSDIFFERENTIATION 205

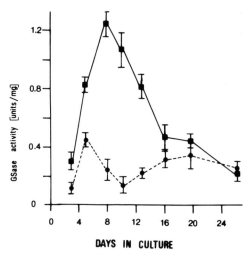

Figure 4. Effects of hydrocortisone on GSase activity. 9-day chick embryo NR cultures were set up at a density of 4.5×10^6 cells per ml. in F medium containing 10% FCS, either with or without 5 μM hydrocortisone continuously present. Glutamine synthetase activity was assayed as described in methods, using purified ovine GSase (Sigma) as a standard. Each point gives the mean and standard error (vertical bar) from at least 6 assays using two or more cultures from the same large batch. All results are given as units of GSase activity per mg protein.

●---●, GSase activity in F medium;

■——■, GSase activity in F medium plus 5 μM hydrocortisone.

ively, forming both lens and pigment cells in the absence of overlying neuronal cells (de Pomerai and Gali, 1981a). We have therefore studied a marker of normal glial differentiation to determine whether there is a reciprocal relationship between normal and foreign differentiation pathways for these cells. The glial enzyme glutamine synthetase (GSase) is precociously inducible by hydrocortisone (HC) in explants and aggregate cultures of embryonic chick NR, but not in sparse monolayer cultures of dissociated NR cells (Linser and Moscona, 1979). However, dense NR cultures containing abundant nuronal as well as glial cells, can be induced by 5 μM HC to express significant GSase activity (Fig. 4; de Pomerai and Soranson, in preparation). Low levels of this enzyme appear to be expressed autonomously (i.e. in the absence of inducer) both in early (5 day) and later (15-20 day) cultures. There is a sharp decline in the levels of HC-induced GSase activity between 10 and 16 days of culture, and by 26 days the enzyme is barely detectable under all culture conditions (Fig. 4). The continuous presence of HC in the culture medium both delays the onset and reduces the extent of later transdifferentiation into lens (8-10 fold lower δ-crystallin levels after 50 days in HC-treated cultures as compared to controls, Fig. 1). Thus at least one medium factor which encourages normal glial differentiation in early cultures is markedly inhibitory to later transdifferentiation. CAT activity also reaches higher levels and is maintained for much longer (up to 30 days) in media containing HC, suggesting that the maintenance of neuronal functions may depend in part on glial differentiation in these cultures (data not shown).

ACKNOWLEDGEMENTS

This work was supported by a grant from the MRC. M. A. H. Gali is supported by a postgraduate studentship from the government of the Republic of Iraq.

REFERENCES

Araki, M., and Okada, T. S., 1978, Effects of culture media on the "foreign" differentiation of lens and pigment cells from neural retina in vitro, Develop., Growth and Differ., 20:71.

Crisanti-Combes, P., Pessac, B., and Calothy, G., 1978, Choline acetyl transferase activity in chick embryo neuroretinas during development in ovo and in monolayer culture, Develop. Biol., 65:228.

Iqbal, K., and Ottaway, J. H., 1980, Improved sensitivity of glutamine synthetase assay and of acyl hydroxamates, Neurochem. Res., 5:805.

Itoh, Y., 1976, Enhancement of differentiation of lens and pigment cells by ascorbic acid in cultures of neural retinal cells of chick embryos, Develop. Biol., 54:157.

Linser, P., and Moscona, A. A., 1979, Induction of glutamine synthetase in embryonic neural retina: localisation in Müller fibres and dependence on cell interactions, Proc. Natl. Acad. Sci. U.S.A., 76:6476.

Moscona, A. A., and Hubby, J. L., 1963, Experimentally induced changes in glutamotransferase activity in embryonic tissue, Develop. Biol., 7:192.

Okada, T. S., 1980, Cellular metaplasia or transdifferentiation as a model for retinal cell differentiation, Current Topics in Develop. Biol., 16:349.

de Pomerai, D. I., and Gali, M. A. H., 1981a, Influence of serum factors on the prevalence of "normal" and "foreign" differentiation pathways in cultures of chick embryo neuroretinal cells, J. Embryol. exp. Morphol., 62:291.

de Pomerai, D. I., and Gali, M. A. H., 1981b, Determination of chick neuroretinal cells in culture: serum factors acting between 12 and 20 days of culture influence the extent of subsequent lens cell formation, Develop., Growth and Differ., 23:229.

de Pomerai, D. I., and Gali, M A. H., 1981c, Alterations in pH and serum concentration have contrasting effects on "normal" and "foreign" pathways of differentiation in cultures of embryonic chick neuroretinal cells, Develop., Growth and Differ., 23: in press.

MICROENVIRONMENTS CONTROLLING THE TRANSDIFFERENTIATION OF VERTEBRATE
PIGMENTED EPITHELIAL CELLS IN IN VITRO CULTURE

Goro Eguchi, Akira Masuda, Yoko Karasawa,
Ryuji Kodama and Yoshiaki Itoh*

Institute of Molecular Biology
Nagoya University
Nagoya 464, Japan

*Biological Laboratory
Aichi Medical University
Nagakute
Aichi 480-11, Japan

SUMMARY

The transdifferentiation of pigmented epithelial cells in vitro is briefly introduced. Several environmental conditions regulating the process have been demonstrated experimentally. On the basis of these data we have suggested regulatory factors in relation to the mechanisms for the initiation of the transdifferentiation of pigmented epithelial cells, focussing particularly on cell surface functions. Finally, we have presented data which contributes to the establishment of a useful and powerful cell culture system which makes it possible for us to analyse the molecular basis of transdifferentiation.

INTRODUCTION

Lens regeneration in the newt is one of the classic problems in the field of developmental biology, and has been widely studied as an exceptionally clear example of tissue type conversion (for reviews see Scheib, 1965; Reyer, 1977; Yamada, 1977). In the newt the lens can be regenerated by cellular metaplasia of the pigmented epithelial cells (PECs) of the dorsal iris. Similar lens regeneration has been demonstrated not only in urodelean species but also in some fishes and even in avian embryos (Scheib, 1965).

On the basis of classic studies it has been thought likely that a strong metaplastic potential of an already specialized cell is not restricted to PECs of urodelean species, but is widely maintained in PECs of vertebrates in general. Therefore, the experimental system of the PEC would seem to be very useful for studying the regulabilit of the differentiated state of cells, and the experimental analysis of cellular metaplasia shown by PECs must provide important basic information for a much deeper understanding of the mechanisms of cel differentiation in normal and pathological states.

In vitro cell culture of uncontaminated PECs has been attempted as one of a range of modern approaches to establish an experimental system more precise and powerful than in vivo systems and also to analyze the phenomena at the cellular and molecular levels (for reviews see Eguchi, 1976, 1979; Okada, 1976, 1980; Yamada, 1977; Clayton, 1978). Eguchi and Okada (1973) first demonstrated that the progeny of single dissociated PECs from chick embryos can transdifferentiate into lens cells. A similar demonstration has been pos sible with the PECs of human fetuses (Yasuda et al., 1978), and the multipotential nature of the differentiation of the adult newt PECs has also been demonstrated (Eguchi, 1979). However, the efficiency of the transdifferentiation in vitro is rather low, particularly in chick embryonic PECs in which repeated culture passages of cells are always required for transdifferentiation.

We have attempted to analyze the environmental factors controlling the transdifferentiation of PECs in vitro in order to establish much more efficient cell culture systems, which allow us to investigate the molecular mechanisms of transdifferentiation of PECs. In this paper, we should like first to review briefly our studies on transdifferentiation of PECs and to discuss the regulatory mechanism of the transdifferentiation of PECs on the basis of the results obtained through our recent approaches.

MODE OF TRANSDIFFERENTIATION OF PECs IN VITRO

PECs of fully grown newts

PECs dissociated from pigmented epithelia of the iris or the retina of adult newts begin to grow actively after a long lag period when cultured in modified Leibovitz L-15 medium supplemented with 10% of fetal calf serum (FCS). Although most PECs lose their melano somes as they grow, many PECs melanize again when they form a typica monolayered epithelial cell sheet. However, in some parts of the cell sheet depigmented PECs continue to proliferate without melanogenesis and eventually pile up to form transparent bodies (lens stru tures) which contain large amounts of major components of lens proteins, α-, β-, and γ-crystallins and possess specific ultrastructure characteristic of the newt lens (Eguchi et al., 1974).

In such cell culture studies the PECs from the dorsal and the ventral halves of the iris were cultured separately, because the lens regenerating potency of the pigmented epithelia in the newt in vivo is strictly localized in the dorsal half of the eye (Sato, 1930, 1951; Mikami, 1941). In cell culture conditions, however, no significant differences in the potency of PECs for the transdifferentiation have so far been observed between the dorsal and the ventral cells (Eguchi et al., 1974; Abe and Eguchi, 1977). The ventral PECs can transdifferentiate into lens cells in the same manner as the dorsal PECs.

It has been clearly demonstrated in both iris and retina PECs of the newt that more than 50% of clonal colonies derived from single PECs differentiated into lens structures within 30 days after culturing, and that some progenies of retina PECs simultaneously differentiate to give lens cells, neuronal cells, and also pigment cells, showing their multipotential nature by this differentiation. The following observations of clonal cultures of newt PECs seem to be particularly important. The cells in some clones maintained a stable melanogenic potential and came into tight contact with each other to form a cohesive cell sheet, while in other clones cells rapidly lost melanosomes as they grew and formed a loose cell sheet. Lens cell differentiation was only achieved in such non-pigmented clones with loose cell-cell contact (Eguchi, 1976, 1979).

PECs of avian embryos and of human fetuses

Lens regeneration, similar to Wolffian regeneration in the newt, is possible in chick embryos at limited stages of development (Scheib, 1965). Retina PECs dissociated from 8- to 9-day-old chick embryos, in which lens regeneration no longer takes place, grew vigorously when dissociated and cultured in a basic culture medium (EF) consisting of 94 parts of Eagle's minimum essential medium (MEM) and 6 parts of fetal calf serum (FCS) without losing their specificity through two or three subcultures. However, in many culture lines depigmented cells appeared, usually in the terminal period of the third subculture and developed into foci consisting of non-pigmented epithelial cells. Transparent piles of cells, which could be identified as lentoid bodies by immunological and electron-microscopic techniques (Okada et al., 1971), were differentiated from these foci 60 to 90 days after primary inoculation of PECs (Eguchi and Okada, 1973). Such a switch of differentiation from retina PECs of chick embryos into lens cells was directly demonstrated by clonal culture. Similar transdifferentiation has been confirmed in retina PECs of Japanese quail embryos (Eguchi, 1976). Thus, the PECs of avian embryos can switch their differentiation into lens cells after repeated passages and DNA replications, when they are dissociated with EDTA and trypsin and cultured in vitro.

inhibitor of melanogenesis (Eguchi, 1976; Eguchi et al., 1979, Eguchi and Itoh, 1981). When freshly dissociated PECs, which were inoculated in a 5.5 cm culture dish containing 3.5 ml of EF at a cell density of 3.5×10^5, began to grow several days after inoculation, the culture medium was replaced by the medium (EFP), containing PTU at a concentration of 0.5 mM. The PECs grew very actively in this medium to form an epithelial cell sheet without melanogenesis. Even when the cultures attained confluence, depigmented PECs in many parts of the cell sheet continued to grow and eventually overlapped onto the previously formed epithelium. From these parts lentoid bodies began to develop within 25 days after inoculation (Fig. 1a), showing a clear contrast to the cultures of PECs maintained in EF (Fig 1b). Such enhancing effects of PTU can be definitely recognized in cultures at clonal cell density.

In a series of experiments, PECs were maintained for 45 days in EFP (for clonal cultures the serum concentration was occasionally raised to 10%). The efficiency of lens differentiation was more than 7 fold higher than that in the control cultures maintained in EF (Eguchi, 1979).

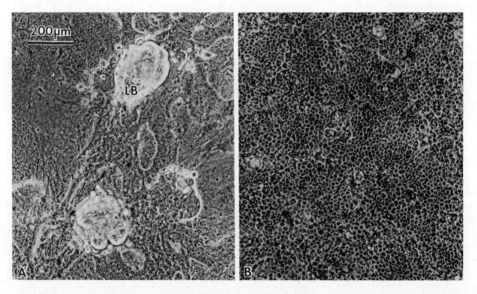

Figure 1. Phase microscopic appearance of cultures of 8- to 9-day old chick embryonic PECs, maintained in vitro for 25 days in the presence or absence of PTU. A: A confluent culture originated from 4×10^5 PECs which were inoculated to a 5.5 cm culture dish and maintained in EFP. Lentoid bodies (LB) are developing from the site where the cell sheet overlaps. B: A confluent culture derived from PECs which were cultured in EF with the same inoculum size as in the culture shown in Fig. 1A.

These effects of PTU seem to be closely correlated with the morphogenetic conditions of dedifferentiated PECs, because the formation of lentoid bodies is achieved only in the parts where cells overlap. To test this possibility, the depigmented epithelial cell sheet, maintained for 20 days in EFP, was punched with a stainless steel tube of 1.0 to 3.0 cm in diameter, and treated with Dispase II (Godo Susei Co. Ltd., Tokyo), a bacterial neutral protease (EC 3.4.24.4) derived from Bacillus polymyxa, at a concentration of 1 500 U/ml in culture medium. This enzyme affects adhesion of PECs to the plastic surface but has no effect on cell-cell contact. Therefore, each disk punched out from the epithelial cell sheet became detached from the plastic surface and formed a tiny vesicle, since detachment of the disk always proceeded centripetally from its periphery. Vesicles thus prepared were cultured on soft agar for 5 days with EFP. Lens cell masses were found to differentiate without exception in all of these vesicles. However, the differentiation of such lens cell masses was found in none of the control vesicles prepared from the pigmented cell sheet of PECs maintained in EF (Table 1 and Fig. 2). Detailed results con-

Table 1. Differentiation of lentoids with antigenic lens specificity obtained by cultivation of vesicles formed from epithelial disks. These were obtained from the non-pigmented progeny of chick embryonic retina PECs cultured under the influence of phenylthiourea.

Origin of vesicles	Pigmented cell sheet formed by PECs cultured in EF for 20 days	Non-pigmented cell sheets formed by PECs cultured in EFP for 20 days
Culture period and medium	5 days in EF	5 days in EFP
Number of vesicles cultured and examined (N)	37	48
Number of vesicles with lentoid differentiation (L)	0	48
Efficiency of lentoid differentiation (L/N x 100)	0 %	100 %

EF: The basic culture medium consisting of 94 parts of Eagle's minimum essential medium (MEM) and 6 parts of fetal calf serum (FCS).
EFP: EF containing phenylthiourea (PTU) at a concentration of 0.5 mM.

cerning the effects of PTU on the transdifferentiation of PECs will be reported elsewhere.

Amplification of the effect of phenylthiourea by modification of culture conditions

The transdifferentiaton of PECs into lens cells was only achieved at the terminal period of the third generation of culture, at the

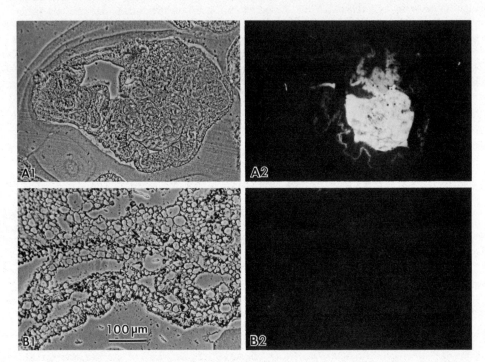

Figure 2. Histological and immunohistochemical appearances of the structures derived from epithelial vesicles of PECs maintained in vitro in the presence or absence of PTU.
A1: A phase microscopic image of a sectioned lentoid differentiated from a vesicle of non-pigmented PECs cultured in EFP. A2: The result of indirect fluorescent antibody staining with anti-δ-crystallin antiserum in the same section as shown in Fig. A1. A large cell mass segregated in the lumen of the vesicle shows intense specific immunofluorescence. B1: A phase microscopic image of irregular epithelial structures derived from vesicles of PECs maintained in EF. All cells of the epithelia contain melanosomes. B2: The result of indirect fluorescent antibody staining of the same section as that shown in Fig. B1. No specific immunofluorescence of δ-crystallin can be seen

earliest, 60 days after the cultures were initiated. Although this prolonged period of cultivation, required for the transdifferentiation of PECs, could be reduced by the introduction of PTU into the culture medium, the efficiency of the transdifferentiation was still rather low. Our collaborative experiments have given rise to a promising finding that the effects of PTU can be greatly increased by testicular hyaluronidase (Boehringer, Mannheim, GmbH, West Germany) (HUase).

Freshly inoculated PECs were transferred to the modified EFP (EFPH) containing HUase at a concentration of 0 to 500 U/ml of the medium several days after culture in EF. Differentiation of lentoid bodies was intensely amplified by HUase in the presence of PTU, while HUase alone suppressed transdifferentiation. Semiquantitative analyses by estimating the number of lentoid bodies developed in a unit area of the culture dish revealed that the efficiency of the transdifferentiation in EFPH (concentration of HUase: 250 to 500 U/ml of medium) is increased more than 50 fold compared with control cultures maintained in EFP (Itoh et al., 1979; Eguchi and Itoh, 1981).

Previous to these attempts, Itoh (1966) clearly demonstrated that in vitro transdifferentiation of neuroretinal cells of chick embryos into both lens and pigment cells can be enhanced by supplementing MEM with dialyzed fetal calf serum (dFCS) in place of FCS and also with ascorbic acid (AsA).

On the basis of his results we have introduced dFCS and AsA to modify the culture conditions of PECs. In a series of experiments, 5×10^5 PECs dissociated from primary cultures maintained in EF for 6 days were seeded and cultured in a fresh culture dish (5.5 cm in diameter) containing 3.5 ml of the medium (EdFPH), which was prepared by replacing FCS of EFPH with dFCS (dialyzed fetal calf serum). In the cultures with EdFPH, these PECs rapidly lost their specificity as pigment cells and grew vigorously to form a highly compact cell sheet, consisting of apparently dedifferentiated epithelial cells, within two weeks of cultivation. These dedifferentiated cells did not express any identifiable specificity when transferred to the same culture condition with EdFPH.

When cells dedifferentiated in EdFPH were dissociated and cultured at extremely high cell density with EdFPH supplemented with AsA at a concentration of 0.5 mM (EdFPHA), the cells began to differentiate very efficiently and synchronously into lens cells within 10 days. Pigment cells have never differentiated in these conditions; the differentiation of lens cells proceeded rapidly and more than 90% of cells came to express lens specificity within two weeks or so. In these culture conditions, the amount of δ-crystallin exceeded 50% of that of total proteins. It was also confirmed immunochemically that the cells differentiated in EdFPHA contain both α-, and β-crystallins. However, the same dedifferentiated population of PECs could exclusively express the specificity of pigment cells, when cultured in EdFPHA in the absence of both PTU and HUase. Some of the quantitative data is summarized in Table 2, which clearly shows that the differentiated state of PECs can be neutralized by culturing them with EdFPH, and that the cell population in this state can readily express the definite specificity of either lens cells or of pigment cells, depending upon the environmental condition.

The detailed results of the experiments described here will be reported elsewhere (Itoh and Eguchi, in preparation).

GENERAL DISCUSSION AND CONCLUSION

As outlined in this paper, although pigmented epithelia of ventral halves of adult newt eyes, 8- to 9-day-old avian embryos, and of human fetuses of 12 weeks after conception never produce lens tissues in vivo, PECs constituting these pigmented epithelia can transdifferentiate into lens cells, when dissociated and cultured in vitro. We know also that transdifferentiation of pigment and lens cells occurs in cultures of neuroretinal cells of both avian embryos and of human fetuses (for review see Okada, 1976, 1980; Clayton, 1978). Therefore, we should like to emphasize that a population of PECs characterised by a "dormant" capacity for transdifferentiation might be preserved in vertebrates in general. It would seem that this capacity of individual multipotent PECs must normally be repressed by the stable tissue architecture.

Although in the eye of the newt in situ the lens can be formed only from the dorsal half of both iris and retina PE, we have found no significant difference in the transdifferentiation capacity between dorsal and ventral PECs when dissociated and cultured in vitro. It was found to be the rule that PEC cells in clones which formed a cohesive cell sheet with close contacts gave rise to pigmented colonies, while those in non-pigmented colonies, with loose cell-cell contacts transdifferentiated into lens even after a short culture period. A similar situation occurs with transdifferentiation of chick PECs which maintained their specificity if cultured on a collagen substratum when they formed a cohesive epithelial cell sheet (Eguchi, 1979; Yasuda, 1979). Sato (1951) has shown that isolated fragments of the newt retinal pigmented epithelium form a lens only when separated from the choroidal substratum and implanted in a lentectomized eye. Thus we may conclude that PECs can be stabilized in part by virtue of their contact with each other and in part by cell-substratum adhesion.

This assumption is also supported by the following observations: PECs of chick embryos, cultured in the presence of PTU, readily overlap onto the cell sheet and lentoid bodies usually differentiate in multi-layered areas. Moreover, the isolated discs derived from de-

Table 2. A result of quantitative analyses of the amount of δ-crystallin and of melanin produced by chick embryonic PECs maintained in various cell culture conditions.

Culture medium** (Days in culture)	Total culture period (days)	Situation of cultures	δ-crystallin*	melanin*
EF : EF (15) (20)	35	Confluent pigmented cell sheet	N.D.	38.3
EF : EFP (6) (20)	26	Confluent cell sheet with lentoid bodies and pigment cells	27.4	3.1
EF : EdFPH : EdFPH (6) (15) (14)	35	Confluent nonpigmented cell sheet	N.D.	trace
EF : EdFPH : EdFPGA (6) (15) (14)	35	Confluent nonpigmented cell sheet with numerous lentoid bodies	560.8	trace
EF : EdFPH : EdFA (6) (15) (14)	35	Confluent pigmented cell sheet	N.D.	21.7

*μg/mg total protein. Method of Laurell (1966) for δ-crystallin, of Oikawa and Nakayasu (1975) for melanin, of Lowry et al. (1951) for total protein.
**Each colon (:) indicates transfer of cultures to the successive generation.
EF: The basic culture medium consisting of 94 parts of Eagle's minimum essential medium (MEM) and 6 parts of fetal calf serum (FCS).
EFP: EF containing phenylthiourea (PTU) at a concentration of 0.5 mM.
EdFHP: The medium consisting of 94 parts of MEM supplemented with 0.5 mM PTU and 250 U/ml testicular hyaluronidase (HUase) and 6 parts of dialyzed fetal calf serum (dFCS).
EdFPHA: EdFPH supplemented with ascorbic acid (AsA) at a concentration of 0.15 mM.
EdFA: EdFPHA lacking in both PTU and HUase.

differentiated PECs following treatment with PTU readily form vesicles. When cultured in soft agar, the lumen fills with cells which transdifferentiate into lens cells. In contrast, pigmented epithelial discs with tight cell-cell contact never differentiated into lens cells, even though they formed similar vesicles following treatment with Dispase. These formed only irregular epithelial structures (cf. Fig. 2) with a persistent lumen.

It is likely that PTU might reduce cell-cell adhesion by affecting the surface properties of PECs so that they are able to form multilayers. In confluent cultures the motility of cells must be repressed in part by cell-cell contact and in part by cell-substratum adhesion. However, in the vesicles formed from epithelial cell sheets, each cell must be free from the restriction of the cell-substratum adhesion, and the cells with less cell-cell contact can readily change their topological position. Thus transdifferentiation might be facilitated by the multilayering following PTU treatment, as compared with the capacity of monolayer cultures of cells. Changes in cell-cell as well as cell-substratum adhesiveness are indicated from the results of culturing day-old chick lens epithelial cells, which differentiate into lentoid bodies (Okada et al., 1973). It has also been suggested that cell surface properties linking cell contact and cell adhesion may regulate the mode of differentiation of lens cells both in vitro and in vivo (Eguchi et al., 1975; Clayton et al., 1976; Clayton, 1979).

PTU is a potent inhibitor of melanogenesis of cultured PECs, which therefore appeared to be dedifferentiating. It was suggested therefore that PTU accelerates the transdifferentiation of PECs by suppressing melanogenesis (Eguchi and Itoh, 1981). That this may not be the case is indicated by recent observations by Masuda. Although naphthylthiourea, an anologue of PTU showed a similar effect to PTU, thiourea which inhibits melanogenesis at a much higher concentration did not enhance transdifferentiation, and another analogue, methyl-thiourea, significantly enhanced both lens and pigment cell differentiation when applied to the cultures of PECs maintained in the presence of HUase (Masuda, unpublished data).

HUase alone has no effect on the transdifferentiation of chick embryo PECs but amplifies the effect of PTU on transdifferentiation. PECs dedifferentiate in culture medium containing PTU and HUase but retain their capacity for differentiation supplemented with dFCS in place of FCS, to either lens cells or pigment cells, depending upon the culture condition. An extremely high efficiency of transdifferentiation of lens cells from PECs can be obtained by manipulation of the culture medium. The amount of δ-crystallin produced in cultures of dedifferentiated PECs which were maintained in EdFHPA for about two weeks exceeded 50% of that of total proteins (cf. Table 2). These results permit us to conclude that the differentiated state of PECs can be neutralized and redifferentiation from such a state can be manipulated.

The importance of the cell surface functions during the process of the transdifferentiation of PECs may be emphasised. HUase must affect the microenvironment surrounding PECs cultured in vitro by digesting the hyaluronates responsible for the maintenance of the physiological states of the cell surface, so influencing cell-cell as well as cell-substratum adhesion of PECs. Furthermore cells grown in cell culture must be affected by dissociating agents. These agents, including proteolytic emzymes, presumably remove some important molecules and change the surface properties of PECs, including the ultrastructural organization of the surface zone, which is thought to be inseparably linked to motility, adhesiveness, growth, and differentiation, itself a necessary step in transdifferentiation.

Finally, there are important effects of ascorbic acid and serum factors, particularly those of the dialyzable components of the serum (Itoh, 1976). Serum factors control the in vitro growth and differentiation of cultures of neuroretinal cell of chick ebmryos, which can transdifferentiate into both lens and pigment cells (de Pomerai and Gali, 1981). However, at present, there is still insufficient evidence for discussion of the mode of action.

In conclusion, it is now evident that a cell population derived from PECs of chick embryos, can be obtained in a neutral state of differentiation, and this population can be directed to differentiate into lens cells or pigment cells by manipulating the culture conditions. Further characterisation and improvement of the system and the growth conditions required and analysis of the molecular basis of the transdifferentiation by this system will surely contribute to studies of the regulability of the differentiated state in normal and abnormal development and of carcinogenesis.

ACKNOWLEDGEMENTS

Some parts of the experimental results introduced in this paper were obtained through collaboration between Professor T. S. Okada, Drs K. Yasuda, K. Watanabe, S. Abe, and one of the authors (G.E.), in the Laboratory of Cell Differentiation and Morphogenesis, Department of Biophysics, University of Kyoto. G.E. is deeply indebted to Professor Okada and his colleagues for their fundamental contribution. I thank sincerely Dr Masamitsu Okamoto, of my laboratory, for his kind advice and assistance throughout our studies.

Our studies were supported in part by a Grant-in-Aid for Cancer Research (Project No. 401539) from the Ministry of Education to G.E. and in part by a research fund from the Mitsubishi Foundation.

REFERENCES

Abe, S., and Eguchi, G., 1977, An anlysis of differentiative capacity of pigmented epithelial cells of adult newt iris in clonal cell culture, Develop. Growth and Differ., 19:309.

Clayton, R. M., 1978, Divergence and convergence in lens cell differentiation: Regulation of the formation and specific content of lens fiber cells, in:"Stem Cells and Tissue Homeostasis," B. I. Lord, C. S. Potten, and R. J. Cole, eds., Cambridge University Press, London and New York, p.115.

Clayton, R. M., 1979, Genetic regulation in the vertebrate lens cell, in:"Mechanisms of Cell Change," J. D. Ebert and T. S. Okada, eds., John Wiley and Sons, New York, p.129.

Clayton, R. M., Eguchi, G., Truman, D. E. S., Perry, M. M., Jacob, J., and Flint, O. P., 1976, Abnormalities in the differentiation and cellular properties of hyperplastic lens epithelium from strains of chickens selected for high growth rate, J. Embryol. exp. Morph., 35:1.

Eguchi, G., 1976, "Transdifferentiation" of vertebrate cells in in vitro cell culture, in:"Embryogenesis in Mammals (Ciba Foundation Symposium 40)," Elsevier, Amsterdam, p.242.

Eguchi, G., 1979, "Transdifferentiation" in pigmented epithelial cells of vertebrate eyes in vitro, in:"Mechanisms of Cell Change," J. D. Ebert and T. S. Okada, eds., John Wiley and Sons, New York, p.273.

Eguchi, G., Abe, S., and Watanabe, K., 1974, Differentiation of lenslike structures from newt iris epithelial cells in vitro, Proc. Natl. Acad. Sci. U.S.A., 71:5052.

Eguchi, G., Clayton, R. M., and Perry, M. M., 1975, Comparison of the growth and differentiation of epithelial cells from normal and hyperplastic lenses of chick: Studies of in vitro cell cultures, Develop. Growth and Differ., 17:395.

Eguchi, G., Hama, T., and Itoh, Y., 1979, Control of transdifferentiation of chick embryonic pigmented epithelial cells into lens cells. I, Zool. Mag. Tokyo, 89:464.

Eguchi, G., and Itoh, Y., 1981, Regulation of transdifferentiation by microenvironmental factors in vertebrate pigmented epithelial cells in vitro, in:"Proccedings of IPCC, Sendai," University of Tokyo Press, Tokyo (in press).

Eguchi, G., and Okada, T. S., 1973, Differentiation of lens tissue from the progeny of chick retinal pigment cells cultured in vitro: A demonstration of a switch of cell types in clonal cell culture, Proc. Natl. Acad. Sci. U.S.A., 70:1495.

Itoh, Y., 1976, Enhancement of differentiation of lens and pigment cells by ascorbic acid in cultures of neural retinal cells of chick embryos, Develop. Biol., 54:157.

Itoh, Y., Hama, T., and Eguchi, G., 1979, Control of transdifferentiation of chick embryonic pigmented epithelial cells into lens cells. II, Zool. Mag. Tokyo, 83:465.

Laurell, C. B., 1966, Quantitative estimation of proteins by electro-

phoresis in agarose gel containing antibodies, Anal. Biochem., 15:45.
Lowry, O. H., Rosebrough, N. J., Farr, A., L., and Randall, R., J., 1955, Protein measurement with the Folin phenol reagent, J. biol. Chem., 193:265.
Mikami, Y., 1941, Experimental analysis of the Wolffian lens-regeneration in adult newt, Triturus pyrrhogaster, Jap. J. Zool., 9:269.
Oikawa, A., and Nakaysasu, M., 1973, Quantitative measurement of melanin as tyrosine equivalents and as weight of purified melanin, Yale J. Biol. Med., 46:500.
Okada, T. S., 1976, "Transdifferentiation" of cells of specialized eye tissues in cell culture, in:"Tests of Teratogenicity in vitro," J. D. Ebert, and M. Marois, eds., North-Holland, Amsterdam, p.91.
Okada, T. S., 1980, Cellular metaplasia or transdifferentiation as a model for retinal cell differentiation, Current Topics in Develop. Biol., 16:349.
Okada, T. S., Eguchi, G., and Takeichi, M., 1971, The expression of differentiation by chicken lens epithelium in in vitro cell culture, Develop. Growth and Differ., 13:323.
Okada, T. S., Eguchi, G., and Takeichi, M., 1973, The retention of differentiated properties by lens epithelial cells in clonal culture, Develop. Biol., 45:318.
de Pomerai, D. I., and Gali, M. A. H., 1981, Influence of serum factors on the prevalence of "normal" and "foreign" differentiation pathways in cultures of chick embryo neuroretinal cells, J. Embryol exp. Morph., 62:291.
Reyer, R. W., 1977, The amphibian eye, development and regeneration, in:"Handbook of Sensory Physiology VII/5: The Visual System in Vertebrates," F. Crescitelli, ed., Springer, Berlin, p.309.
Sato, T., 1930, Beitrage zur Analyse der Wolffischen Linsen-regeneration II, Wilhem Roux Arch. EntwMech. Org., 122:451.
Sato, T., 1951, Uber die Linsenbildende Fahigkeit des Pigment-epithels bei Diemyctylus pyrrhogaster I. Pigmentepithel aus Dorsalen Augen bereich, Embryologia, 1:21.
Scheib, D., 1965, Recherches recentes sur la regeneration du crystallin chez les vertebres. Evolution du probleme entre 1931 et 1963, Ergebn. Anat. EntwGesch., 38:45.
Yamada, T., 1977, Control mechanisms in cell-type conversion in the newt lens regeneration, Monographs in Developmental Biology, Vol. 13, A. Wolsky, ed., S. Karger, Basel, Munchen, Paris, London, New York, Sidney.
Yasuda, K., Okada, T. S., Hayashi, M., and Eguchi, G., 1978, A demonstration of a switch of cell type in human fetal eye tissues in vitro: Pigmented cells of the iris or the retina can transdifferentiate into lens, Expl. Eye Res., 26:591.
Yasuda, K., 1979, Transdifferentiation of "lentoid" structures in cultures derived from pigmented epithelium was inhibited by collagen, Develop. Biol., 68:618.

CAN NEURONALLY SPECIFIED CELLS TRANSDIFFERENTIATE INTO LENS?

T. S. Okada, K. Yasuda, H. Kondoh, S. Takagi, and K. Nomura

Institute for Biophysics
University of Kyoto
Kyoto 606
Japan

Both neural and pigmented retina cells readily transdifferentiate into lens, when the dissociated cells are maintained in cell culture for about 3-5 weeks. This unique ability is widely distributed in various vertebrate embryos including birds and mammals, which cannot usually regenerate lost parts of the eye in situ (cf. a recent review by Okada, 1980).

In the case of the formation of lens from pigmented retina the cellular origin of the lens cells is unequivocal, since pigmented retina consists of a homogeneous population of pigment cells and a direct proof of transdifferentiation can be obtained by clonal cell culture of a single living black cell (Eguchi and Okada, 1973). The situation is more complex and intriguing in the case of cultures of neural retina (NR), because this tissue is destined to develop into diverse cell types in normal development.

In avian embryos of 3.5 to 9 days of incubation, which are now widely used as materials for the transdifferentiation studies in cell cultures, the NR is well defined as a separate tissue. However, it is probable that lens and pigment cells in cell culture of NR are derived from multipotential progenitor cells which are included in this tissue but have not expressed any differentiated phenotypes. If this is the case, then lens (and pigment cells) differentiation in cultures of NR does not belong to the category of a true transdifferentiation event, but is a case of differentiation.

When cells of NR of 3.5-day-old chick embryos were cultured, at least two distinct cell types, N-cells and E-cells, appeared in the early stages before the differentiation in vitro of lens and pigment cells. Among these two cell types, N-cells appeared to be neuronal from their morphology. In fact, changes in the number of N-cells closely parallels the activity of choline-acetyltransferase (CAT), an enzyme which serves as a marker for the neuronal phenotype of NR (Nomura et al., 1980). The specific subject of the present communication is to examine: (1) the expression of neuronal phenotypic markers in N-cells and (2) lens differentiation from N-cells to determine the occurrence of a <u>true</u> transdifferentiation in this system.

RESULTS

<u>1) Continuous Observation of N-cell Clusters in Culture.</u>

By 8-10 days, cultures of 3.5-day-old chick embryonic NR consisted of small N-cells, often with bipolar processes, and larger E-cells. Many N-cells were assembled into clusters, which were interconnected by very elongated processes (putative axons). The locations of selected N-cell clusters were marked and further changes in marked clusters were observed by daily photography and cinematography (Okada et al., 1979; Okada, 1980). In some clusters, an apparent change of N-cell clusters into lentoid bodies was observed. The change did not start synchronously within a single culture plate, but, once started, it was very rapid, taking 2-3 days for the differentiation of lentoid bodies.

<u>2) Identification of Neuronal Phenotypes in N-cells.</u>

a) <u>Staining with merocyanine-540 (MC).</u> A fluorescent dye, merocyanine-540, is known to be a specific probe for detecting cells with highly excitable membrane (Easton et al., 1978). When early cultures were stained with this dye and observed under a fluorescent microscope, only N-cells were fluorescent. In late cultures, many "mosaic" clusters appeared which consisted of MC-stained cells, and lentoid cells which were positive in immunofluorescence using anti-δ-crystallin serum.

b) <u>Separation of N-cell and E-cell fractions.</u> Cultures at 8-10 days were harvested and the cell suspension was separated on a Percoll gradient. In two separate fractions, the activity of CAT was much higher in the fraction which contained many smaller N-cells as compared with the other fraction containing a majority of E-cells.

c) <u>Separation of MC-stained and non-stained cell fractions.</u> The cell suspension was prepared from 8-10 day cultures previously stained with MC. Cells stained with MC were separated from non-stained cells by using a fluorescence activated cell sorter (FACS). The activity of CAT was determined in these separated fractions and it was found that the activity, averaged per individual cell, was about 13.5 times higher in the fraction with MC-stained cells than in the other fraction.

3) Culture of Cells of the Isolated Fraction.

The fractions separated by FACS were cultured for 10-11 days. At the end of this period, lentoid differentiation had occurred in both fractions. No quantitative data have been obtained due to a limited supply of the culture samples to be used for the measurement both of δ-crystallin content, and of CAT activity.

4) Chimeric Cultures of Chick and Quail Cells.

In our present technique, unfortunately, cells of the separated fractions did not grow well in further cell culture and the differentiation of lentoid bodies was poor in quantity. It is highly probable that both cell growth and the expression of "altered" differentiative phenotypes require cellular interactions between different cell types derived from NR and contained in a single culture plate. In particular, the maintenance of N-cells seems to be greatly dependent on the presence of E-cells. It is necessary therefore, to recombine the separated fractions again for further cell culture, in order to obtain extensive cell growth and differentiation. If we can use marked cells of either fraction, it would be possible to identify the origin of lentoid bodies formed in the recombination cultures.

A recombination of chick and quail cell fractions was therefore carried out. It is well known that the nuclear structure is conspicuously different among these two species (LeDouarin, 1973). More fortunately, δ-crystallin differs in electrophoretic mobility between these two species, though totally indistinguishable immunologically.

The N-cell fraction was obtained from 8-10 day cultures of 3.5-day-old quail embryonic NR, while the E-cell fraction was similarly prepared from chick NR cultures. In this experiment, a second fractionation was performed on the N-cell fraction, to separate it into three sub-fractions, N_1, N_2, and N_3. Each of these subfractions was recombined with chick E-fraction for further cell cultures. After 10 days, the cultured cells were harvested and homogenized. δ-crystallin was immuno-precipitated from the saline extracts with specific anti-δ, and the species-specificity of δ-crystallin was identified by means of electrophoresis in SDS gel. In all recombination cultures, both chick and quail δ-crystallin bands were found (Fig. 1). In the case of quail N_2- plus chick E-cell fraction, the formation of lentoid bodies was most abundant and the share of quail δ-crystallin was largest here. N_2 is a highly purified fraction consisting of mainly N-cells, stainable with MC, and with the highest activity of CAT amongst the three subfractions derived from the N-cell fraction.

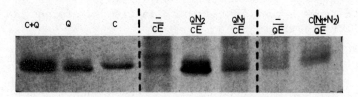

Figure 1. Differences in electrophoretic mobilities betweeen chick E and quail crystallin. In all samples, δ-crystallin was precipitated from the extracts by using a specific antiserum against chick δ-crystallin and applied to the electrophoresis in SDS. C + Q: a mixture of chick and quail δ-crystallin with two bands. Q: quail δ-crystallin. C: chick δ-crystallin. -/CE: cultures of the separated chick E-fraction. QN_2/CE: chimeric cultures of quail N_2-fraction and chick E-fraction. QN_1/CE: chimeric cultures of quail N_1- and chick E-fractions. -/QE: cultures of separated quail E-fraction. $C(N_1 + N_2)/QE$: chimeric cultures of chick N_1- and N_2-fractions combined with quail E-fraction. Note the presence of a very heavy band of quail δ-crystallin in QN_2/CE. In case of $C(N_1 + N_2)/QE$, a band of chick δ-crystallin is clearer than that of quail δ-crystallin.

DISCUSSION AND CONCLUSION

It has been shown that non-neuronal cells in cell cultures of NR (E-cells) can be a progenitor of lentoid and pigment cells (Okada, 1977; Moscona and Degenstein, 1980). The series of experiments reported here provide circumstantial evidence for the differentiation of lentoid bodies from the N-cells found in the early culture period. N-cells probably correspond to matrix cells and to very immature neuroblasts of the developing NR in situ. N-cells, however, are differentiated in cell culture, in the sense that they are distinct from E-cells. Stainability with MC, and the activity of CAT suggest the neuronal characters of these cells. In other words, N-cells have expressed some of the phenotypic markers of neuronal cells. Thus, we can state that a true transdifferentiation occurs in the system of cell culture of NR; namely lentoid differentiation from such cells as are immature but have already expressed neuronal phenotypes.

Recently, several important studies have been carried out which demonstrate an unexpected flexibility in the differentiation of nervous tissue cells. Instability in the choice of neuro-transmitter molecules of autonomic neurones is well known (Patterson et al.,

1975; LeDouarin et al., 1978). A multipotentiality in the differentiation of cell lines derived from neurotumors is described in this symposium by Sueoka (cf. Imada and Sueoka, 1978). Pigment cell differentiation occurs in cultured ganglion cells, though probably not from neuronal cells but rather from non-neuronal cells included in the ganglia (Nicholas and Weston, 1977). Transdifferentiation of immature neuronal cells into lens is a very dramatic change in the pathway of differentiation. Studies to elucidate a "key" to lead to this change as well as molecular events underlying the process should make an important contribution to the understanding of the mechanisms of cell differentiation in general.

REFERENCES

Easton, T. G., Valinsky, J. E., and Reich, E., 1978, Merocyanine 540 as a fluorescent probe of membranes: Staining of electrically excitable cells, Cell, 13:475.

Eguchi, G., and Okada, T. S., 1973, Differentiation of lens tissue from the progeny of chick retinal pigment cells cultured in vitro: A demonstration of a switch of cell types in clonal cell culture, Proc. Natl. Acad. Sci. U.S.A., 70:1495.

Imada, M., and Sueoka, N., 1978, Clonal sublines of rat neurotumor RT4 and cell differentiation. I. Isolation and characterization of cell lines and cell type conversion, Develop. Biol., 66:97.

LeDouarin, N., 1973, A biological cell labelling technique and its use in experimental embryology, Develop. Biol., 30:217.

LeDouarin, N., Teillet, M. A., Ziller, C., and Smith, J., 1978, Adrenergic differentiation of cells of the cholinergic ciliary and Remak ganglia in avian ebmryos after in vivo transplantation, Proc. Natl. Acad. Sci., U.S.A., 75:2030.

Moscona, A. A., and Degenstein, L., 1980, Lentoids in aggregates of embryonic neural retina cells, Cell Differ., 10:39.

Nicholas, D. H., and Weston, J. A., 1977, Melanogenesis in cultures of peripheral nervous tissue. I. The origin and prospective fates of cells giving rise to melanogenesis, Develop. Biol., 60:217.

Nomura, K., Takagi, S., and Okada, T. S., 1980, Expression of neuronal specificities in "transdifferentiating" cultures of neural retina, Differentiation, 16:141.

Okada, T. S., 1977, A demonstration of lens forming cells in neural retina in clonal cell culture, Develop., Growth and Differ., 19:47.

Okada, T. S., 1980, Cellular metaplasia or transdifferentiation as a model for retinal differentiation, Curr. Topics in Dev. Biol., 16:349.

Okada, T. S., 1981, Phenotypic expression of embryonic neural retinal cells in cell culture, Vision. Res., 21:83.

Okada, T. S., Yasuda, K., and Nomura, K., 1979, The presence of multipotential progenitor cells in embryonic neural retina as revealed by clonal cell culture, in:"Cell Lineage, Stem Cells and Cell Determination," N. LeDouarin, ed., Elsevier/North Holland, Amsterdam, p.335.

Patterson, P. H., Reichardt, L. F., and Chum, L. L. Y., 1975, Biochemical studies on the development of primary sympathetic neurons in cell culture, Cold Spr. Harb. Symp. quant. Biol., 40:389.

THE EFFECTS OF N-METHYL-N'-NITRO-N-NITROSOGUANIDINE ON THE

DIFFERENTIATION OF CHICKEN LENS EPITHELIUM IN VITRO:

THE OCCURRENCE OF UNUSUAL CELL TYPES

Ruth Clayton and Charles Patek

Institute of Animal Genetics
University of Edinburgh
Edinburgh   EH9 3JN

INTRODUCTION

All vertebrate embryo retina cells so far tested are able to transdifferentiate into lens cells in cell culture. (Recent reviews include Clayton, 1978; Eguchi, 1979; Okada, 1980.) This potential is associated with a molecular overlap between these tissues, lens cyrstallin mRNA being normally expressed in both pigmented and neural retina of the embryo at low levels (Jackson et al., 1978). The rate of increase of crystallin mRNA with transdifferentiation is related to the initial level (Clayton et al., 1979; Thomson et al., 1981). Although retina mRNAs are also expressed in lens at low levels (Jackson et al., 1978), lens cells do not transdifferentiate to retina cell types under any of the conditions which facilitate transdifferentiation of other tissues to lens cells.

Resistance to transdifferentiation is also shown by amphibian ventral iris in situ, the cells of which do not transdifferentiate to lens in vivo, but which show transdifferentiation potential in cell culture (Eguchi, 1979). Exposure in vivo to N-methyl-N'-nitro-N-nitrosoguanidine (MNNG), an alkylating agent which is a mutagen and a carcinogen, permits the expression in situ of this potential for transdifferentiation into lens cells (Eguchi and Watanabe, 1973). Tumours were not observed but the potential for lens formation was regularly obtained and persistent. The regularity of the effect points to an epigenetic mechanism, related to the normal unexpressed potential of iris cells. Mutations, which would presumably occur, would be random and therefore probably not relevant.

We have investigated the possibility that exposure to MNNG might similarly permit lens cells to disclose whether their content of retina mRNA species might also confer a potential for transdifferentiation events.

MATERIALS AND METHODS

Day old chicken lens epithelium (LE) from a slow growing (egg-laying) strain (RN) were seeded at 0.8×10^5 cells/ml and grown as described previously (Clayton et al., 1980). The cultures were treated for 1 hour with subcytotoxic levels of MNNG (0.75 or 7.5 µg/ml) when plated out on the first day of culture (zero hours of culture), washed, and the medium renewed every 3 days. After 28 days they were harvested and subcultured at the same initial cell density. Neural retina (NR) cells from 8 day chick embryos were grown and treated with MNNG at 7.5 µg/ml, either on days 9 and 11 or on days 19 and 21, as described previously (Clayton et al., 1980) and the cultures followed for 50-60 days.

RESULTS

By day 28, primary LE control cultures contained normal epithelial cells (0.9×10^5 cells/ml) and differentiated lentoids (bodies composed of lens fibre cells, Fig. 1A). The secondary control cultures were similar, with a comparable cell number, but also contained a small number of fibroblast-like cells (Fig. 1B). Lentoids were slightly reduced in size and number. By day 28, primary cultures treated with MNNG at 7.5 µg/ml contained fewer cells than the controls (0.3×10^5/ml), and a small proportion of the cells were fibroblast-like (Fig. 1B). These cultures failed to differentiate into lentoids. The secondary cultures from these MNNG treated cells differed in several ways from control secondary cultures. Their rate of cell growth was three times higher than primary and secondary control cultures. Their rate of cell growth was three times higher than primary and secondary control cultures, and eight times higher than that of primary MNNG-treated cultures. By day 14, MNNG-treated secondary cultures were comprised of 4 distinct morphological cell types, (a) a small proportion of epithelial cells (Fig. 1A) (b) a large majority of fibroblast-like cells (Fig. 1B) (c) a lesser number of black pigmented cells (Fig. 1C) and (d) a small but significant number of 'neurone-like' cells (Fig. 1D). No lentoids were obtained. The black 'pigment epithelium-like' and the 'neurone-like' cells were never seen in control cultures, nor in secondary cultures derived from cells treated with MNNG on the 7th day of primary culture, which contained only epithelial and a high proportion of fibroblast-like cells (Patek and Clayton, these proceedings), nor from MNN treated cells from two other genotypes tested (results not shown). Pigmented and neurone-like cells were however repeatably obtained fro

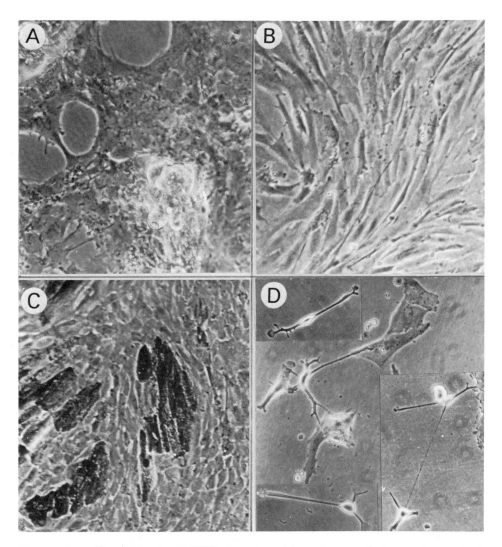

Figure 1. The effect of MNNG treatment on day 1 of the primary culture (7.5 μg/ml) on the in vitro differentiation of chicken lens epithelial cells in primary and secondary culture. (A) day 28 primary control cultures with lentoid bodies (B) day 28 MNNG-treated primary cultures with 'fibroblast-like' cells and no lentoids (C) day 14 secondary MNNG-treated cultures with black 'pigment epithelium-like' cells (D) day 14 secondary MNNG-treated cultures with 'neurone-like' cells.

cultures of RN genotypes treated at 1 hour of culture.

The action of MNNG was dose-dependent. Cells treated with a lower level (0.75 µg/ml) of MNNG were morphologically indistinguishable from control cultures, exhibited similar growth rates, and lentoids were differentiated. A correlation between the presence of fibroblast-like cells, high actin levels and depressed crystallin levels is reported in Patek and Clayton, these proceedings. The effects of MNNG treatment of NR cultures are shown in tabular form in Table 1. The morphological changes in these cultures have been reported previously (Clayton et al., 1980; Clayton, 1982) and will be described in further detail elsewhere (Patek and Clayton, in preparation).

DISCUSSION

Fibroblast-like cells appear in long-term cultures of bovine and rat LE cells (Tassin et al., 1976; Rink and Vornhagen, 1979), and natural crystallin loss from bovine and rat LE cultures in such cultures is also well documented (van Venrooij et al., 1974; Courtois et al., 1978; Hamada et al., 1979). Actin is normally present at low levels in the lens (Ramaekers et al., 1980). A progressive increase in actin synthesis and accumulation occurs in successive subcultures of MNNG treated LE cells (Patek and Clayton, these proceedings). Crystallin loss, the synthesis and accumulation of high levels of actin, and fibroblastic morphology are also found in SV40 transformed hamster LE cells in vitro (Bloemendal et al., 1980). Our observations indicate that an effect of MNNG on cultures appears to be to both increase and accelerate the appearance of fibroblast like cells, crystallin loss and the increase of actin. The finding that crystallins are still synthesised in these treated cells, although accumulated levels are low or undetectable (Patek and Clayton, these proceedings), suggests either a rapid turnover, or a relatively low rate of synthesis, or both.

Although further work is required to establish conclusively the precise nature of these 'neurone-like' and 'black pigment epithelium-like' cells, their morphology suggests the possibility, which requires exploration, that they may have higher levels of neural and pigmented retina mRNAs respectively than are found in the normal lens. Their appearance in cultures of lens epithelium is without precedent. Cytochalasin B can cause arborisation of lens epithelial cells (Mousa and Trevithick, 1977), but this process is immediate and reversible. The delayed appearance of stable neurone-like cells, and of epithelial cells containing pigment has never been observed by us nor recorded by other laboratories, under any culture conditions used hitherto, and their appearance in secondary culture seems therefore to be linked to the exposure to MNNG many cell divisions earlier. Furthermore their appearance is restricted to treatment of N-r genotype cells

Table 1. Transdifferentiation of day 8 chicken neural retina cells in primary cell culture following exposure to MNNG.

	NEURONES	NEUROEPITHELIUM	LENS CELLS	PREPIGMENT CELLS	PIGMENTED CELLS
Control cultures	6-17 days, then lost naturally.	Morphology normal	25-30 days	28-30 days. These cells become pigmented.	33-38 days
MNNG for 1 hour on days 9 and 11	Between 6-12 or 13 days, then lost. New neurone-like cells reappear at 30-36 days.	Morphology normal, but by 30 days the cell count is half that of controls.	32-40 days	None	None
MNNG for 1 hour on days 19 and 21.	6-17 days, then lost naturally. New neurone-like cells reappear at 35-50 days.	Morphology normal, but by 30 days the cell count is 75 % that of controls.	27-32 days. α-crystallin reduced in amount.	36-38 days, but about 50 % of the number in controls. Only 20 % of these become pigmented.	43-48 days, but cells exhibit morphological abnormalities.

Fibroblast-like cells are only seen after MNNG treatment on days 9 and 11. They begin to appear at about 30 days of culture.

only, with MNNG and 0.75 µg/ml at 0 hours of culture, and they do not appear if cultures are treated with the higher concentration of MNNG nor if cultures are treated at 7 days (Patek and Clayton, these proceedings). An antigen with neural retina specificity was obtained in spontaneously transformed rat LE cultures by Hamada et al. (1979), although these cells did not exhibit neurone-like morphology.

The effects of MNNG on transdifferentiating NR cultures are also delayed by several cell divisions and appear also to be determined by the state of differentiation of the culture at the time of treatment. Early treatment abolishes the pigment pathway and delays lentoid formation, while later treatment, when committment to these pathways is presumably at a more advanced stage (Thomson et al., 1979) leads only to delay of pigment transdifferentiation and a slight modification of lentoid composition.

The apparently specific nature of the effects on differentiation of this carcinogen implies that the mechanism is both epigenetic and non-random and is a characterisitic of the cell population rather than the chemical agency. Other reports suggest that carcinogens applied to a tissue at a critical stage may modify differentiation rather than lead to tumour formation. Treatment of the regeneration blastoma of an amputated newt limb with MNNG led to supernumerary limb formation, but tumours were not seen over a 5 month period (Tsonis and Eguchi, 1981). This observation is in agreement with that of Breedis (1952) who also obtained accessory limb from newts using methylcholanthrene and other carcinogens. Exposure of the human female foetus to stilboestrol during the 18th week of gestation may lead to eventual development of carcinoma of the cervix and vagina but can also lead to specific non-malignant anomalies of differentiation of the same tissues (Herbst et al., 1977).

Even if malignancies are produced, they may be related closely to the stage of differentiation of a target tissue. Exposure of foetal rats to ethylnitrosourea at 15 days of gestation leads to tumours of predominantly neuronal characteristics but at 18 days leads to tumours of predominantly glial characterisitics. All foetal tissues are, presumably, exposed to the carcinogen, but the effects over this period are restricted, the periods of susceptibility apparently corresponding to the period of rapid cell division in the CNS, when neurones differentiate before glia (Rajewsky et al., 1977). Similarly the effects of ethyl nitrosourea on the oppossum embryo are dependent on the stage of differentiation treated (Jurgelski et al., 1979).

Specific shifts in differentiation following MNNG treatment were found for ventral iris by Eguchi and Watanabe (1973) and for gut cells in culture by Fukamachi and Takayama (1979). Other agencies have been found to permit specific epigenetic shifts of differentiation. Muscle cells have been obtained from fibroblasts in vitro by treatment with 5-azacytidine (Taylor and Jones, 1979) and for cells from fibroblasts following treatment with 3-diazadenosine (Chiang, 1981).

A disturbance of regulatory mechanisms, rather than permanent genetic damage, and a definition of possible responses to such disturbance by the cells themselves is also indicated by examples of reversibility of the malignant state; for example the normal differentiation of teratocarcinoma cells (Mintz and Illmensee, 1975) or the transdifferentiation of normal lens cells from retinoblastoma cells (Okada, 1978). A similar argument applies in those cases where transformed cell types, including Friend cells, lymphoblastomas and neuroblastomas, show partial or considerable normal differentiation after treatment with such agencies as DMSO, 5BuDR, DbcAMP (Harrison, 1977; Reuben et al., 1976; Symonds and Sachs, 1979; Ross et al., 1975). Indeed neuronal differentiation of neuroblastoma cells has followed exposure to the carcinogen, MNNG (Yoda and Fujimura, 1979).

Preliminary evidence suggests that one effect of MNNG on LE cells in vitro is to induce selective quantitative shifts amongst the translatable mRNAs (Clayton et al., 1980). Similar results were obtained with thioacetamide treated liver cells by Chakrabarty and Schneider (1978). This might plausibly permit the types of change observed, including selective repression or loss of certain crystallins (Patek and Clayton, these proceedings), the increase in actin, and the appearance of unusual cell types, whose morphology opens up the possibility that they may express increased levels of the retina mRNAs which are normally found in the intermediate abundance class of lens mRNA. These possibilities are now open to further test.

The presence in a tissue of naturally occurring low levels of "inappropriate" gene products (such as chorionic gonadotrophin in tissues of the gastrointestinal system), would appear to be linked to the capacity of these tissues to synthesise these products at high levels in tumours arising in situ (Baylin and Mendelsohn, 1980). It has been suggested (Clayton, 1982) that this change in differentiation, following some unknown oncogenic stimulus, is a phenomenon similar to those outlined here.

ACKNOWLEDGEMENTS

We are grateful to the C.R.C. and the M.R.C. for their support, and to Ross Breeders Limited, Newbridge, Dumfries for the supply of day old chickens. We thank Mr S. Williamson for technical assistance.

REFERENCES

Baylin, S. B., and Mendelsohn, G., 1980, Ectopic (inappropriate) hormone production by tumours: mechanisms involved and the biological and clinical implications, Endocr. Rev., 1:45.

Bloemendal, H., Lenstra, J. A., Dodemont, H., Ramaekers, F. C. S., Groeneveld, A. A., Dunia, I., and Benedetti, E. L., 1980, SV40-transformed hamster lens epithelial cells: A novel system for the isolation of cytoskeletal messenger RNAs and their translation products, Expl. Eye Res., 31:513.

Breedis, C., 1952, Induction of accessory limbs and of sarcoma in the newt (Triturus viridescens) with carcinogenic substances, Cancer Res., 12:861.

Chakrabarty, P. R., and Schneider, W. C., 1978, Increased activity of rat liver messenger RNA and of albumin messenger RNA modulated by thioacetamide, Cancer Res., 38:2043.

Chiang, P. K., 1981, Conversion of 3T3-L1 fibroblasts to fat cells by an inhibitor of methylation: Effect of 3-diazadenosine, Science, 211:1164.

Clayton, R. M., 1978, Divergence and convergence in lens cell differentiation: Regulation of the formation and specific content of lens fibre cells, in:"Stem Cells and Tissue Homeostasis," B. I. Lord, C. S. Potten, and R. J. Cole, eds., Cambridge Unversity Press, Cambridge, p.115.

Clayton, R. M., 1982, Cellular and molecular aspects of differentiation and transdifferentiation of ocular tissues in vitro, in: "Differentiation In Vitro," M. M. Yeoman and D. E. S. Truman, eds., Cambridge University Press, Cambridge, p.83

Clayton, R. M., Thomson, I., and de Pomerai, D. I., 1979, The relatship between crystallin mRNA expression in retinal cells and their capacity to redifferentiate into lens, Nature, 282:628.

Clayton, R. M., Bower, D. J., Clayton, P. R., Patek, C. E., Randall, F. E., Sime, C., Wainwright, N. R., and Zehir, A., 1980, Cell culture in the investigation of normal and abnormal differentiation of eye tissues, in:"Tissue Culture in Medical Research II," R. J. Richards, and K. T. Rajan, eds., Pergamon Press, Oxford and New York, p.185.

Courtois, Y., Simmonneau, L., Tassin, J., Laurent, J., Laurent, M., and Malaise, E., 1978, Spontaneous transformation of bovine lens epithelial cells, Differentiation, 10:23.

Eguchi, G., 1979, Transdifferentiation in pigment epithelium cells of the vertebrate eye in vitro, in:"Mechanisms of Cell Change," J. D. Ebert, and T. S. Okada, eds., John Wiley, New York, p.273.

Eguchi, G., and Watanabe, K., 1973, Elicitation of lens formation from the 'ventral iris' epithelium of the newt by a carcinogen, N-methyl-N'-nitro-N-nitrosoguanidine, J. Embryol. exp. Morph., 30:63.

Fukamachi, H., and Takayama, S., 1979, Promotion of epithelial keratinization by N-methyl-N'-nitro-N-nitrosoguanidine in rat forestomach, Experientia, 35:666.

Hamada, Y., Watanabe, K. Y., Aoyama, H., and Okada, T. S., 1979, Differentiation of rat lens epithelial cells in short- and long-ter cultures, Develop., Growth and Differ., 21:205.

Harrison, P. R., 1977, The biology of the Friend Cell, in:"International Review of Biochemistry, Biochemistry of Cell Differentiation, II," J. Paul, ed., University Park Press, Baltimore.

Herbst, A. L., Scully, R. E., Robbey, S. J., Welch, W. R., and Cole, P., 1977, Abnormal development of the human genital tract following prenatal exposure to diethyl stilboestrol, Cold Spring Harbour Conferences on Cell Proliferation, 4:399.

Jackson, J. F., Clayton, R. M., Williamson, R., Thomson, I., Truman, D. E. S., and de Pomerai, D. I., 1978, Sequence complexity and tissue distribution of chick lens crystallin, Develop. Biol., 65:383.

Jurgelski, W., Hudson, D., and Falk, H. L., 1979, Tissue differentiation and susceptibility to embryonal tumor induction by ethylnitrosourea in the oppossum, Natl. Cancer Inst. Monogr., 51:123.

Lawley, P. D., and Thatcher, C. J., 1970, Methylation of deoxyribonucleic acid in cultured mammalian cells by N-methyl-N'-nitro-N-nitrosoguanidine, Biochem. J., 116:693.

Mintz, B., and Illmensee, K., 1975, Normal genetically mosaic mice produced from malignant teratocarcinoma cells, Proc. Natl. Acad. Sci. U.S.A., 73:3585.

Mousa, C. Y., and Trevithick, J. R., 1977, Differentiation of rat lens epithelial cells in tissue. II. Effects of cytochalasins B and D on actin organisation and differentiation, Develop. Biol., 60:14.

Okada, T. S., 1978, Redifferentiation and transdifferentiation of retinoblastoma cells, in:"Proceedings of the VIth International Congress on Eye Research," Osaka, Japan.

Okada, T. S., 1980, Cellular metaplasia or transdifferentiation as a model for retina cell differentiation, Current Topics in Develop. Biol., 16:349.

Rajewsky, M. F., Augenlicht, L. H., Biessmann, H., Both, R., Hilser, D. F., Laerum, O. D., and Lomakena, L. Y., 1977, Nervous-system-specific carcinogenesis by ethylnitrosourea in the rat: Molecular and cellular aspects, Cold Spring Harbor Conferences on Cell Proliferation, 4:709.

Ramaekers, F. C. S., Osborn, M., Schmid, E., Weber, K., Bloemendal, H., and Franke, W. W., 1980, Identification of the cytoskeletal proteins in lens forming cells. A special epithelioid cell type, Expl. Cell Res., 127:309.

Rink, H., and Vornhagen, R., 1979, Crystallins of lens epithelium cells during ageing and differentiation, Ophthal. Res., 11:355.

Ross, J., Olmstead, J. B., and Rosenbaum, J. L., 1975, The ultrastructure of mouse neuroblastoma cells in tissue culture, Tissue and Cell, 7:107.

Rueben, R. C., Wilfe, R. L., Breslaw, R., Rifkind, R. A., and Marks, P. A., 1976, A new group of potent inducers of differentiation in murine erythroleukaemia cells, Proc. Natl. Acad. Sci. U.S.A., 73:862.

Symonds, G., and Sachs, S., 1979, Activation of normal genes in malignant cells: activation of chemotaxis in relation to other stages of normal differentiation in myeloid leukaemia, Som. Cell Genet., 5:931.

Tanaka, M., Levey, J., Terada, M., Breslow, R., Rifkind, R., and Marks, P. A., 1979, Induction of erythroid differentiation in murine virus infected erythroleukemia cells by highly polar compounds, Proc. Natl. Acad. Sci. U.S.A., 72:1003.

Tassin, J., Simmonneau, L., and Courtois, Y., 1976, Epithelial lens cells: a model for studying in vitro ageing and differentiation, in:"Biology of the Epithelial Lens Cell in Relation to Development, Ageing and Cataract," Y. Courtois, and F. Regnault, eds., INSERM, Paris, p.145.

Taylor, S. M., and Jones, P. A., 1979, Multiple new phenotypes induced in 10T½ and 3T3 cells treated with 5-azacytidine, Cell, 17:771.

Thomson, I., Yasuda, K., de Pomerai, D. I., Clayton, R. M., and Okada, T. S., 1981, The accumulation of lens-specific protein and mRNA in cultures of eye cup from 3½ day chick embryos, Expl. Cell Res., 135:445.

Tsonis, D. A., and Eguchi, G., 1981, Effects of carcinogens on limb regeneration, Differentiation, 20:52.

Van Venrooij, W., Groenveld, A. A., Bloemendal, H., and Benedetti, E. L., 1974, Cultured calf-lens epithelium I. Methods of cultivation and characterisitics of the culture, Expl. Eye Res., 18:517.

Yoda, K., and Fujimura, S., 1979, Induction of differentiation in cultured mouse neuroblastoma cells by N-methyl-N'-nitro-N-nitrosoguanidine, Biochem. Biophys. Res. Commun., 87:128.

SUMMARY OF DISCUSSION ON REVERSIBLE MALIGNANCY,

TRANSDIFFERENTIATION AND RELATED TOPICS

Ratcliffe's paper on ectopic hormones gave rise to some discussion about whether the tumours taken from lungs arose from cells of lung tissues or by metastasis of cells from tissues outside the lungs. The question was put by Iscove and Okada pointed out that if teratocarcinoma cells are injected into mice intravenously many of them become trapped in the lung, suggesting that here was a locality in which metastasising cells might accumulate. Both Ratcliffe and Woodruff explained that there was no certain proof that the cells were derived from lung, but that those producing ectopic hormones had a strong morphological resemblance to tumour cells which did not produce hormones and which were generally regarded as lung tumours. They were found in cases where no other tumours were known elsewhere in the patient and were generally bronchogenic carcinomas rather than tumours arising out of the alveolar cells. The clinician was not always in a position to investigate all of the aspects which a scientist might wish to. In reply to questions from Pessac, Ratcliffe reported that there was evidence that a single cell of a tumour might produce more than one peptide hormone, for example calcitonin and ACTH. No lung tumour was known which produced metenkephalin, though this had been found in adrenal medullary tumours. Some tumours in culture produced ectopic hormones when there had been no clinical evidence of such hormone production by the tumour in situ. Bird enquired whether it was likely that hormone production gave the tumour cell any selective advantage over cells that did not produce hormones, but both Ratcliffe and Woodruff doubted whether this was likely; certainly hormone production frequently made the prognosis for the patient worse than would otherwise have been the case.

Clayton proposed three possible ways in which the ectopic hormone production might arise: (1) A cell has changed its differentiation expression either to produce an entirely new gene product or to produce something in abundance which had hitherto been synthesised at very low levels; (2) a cell had migrated out from a tissue such as the pituitary, lodged in the lung and turned into a tumour; or (3) cells of neural crest origin which migrate and behave like stem cells, giving rise to the tumours in tissues such as lungs but secreting products appropriate to other tissues. Clayton believed that experiments

such as Le Douarin's had shown that neural crest cells would not be
expected to occur in lungs. It also seemed improbable that cells had
migrated from the pituitary to the lung, so that this left transdifferentiation as the most likely cause of ectopic hormone production.

Ratcliffe said that among the possibilities to consider was a
model (of Baylin and Mendelsohn) which involved some cells of the lineage that gives rise to the lung remaining in an immature state, producing foetal antigens such as carcinoembryonic antigen and chorionic
gonadotrophin: maturing to endocrine type cells (K cells) while other
lineages became the major cell types of normal lung. If there were
a maturational arrest, endocrine line cells might ultimately be the
origin of the lung tumours. This early cell type could undergo a
block at some stage of further differentiation, proliferating, and
producing small amounts of several hormones which appear to go together, such as group 1 peptides, ACTH, the calcitonins, etc. It
is know from studies of smoking dogs that ACTH may be produced in the
lung before malignancy occurs, and that lung produces low levels of
chorionic gonadotrophin. Birnie agreed that this explanation re-
sembled one theory of the origin of leukaemia. Harrison said that
the interpretation of ectopic hormone production required a knowledge
of the nature of the hormone-producing cell, of whether it had differentiation markers and antigens characterisitic of lung cells, of endocrine cells, of neither or of both. It was also important to know
the hormone production per cell in the tumour and in the normal lung
cell. It was possible that the hormone syndromes might in some cases
be due to over-production of the hormone by a large mass of tumour
cells, each producing only a little hormone. Ratcliffe replied that
in certain cases the rate of production, storage and secretion in
cells of a lung tumour such as a bronchocarcinoma was as great as in
the pituitary itself, while in other cases a large tumour contained
little, indicating that the level of hormone per cell was only a
hundredth or a thousandth of that in a cell in the pituitary.
Woodruff pointed out that lung tumours which do and those which do
not produce hormone are morphologically similar. Ratcliff agreed
with Harrison that the lung and haemopoietic systems were similar in
being self-renewing, so that maturation arrest at different stages
might occur in lineages in both systems, but that little was known
about the 'stem cells' which led to the renewal in the lung. Pessac
asked whether the number of peptide hormones produced in a known cell
might indicate its possible position in lineages. Ratcliffe replied
that there is evidence that one cell can produce more than one peptide
(e.g. calcitonin and ACTH). In culture lung tumour cells may produce
more hormone than they do in situ. Pessac referred to neurone-like
characteristics of carcinoma cells in culture suggesting an origin
from neuronal stem cells.

The relationship between 'stem cells' and 'transdifferentiation'
was later discussed extensively, following a comment by Paul. We ten

to envisage normal differentiation as following a travelling pathway, alternative decisions being made at each branch point. Hadorn's classical transdetermination experiments were interpreted as exhibiting a phenomenon in which the normal branching rules were breached and cells moved from one branch directly to another. Transdifferentiation presumably also implied the same kind of phenomenon. However, Sueoka's AC cells seem to behave like stem cells with a limited number of alternative decisions. Was it not possible that Eguchi's experiments could be interpreted in a similar way, with the pigmented epithelial cells, resembling stem cells? Okada replied that multipotential cells are usually transitory and proceed through stages of differentiation, while Sueoka had succeeded in making permanent cultures of cells behaving like stem cells and was dealing with differentiation not transdifferentiation. When cells can be identified as belonging to a particular cell type and then they undergo changes to produce another distinct cell type, different from the characteristics of the starting cells, then this provided an operational definition of 'transdifferentiation'. Some people might like to refer to the starting cell type as 'stem cells with some differentiative phenotype' but it seemed contradictory to call any cells with definitive phenotypes 'stem cells'. Clayton made the point that there were many branching points between the zygote and the sequences leading to pigmented cells of the retina or iris, derived from the optic vesicle, and the lens fibre cell derived from the ectoderm of the head. During transdifferentiation there was no reason to suppose that the cells went backwards past these branch points and then proceeded along the alternative pathway. The change was more or less directly from one differentiated state to the other. De Pomerai believed that a cell which was differentiated to the extent of expresssing pigment could not reasonably be regarded as a stem cell. Moscona said that the crucial question was whether a phenotype which under normal conditions would not undergo any further phenotypic change, which had undergone a committed programme of genomic expression, could be made to switch into a completely different programme. If this were so, what was the mechanism of the switch? In the experiments he had described glial cells, which under normal conditions were completely stable, could be switched to do something completely different. He believed that cell contact and communication were important in regulating the stability of the original phenotype, and also in bringing about the switch to the new phenotype.

Yaffe enquired whether it was possible to maintain a cell type in culture for many generations, demonstrating that it could breed true, and then, by some manipulation, cause an alteration into a new and distinct cell type. Okada said that this was essentially what they had done in 1979 for neural retinas, growing a retina cell so that for ten to fifteen generations it replicated as a pigmented cell, and then some pigment characteristics were lost and the cells changed into lens. De Pomerai said changes to stable lens of pigment cell differentiation could occur after many cell divisions from the progeny

of a single neural retina cell. These pigment cells were stable for
numbers of cell divisions but they too could change to lens. Single
pigment retina cells could also give rise after 10 or 15 divisions
to lens cells. A formal demonstration was possible since it was known
that on a collagen substrate the retinal pigment cells would replicate
for several generations and would not transdifferentiate, while on
transfer to a plastic substrate they would undergo transdifferentiation. Clayton referred to two relevant experiments: in Moscona's
all of the cells at the centre of the aggregates showed immunofluorescence for crystallin and if this had been due to selective growth
of a minority population of stem cells, this would imply a very great
loss of cells from the original population and there was no evidence
for this. In Eguchi's experiments he had also showed a massive transformation from pigment cells under certain conditions and again there
was no evidence of loss of the original cell population. The fact
that such transdifferentiations had only been found in comparatively
few instances could be explained by (1) the limited number of systems
examined; (2) the fact that ideal conditions for the transdifferentiation had not been found; (3) that changes may have occurred but
passed without detection. The change from a black pigmented cell to
a transparent lentoid is more conspicuous than some which might occur.
The conversion of fibre cells to muscle cells and heart cells to fat
cells have also been reported. In looking for fully differentiated
cells which might undergo transdifferentiation there was no reason
to avoid studying dividing cells, because these can be fully differentiated, as, for example, the basement cells of the skin, as shown
by regional characteristics of skin grafts. McDevitt drew attention
to the fact that iris cells in situ do not divide whereas, when set
up in culture, they do undergo division. Harrison asked for clarif-
ication of the essential difference between the cloning experiments
described by those working on transdifferentiation of ocular tissues
and the experiments carried out on the haemopoietic system. Bone
marrow cells can be cloned and under suitable conditions give rise
to macroscopic colonies containing various haemopoietic cells (granulocytes, megakaryocytes, erythroid cells). These macroscopic colonies can be disaggregated and recloned, yeilding secondary colonies
also containing various haemopoietic cells. Which type of cell appears depends upon growth conditions, such as the presence of hormone
and growth factors. The experiments have been interpreted as indicating the presence of a bi- or tri-potential cell which is maintained
in low numbers in the macroscopic colonies and which gives rise to
different cell types on cloning, if the experimental conditions are
changed. There seemed to be a strong analogy between such experiment
and the findings of Eguchi and Okada. It was only their cloning experiments which provided any evidence that pigmented cells, rather
than some bi-potential precursor cell, could give rise to lens cells.
Now the cloning efficiencies were very low and large numbers of cells
were plated out. Could it not be that one of the cells cloned is a
bi-potential cell, which because of the initial culture conditions,
gives rise mainly to pigmented cells, but the remaining precursor

cells differentiate at a later stage to produce lens cells. Okada replied that when cells were cloned from neural epithelium there was a progressive loss of pluripotent cells, so that in time the cells had only one potentiality. Thus they were unlike stem cells in the strict sense and he preferred to call them simply progenitor cells. Buckingham, commenting on the role of cell division during transdifferentiation, said that for a molecular biologist changes or modulations of gene expression were most easily envisaged as occurring during mitoses when rearrangements of chromatin structure may take place. Eguchi's data seemed to show that a clone of pigment cells ceased to show pigmentation at about the 8-cell stage, whereas the lens type of morphology occurred after further division when several hundred cells were present. Moscona's system showed that the transformation from glial cells to lens was also accompanied by cell division, though the DNA inhibitor studies suggested that this was not essential. Was it possible to rule out cell division in this case? Moscona replied that he had used post-mitotic cells from 16· days embryo for 2 days in monolayer in the presence of mitosis inhibitor and then reaggregated, when the cells rapidly convert to lens, but complete inhibition of DNA synthesis had not been achieved and so it was not possible to give an unambiguous answer. Eguchi commented that pigment epithelial cells can stably maintain their specificity throughout a number of passages in cell culture. In particular, chick embryo pigment epithelial cells transdifferentiate only when they have been in culture for more than two months. Transdifferentiation of pigment epithelial cells was demonstrated in a clonal culture starting from a single cell, but it was not known how many cell replications are required for the transdifferentiation in vitro. At high initial cell densities there is very little cell proliferation.

De Pomerai pointed to the importance of a critical cell mass in the process of transdifferentiation but Moscona observed that this could be achieved by reaggregating cells. Courtois asked what information was available about the internal or external signals required to induce transdifferentiation or to prevent it. De Pomerai mentioned the role of normal tissue architecture in preventing it. Disruption was necessary to achieve transdifferentiation. Of factors in the medium, high glucose concentrations prevented it. Courtois asked about the possible role of lentropin, which Beebe showed increased the rate of differentiation of chick lens epithelium into fibres, and for an explanation of why only the dorsal part of the iris seemed to be involved in lens regeneration in amphibians. McDevitt replied that the role of neural retina in lens regeneration had long been established, but there had been no plausible explanation of why ventral iris normally will not regenerate a lens while dorsal iris does so. The fact the MNNG could cause the ventral iris to give rise to lens was more likely to be to its disruptive effect than to its mutagenic effect in view of the time scale involved. No-one appeared to have studied the effect of lentropin on regeneration of lens either in vivo or in vitro. One of the problems of regeneration was that once depigmentation of the iris had begun some cells will revert to being pigment cells. On a further lentectomy the corresponding cells, on depigmentation, may go on and transdifferentiate. In reply to questions from Moscona, McDevitt agreed that closely related amphibian species differ in their capacity to form a lens from iris and that when iris from a regenerating species is transplanted into a non-regenerating species, regeneration will occur, so the capacity is intrinsic to the iris and not due to humoral, extrinsic factors. Eguchi reported that if the iris was inverted in the eye and allowed to grow the part which was now dorsal, and which had been ventral, will regenerate.

In a discussion of Clayton's paper, Courtois asked when specific crystallin mRNA was first detected. It had not been looked for earlier that at 3½ days. δ-crystallin synthesis had commenced then but the eye cup had not really differentiated. In reply to questions from Moscona, Clayton reported that no-one had sought for crystallin mRNA outside the eye, though crystallin-like antigens had been found outside the eye by a number of workers, most recently by Barabanov who also obtained lentoids from mouth ectoderm. One possibility was that there might be a family of related genes, some of which were expressed in the eye and other in tissues outside the eye. This question was not yet resolved.

Relating to Sueoka's paper, Birnie enquired whether he had had an opportunity to compare nuclear and cytoplasmic RNA in his cell lines. Sueoka replied that they were concentrating on cloning marker genes such as those for S100 and protein C2. In general the transformed cell lines retained the complexity of the original tissue with respect to nuclear RNA but not to cytoplasmic RNA. Moscona suggested that the concept that differentiation represented irrevocable committment should be reconsidered in the light of recent information of events modifying of genomic information, and the possibility that a differentiated cell type could be transformed into another should be tested before being excluded.

McDevitt pointed out that not all pigmented cells of the iris give rise to a lens after lentectomy, but if the new lens is again removed some cells which remained as pigment cells will now differentiate to lens cells.

INTRODUCTORY REVIEW:

STRATEGIES OF REGULATION: SIGNALS, RECEPTORS AND EFFECTORS

D. E. S. Truman and R. M. Clayton

Department of Genetics
University of Edinburgh
Edinburgh   EH9 3JN

While the pattern of behaviour of a differentiating cell is in part influenced by intrinsic factors, as is evident in the mosaic development of molluscs, for example, there are many instances which reveal the profound importance of factors external to the cell in the regulation of differentiation. The regulative nature of development in vertebrates tells us that the fate of a cell depends to a degree on the activities of neighbouring cells. Direct evidence of the influence of one cell on another comes from studies of epithelial-mesenchymal interaction, pioneered by Grobstein (1964). Of the possible modes of cell communication, the best characterised are systems of hormonal regulation of development. To add to these instances we have a growing list of growth factors and of proteins which have a specific directive influence on development, such as the macrophage-granulocyte inducer, MGI (Sachs, 1978). The developmental significance of epidermal and nerve growth factors has been reviewed by Gospodarowicz (1981). Another growth factor is described in these proceedings by Courtois et al.

The sequence of events by which genetic information in the nucleus is made manifest in the form of specific proteins in the cytoplasm, though not understood in all the subtleties of its regulation, is known in considerable detail. In contrast, our knowledge of the means whereby information from the outside of the cell is conveyed to the genes in the nucleus is meagre. The session of the conference on signals, receptors and effectors provided examples of some of the systems which may add to our knowledge of this line of information flow.

The study of hormonal regulation has given us some details of systems which represent one way of communicating to the genetic material from outside the cell. It will be appreciated that not all instances of hormonal regulation of cellular metabolism involve intervention with gene activity. For instance, the regulation of phosphorylase activity by glucagon is brought about by the modification of the structure of a whole cascade of enzymes, but all these changes are post-translational modifications and so do not directly implicate gene activity (Krebs and Beavo, 1979; Fletterick and Madsen, 1980). Other instances of hormonal regulation certainly do involve modification of gene expression. Among the most thoroughly researched of these examples is the control by oestrogen of the synthesis of vitellogenin in Xenopus liver, of casein synthesis in the rodent mammary gland and of α-amylase synthesis in barley aleurone cells. These and other instances are reviewed by Ho (1979).

The hypothesis of oestrogen action which has gained wide acceptance and which has good experimented support may be summarised as follows: oestrogens in the blood stream enter all cells, but in target tissues are firmly bound to receptors and so become accumulated against a concentration gradient. These receptors, which are proteins, are located in the cytoplasm before they bind hormones, but after the hormone molecules have become bound to the receptor, the hormone receptor complex becomes translocated to the nucleus. The complex interacts with the chromatin so as to lead to the greatly enhanced transcription of certain specific genes and hence there is an increased accumulation of specific mRNA molecules. This, in turn, is followed by synthesis of certain proteins. While the genes, their messengers and the proteins are characterised in detail, much less is known about the receptors, and the nature of their interaction with chromatin and the mechanism of stimulation of transcription is the subject of much speculation and controversy. The field has been reviewed many times, for example by Gorski and Gannon (1976), O'Malley et al. (1977), Green (1980), Thrall et al. (1978).

The model outlined above can be used as a guide to constructing hypotheses about the mechanism of action of other factors which influence gene expression, but if we do we should be aware of some of the further complexities of the system. Not only is protein synthesis stimulated by enhanced transcription, but mRNA accumulation is increased by stabilization of the mRNA during hormonal action. This has been shown for ovalbumin message (Palmiter and Casey, 1974) and also for the casein message in the mammary gland under the influence of prolactin (Guyette et al., 1979). Another aspect of the response of the genes for the egg white proteins is that there is a delay in the response of the genes, which varies from gene to gene and also depending upon which hormones are responsible for the stimulation. This delay has been interpreted as an indication of the need to convey a signal from the acceptor site on the chromatin at which the hormone receptor complex acts to the site of the gene which is expressed (Palmiter et al., 1976).

The nature and molecular consequences of the binding of the hormone-receptor complex to the chromatin remain unclear. Many of the studies of the binding of the hormone receptor complex suggest that there are a large number of acceptor sites in the nucleus - perhaps some thousands of sites. However, there may be a more limited number of sites of higher affinity responsible for the more specific response (Yamamoto and Alberts, 1976; Katzenellenbogen, 1980). There is evidence that the response of the chromatin to the hormonal treatment may be attributed to the non-histone proteins of the chromatin. Experiments involving the reconstitution of chromatin using DNA, histones and non-histone proteins from oviducts of birds treated with oestrogen and those from which the hormones had been withdrawn indicated that the response to hormones involved changes in the non-histone protein fraction (O'Malley et al., 1977). A further indication of a changed state of the chromatin following hormone treatment was the observation of Garel and Axel (1977) that the actively transcribed ovalbumin gene is more sensitive to digestion by DNAse I than is the gene in its inactive state.

The distinctive feature of target tissues which respond to oestrogen is that their cytoplasms contain the proteins which bind the hormone and then translocate to the chromatin. There is evidence from other steroid hormone systems that competence to respond to the hormone is dependent upon a protein receptor. In the case of tissues sensitive to testosterone there is genetic evidence of the role of the receptor: absence of the receptor leads to failure to respond as in the Tfm mutation in mouse and in man (Ohno, 1976). While this provides a molecular mechanism for competenece of one type, it is by no means clear that all the embryological phenomena embraced by the term "competence" are of the same type, and the steroid receptor model may not be a sufficient model for all such cases.

If hormonally regulated gene expression can provide us with a model for the response of a cell to signals, can it also provide a model for the signals themselves? In one respect, it certainly can. The notable feature of the hormones is their chemical diversity, including steroids, catecholamines, small peptides and comparatively large proteins. Both the signals and receptors involved in embryonic development and differentiation probably have at least this degree of diversity.

The classical example of signalling between parts of a developing embryo is primary embryonic induction. Attempts to identify the nature of the signal in this system have had a very confused history, but there is some evidence that proteins may act on the conveyors of information. When one looks at the secondary embryonic inductions, that is the epithelial-mesenchymal interactions which play an important role in differentiation and morphogenesis of many organs, there is still ambiguity about the nature of the signal between the tissues. While the classical experiments of Grobstein, using filters interposed between the interacting tissues, were originally interpreted as indicating that diffusible substances conveyed signals between the tissues, it is now believed that in some systems, such as the kidney system the growth of minute cellular processes through the pores of the filters may be necessary if one tissue is to affect another (Saxen et al., 1976). In the pancreas sytem, however, a diffusible factor is relatively well characterised (Rutter et al., 1978).

Associated with the role of cell contact in development, we should consider the nature of non-diffusible components in the intercellular spaces. There is growing evidence that macromolecules secreted into the intercellular space can have permissive or directive effects on cellular differentiation. In the developing cornea contact with collagen appears to be all that is required to initiate primary stroma production (Meir and Hay, 1975). Lash and Vasan (1978) and Lash and Cheney (1982) have shown that during chondrogenesis the production of cartilage-specific products is stimulated by proteoglycons or by collagens from a variety of sources. Type II collagen, which is the type synthesised by the in vivo inducer, the notochord, was the most effective inducer of chondrogenesis. Dessan et al. (1980) have described the sequence of syntheses during chondrogenesis in limb buds, in which type I collagen and fibronectin increase in the early condensation stages, while during cartilage differentiation type II collagen accumulates and type I collagen and fibronectin disappear. If type II collagen is both an inducer of cartilage differentiation and a product in the intercellular space during that differentiation, then we have an example of positive feed-back loop which would lead to a stabilization of the state of differentiation. Such a cycle in which the intercellular material is both inducer and product could also explain the necessity of removing cells from their aggregated state, and hence contact with their normal intercellular environment, if differentiation is to be modified, as in transdifferentiation. While a positive feedback system would be effective in driving a cell firmly in a particular direction of differentiation, there would need to be some negative feedback system to prevent an excessive development of the tissue derived from that cell type.
Such a negative feedback might be provided by the action of the substances characterised as chalones. The concept of a chalone is that it is a substance specifically produced by an organ which has an inhibiting effect on the growth of that organ. As the organ grows the chalone accumulates and prevents further growth: excision of part of the organ would reduce the amount of chalone and permit growth to resume. Attempts to isolate, purify and characterise chalones have indicated that they are proteins, but knowledge of their mechanism of action is limited (Bullough, 1975). Direct contact between cells also provides a mechanism of regulation of cell division (Glaser, 1980).

The importance of the cell surface in the mediation of signallin between the cell and its environment is the subject of a literature too vast to be reviewed here, but a few examples are mentioned in order to indicate some areas of interest. The role of the cell surface in the interactions of cell recognition has been emphasised for many years (for reviews see Curtis, 1979; Yamada et al., 1980; Frazier and Glasser, 1979). Recent work has added more detail to our knowledge of the role of cell surface carbohydrates in the recognition process (e.g. Stanley and Sudo, 1981) and in primary embryonic induction (Barbieri et al., 1980). The action of epidermal growth factor involves interaction with cell surface receptors (Schechter et al., 1978) and there is evidence that the response of rat mammary epithelium to epidermal growth factor is mediated by an effect on the synthesis of type IV collagen (Salomon et al., 1981).

Communication of some sort of signal from the surface of the cell to the nucleus and its genetic material is implied by many developmental phenomena. While the steroid hormone system has placed emphasis on diffusible globular proteins being transferred from cytoplasm to nucleus, the possibility remains that there may be a variety of types of communication from cell surface to gene, and there is accumulating evidence that cytoskeletal elements might have a part to play in influencing the nucleus. The structure and functions of the cytoskeleton have been reviewed by Lazarides (1980, 1981). The possibility that microtubules may mediate between the cell surface receptors and events in the cell cycle is reviewed by Crossin and Carney (1981) who put forward supporting data based on the action of microtubule inhibitors in chick, mouse and human fibroblasts leading to the intiation of DNA synthesis in these cells.

Cyclic AMP has been implicated in a wide variety of roles in cells since it was first found to be involved as a 'second messenger' in some hormonal system (Sutherland, 1972). An interaction between cAMP and the cytoskeleton has been found with the discovery of a cAMP-dependent protein kinase which can bring about phosphorylation of the proteins of cytoskeletal elements (O'Connor et al., 1981). While our knowledge of the chemistry of cytoskeletal elements continues to expand rapidly, the most tantalising question concerning these organelles remains the nature of the forces which causes them to occupy particular positions in the cell. If the location of the cytoskeletal elements determines the shape of a cell and hence, frequently, the morphology of the tissue in which it takes part (as, for example in the movements of neurulation (Schroeder, 1970)), then we are driven to seek what it is that controls the location of the cytoskeletal elements. The microtubule organising centre (MOTC) is being investigated and characterised and our knowledge of this structure leads on to the question of what controls the distribution of MTOCs (Stearns and Brown, 1979).

The steroid hormones, their receptors and the response of specific genes to their actions represent the best understood system of intercellular communication. Amongst the least understood in molecular terms are these systems in which cellular behaviour is dependent upon cell protein and for which gradients of positional information have been postulated. There are many gradients in living organisms, of concentrations of substances and of metabolic activity, and it is not easy to distinguish cause and effect in those systems in which gradients have been detected. Whatever the form of the gradient, its information will presumably have to be conveyed via the cell surface. Trisler et al. in these proceedings describe evidence of glycoprotein gradients conveying positional information in the retina. It has been suggested that gap junctions provide a channel for the communication of postitional information (Wakeford, 1979; Loewenstein, 1979).

REFERENCES

Barbieri, F. D., Sanchez, S. S., and Del Pino, E. J., 1980, Changes in lectin-mediated agglutinability during primary embryonic induction in the amphibian embryo, J. Embryol. exp. Morph., 57:95.

Bullough, W. S., 1975, Mitotic control in adult mammalian tissue, Biol. Rev., 50:99.

Crossin, K. L., and Carney, D. H., 1981, Evidence that microtubule depolymerization early in the cell cycle is sufficient to intiate DNA synthesis, Cell, 23:61.

Curtis, A. S. G., 1979, Cell recognition and intercellular adhesion, in:"Biochemistry of Cellular Regulation, Vol. IV. The Cell Surface," P. Knox, ed., C.R.C. Press, Boca Raton, Fla., p.151.

Fletterick, R. J., and Madsen, N. B., 1980, The structures and related functions of phosphorylase a, Ann. Rev. Physiol., 43:251.

Frazier, W., and Glaser, L., 1979, Surface components and cell recognition, Ann. Rev. Biochem., 48:491.

Garel, A., and Axel, R., 1977, The structure of the transcriptionally active ovalbumin genes in chromatin, Cold Spring Harbor Symp. quant. Biol., 42:701.

Glaser, L., 1980, From cell adhesion to growth control, in:"The Cell Surface: Mediator of Developmental Processes," S. Subtelny and N. K. Wessells, eds., Academic Press, New York, p.79.

Gorski, J., and Gannon, F., 1976, Current models of steroid hormone action: a critique, Ann. Rev. Genet., 11:239.

Gospodarowicz, D., 1981, Epidermal and nerve growth factors in mammalian development, Ann. Rev. Physiol., 43:251.

Green, C. D., 1980, The regulation of gene expression by steroid hormones in animal cells, in:"Biochemistry of Cellular Regulation: Vol. 1. Gene Expression," M. J. Clemens, ed., C.R.C. Press, Boca Raton, Fla., p.59.

Grobstein, C., 1964, Cytodifferentiation and its control, Science, 143:643.

Guyette, W. A., Matusik, R. J., and Rosen, J. M., 1979, Prolactin-mediated transcriptional and post-transcriptional control of casein gene expression, Cell, 17:1013.

Ho, T.-H. D., 1979, Hormonal control of gene expression, in:"Physiological Genetics," J. G. Scandalios, ed., Academic Press, New York, p.110.

Katzenellenbogen, B. S., 1980, Dynamics of steroid hormone receptor action, Ann. Rev. Physiol., 42:17.

Krebs, E. G., and Beavo, J. A., 1979, Phosphorylation-dephosphorylation of enzymes, Ann. Rev. Biochem., 48:923.

Lash, J. W., and Vasan, N. S., 1978, Somite chondrogenesis in vitro stimulation by exogenous extracellular matrix components, Develop. Biol., 66:151.

Lash, J. W., and Cheney, C. M., 1982, Diversification with embryonic chick somites: in vitro analysis of proteoglycan synthesis, in: "Differentiation in Vitro," M. M. Yeoman and D. E. S. Truman, eds., Cambridge Universtiy Press, Cambridge, p.193.

Lazarides, E., 1980, Intermediate filaments as mechanical integrators of cellular space, Nature, Lond., 283:249.

Lazarides, E., 1981, Intermediate filaments - chemical heterogeneity in differentiation, Cell, 23:649.

Loewenstein, W. R., 1979, Junctional intercellular communication and the control of growth, Biochim. Biophys. Acta., 560:1.

Meier, S., and Hay, E. D., 1975, Stimulation of corneal differentiation by interaction between cell surface and extracellular matrix. I. Morphometric analysis of transfilter "induction", J. Cell Biol., 66:275.

O'Connor, C. M., Gard, D. L., and Lazarides, E., 1981, Phosphorylation of intermediate filament proteins by cAMP-dependent protein kinase, Cell, 23:135.

Ohno, S., 1976, Major regulatory genes for mammalian sexual development, Cell, 7:315.

O'Malley, B. W., Towle, H. C., and Schwartz, R. J., 1977, Regulation of gene expression in eukaryotes, Ann. Rev. Genetic., 11:239.

Palmiter, R. D., and Carey, N. H., 1974, Rapid inactivation of ovalbumin messenger ribonucleic acid after acute withdrawal of estrogen, Proc. Natl. Acad. Sci. U.S.A., 71:2357.

Palmiter, R. D., Moore, P. B., Mulvihill, E. R., and Emtage, S., 1976, A significant lag in the induction of ovalbumin messenger RNA by steroid hormones: a receptor translocation hypothesis, Cell, 8:557.

Rutter, W. J., Pictet, R. L., Harding, J. D., Chirgwin, J. M., MacDonald, R. J., and Przybyla, A. E., 1978, An analysis of pancreatic development: role of mesenchymal factor and other extracellular factors, in:"Molecular Control of Proliferation and Differentiation," J. Papaconstantinou and W. J. Rutter, eds., Academic Press, New York, p.205.

Sachs, L., 1978, Control of normal cell differentiation and the phenotypic reversion of malignancy in myeloid leukaemia, Nature, Lond., 274:535.

Salomon, D. S., Liotta, L. A., and Kidwell, W. R., 1981, Differential response to growth factor by rat mammary epithelium plated on different collagen substrata in serum-free medium, Proc. Natl.

Acad. Sci. U.S.A., 78:382.

Saxen, L., Lehtonen, E., Karkinen-Jaaskelainen, M., Nordling, S., and Wartiovaara, J., 1976, Are morphogenetic tissue interactions mediated by transmissible signal substances or through cell contacts?, Nature, Lond., 259:662.

Schroeder, T. E., 1970, Neurulation in Xenopus laevis. An analysis and model based upon light and electron microscopy, J. Embryol. exp. Morph., 23:427.

Schechter, Y., Hernaez, L., and Cuatrecasas, P., 1978, Epidermal growth factor: biological activity requires persistent occupation of high affinity cell surface receptors, Proc. Natl. Acad. Sci. U.S.A., 75:5788.

Stanley, P., and Sudo, T., 1981, Microheterogeneity among carbohydrate structures at the cell surface may be important in recognition phenomena, Cell, 23:763.

Stearns, M. E., and Brown, D. L., 1979, Purification of cytoplasmic tubulin and microtubule organising center proteins functioning in microtubule initiation from the alga Polytomella, Proc. Natl. Acad. Sci. U.S.A., 76:5745.

Sutherland, E. W., 1972, Studies on the mechanism of hormone action, Science, 177:401.

Thrall, C. L., Webster, R. A., and Spelsberg, T. C., 1978, Steroid receptor interaction with chromatin, in:"The Cell Nucleus, Vol. VI, Chromatin, Part C.," H. Busch, ed., Academic Press, New York, p.461.

Trisler, G. D., Schneider, M. D., Mostral, J. R., and Nirenberg, M., 1982, Molecules that define a dorsal-ventral axis of retina can be used to identify cell position, These proceedings.

Wakeford, R. J., 1979, Cell contact and positional communication in hydra, J. Embryol. exp. Morph., 54:171.

Yamada, K. M., Oldern, K., and Hahn, L. H. E., 1980, Cell surface protein and cell interactions, in:"The Cell Surface: Mediator of Developmental Processes," S. Subtelney and N. K. Wessells, eds., Academic Press, New York, p.43.

Yamamoto, K. R., and Alberts, B. M., 1976, Steroid receptors: elements for modulation of eukaryotic transcription, Ann. Rev. Biochem., 45:721.

COMMITMENT, DNA SYNTHESIS AND GENE EXPRESSION

IN ERYTHROLEUKEMIA CELLS

Richard A. Rifkind, Elliot Epner and Paul A. Marks

Memorial Sloan-Kettering Cancer Center
New York
NY 10021

Evidence has emerged from a number of systems that commitment, characterized by restriction of cell divisions to a small number ("terminal" cell division) and the synthesis of those proteins characteristic of the differentiated state, may require DNA synthesis (Rutter et al., 1973; Holtzer et al., 1972; Marks and Rifkind, 1978). We have studied this problem with murine erythroleukemia cells (MELC), which can be induced to differentiate and initiate terminal cell division by a variety of agents (Marks and Rifkind, 1978). Induced MELC express many characteristics indistinguishable from normal erythroid differentiation, including accumulation of α and β globin mRNA (Ross et al., 1972; Ohta et al., 1976), α, $β^{maj}$, and $β^{min}$ globins, hemoglobins (Boyer et al., 1972; Ostertag et al., 1972), and spectrin (Eisen et al., 1977), and they display terminal cell division (Fibach et al., 1977; Gusella et al., 1976). Several studies provide evidence that this program of induced differentiation depends upon an action of the inducer which is cell cycle-related (McClintock and Papaconstantinou, 1974: Levy et al., 1975; Harrison, 1977). Consistent with this hypothesis are the findings we (Gambari et al., 1979) and others (Geller et al., 1978) have shown that G_1 or G_2 MELC, fractionated by unit gravity sedimentation, are committed to differentiate more rapidly than S phase of unfractionated cells cultured under similar conditions. More recently, Conkie et al. (1981), have provided evidence from work with a temperature sensitive variant of MELC, that induced differentiation requires progression through the cell cycle. In our own laboratory, we find that inducer-mediated events during early S appear to be required for subsequent expression of α and β globin structural genes in MELC culture with the inducer, hexamethylene bisacetamide (HMBA) (Gambari et al., 1978). Using cells synchronized by centrifugal elutriation, in early S, mid-S, and late S/G_2, the progress of the cells through the cell division cycle was determined by flow microfluorometry, while the appearance of newly synthesized globin mRNA was determined by measuring the amount of ^{3}H-uridine-labelled mRNA that hybridizes to an oligo(dT)globin cDNA cellulose column (Gambari et al., 1979). Early S cells cultured with HMBA first show accumulation of newly synthesized globin mRNA in the subsequent G_1. If HMBA is added to cells synchronized in mid-S or in late S/G_2, the onset of accumulation of newly synthesized globin mRNA is delayed until the cells have traversed a full cell cycle, including early S, in the presence of inducer. These studies suggest that the expression of globin genes is regulated by inducer-related events in the early portion of the S phase.

The nature of this cell cycle, early S-phase, inducer-related event is not yet known. However, investigations with a number of other cell systems suggest that genes replicating in early S play a role in regulating both proliferation and differentiation (Gambari et al., 1979; Brown and Schildkraut, 1979; Oslcy ct al., 1977; Balazs et al., 1973; Stambrook, 1974; Ho and Armentrout, 1978; Tobia et al., 1970, 1971; Lima-de-Faria and Jaworska, 1968; Pardue and Gall, 1970). Brown and Schildkraut (1979) report that unifilar substitution of bromodeoxyuridine for thymidine in DNA during early S phase inhibits growth and differentiation of MELC subsequently exposed to Me_2SO. Likewise, incorporation of bromodeoxyuridine into early but not late replicating DNA inhibits growth of HeLa (Mueller and Kajiwara, 1964) and hamster cells (Chang and Baserga, 1977), fusion of myoblasts (Lough and Bischnoff, 1976) and the synthesis of a number of enzymes in mouse L cells (Kasupski and Mukherjee, 1977). These studies focus attention on events during early S as a target for study of the molecular mechanisms involved in inducer-mediated differentiation of MELC.

Previous studies in this laboratory as well as others (Balazs et al., 1973; Stambrook, 1974; Ho and Armentrout, 1978: Tobia et al., 1970, 1971: Lima-de-Faria and Jaworska, 1968) provide evidence that the S phase is ordered temporally with respect to replication of specific DNA sequences. Using bromodeoxyuridine density labelling (to isolate newly synthesized DNA from total DNA) and appropriate radioactive DNA probes, it was shown that in HeLa and SD3T3 cells ribosomal RNA genes replicate throughout S (Balazs et al., 1973) and late in S in Chinese hamster cells (Stambrook, 1974). Replication of HeLa 5S RNA genes occurs primarily in early S (Ho and Armentrout, 1978). Recently, we have studied the pattern of replication of the α and β globin and rRNA genes in MELC (Epner et al., 1981). Synchronous populations of cells were obtained by centrifugal elutriation. Newly synthesized, bromodeoxyuridine-substituted, DNA was isolated by centrifugation to equilibrium in neutral CsCl gradients (for double stranded DNA) or alkaline CsCl gradients (for single stranded, Bromodeoxyuridine-substituted DNA). The results indicate, on the one hand, that α and β globin gene sequences are located in regions of the MELC genome that replicate early in S phase during exponential growth. Ribosomal RNA genes, on the other hand, appear to replicate in early, middle, and late S phase.

These results, taken together with those studies which suggest that inducer-mediated events occurring during early S phase are critical to subsequent expression of globin genes, as well as other studies that implicate cell cycle-related events in the activation of globin genes during normal erythropoiesis (Groudine and Weintraub, 1981; Rifkind et al., 1978), suggest a correlation between replication of the globin gene sequences and inducer effects leading to the expression of these sequences. Unique sequence genes that are transcribed (or have the potential of being transcribed) may be preferentially replicated early in S phase. Whether early S phase has special characteristics with respect to the modification of DNA and/or chromatin structures critical for transcription remains to be determined.

ACKNOWLEDGEMENTS

These studies were supported, in part, by Grants CA-13696, CA-31768, and Core Grant CA-08748-16 from the National Cancer Institute, and Grant CH-68A from the American Cancer Society.

REFERENCES

Balazs, I., Brown, E., and Schildkraut, C., 1973, The temporal order of replication of some DNA cistrons, Quantitative Biol., 38:239.

Boyer, S. H., Wuu, K. D., Noyes, A. N., Young, R., Scher, W., Friend, C., Preisler, H., and Bank, A., 1972, Hemoglobin biosynthesis in murine virus-induced leukemic cells in vitro: structure and amounts of globin chains produced, Blood, 40:823.

Brown, E., and Schildkraut, C., 1979, Perturbation of growth and differentiation of Friend murine erythroleukemia cells by 5-bromo-deoxyuridine incorporation in early S phase, J. Cell Physiol., 99:261.

Chang, H. L., and Baserga, R., 1977, Time of replication of genes responsible for a temperature-sensitive function in a cell cycle-specific ts mutant from a hamster cell line, J. Cell Physiol., 92:333.

Conkie, D., Harrison, P. R., and Paul, J., 1981, Cell cycle dependence of induced hemoglobin synthesis in Friend erythroleukemic cells temperature-sensitive for growth, Proc. Natl. Acad. Sci. U.S.A., 78:3644.

Eisen, H., Nasi, S., Georopoulos, C. P., Arndt-Jovin, D., and Ostertag, W., 1977, Surface changes in differentiating Friend erythroleukemic cells in culture, Cell, 10:689.

Epner, E., Rifkind, R. A., and Marks, P. A., 1981, Replication of α and β globin DNA sequences occurs during early S phase in murine erythroleukemia cells, Proc. Natl. Acad. Sci. U.S.A., 78:3058.

Fibach, E., Reuben, R. C., Rifkind, R. A., and Marks, P. A., 1977, Effect of hexamethylene bisacetamide on the commitment to differentiation of murine erythroleukemia cells, Cancer Res., 37:440.

Gambari, R., Terada, M., Bank, A., Rifkind, R. A., and Marks, P. A., 1978, Synthesis of globin mRNA in relation to the cell cycle during induced murine erythroleukemia differentiation, Proc. Natl. Acad. Sci. U.S.A., 75: 3801.

Gambari, R., Marks, P. A., and Rifkind, R. A., 1979, Murine erythroleukemia cell differentiation: Relationship of globin gene expression and of prolongation of G_1 to inducer effect during G_1/ early S, Proc. Natl. Acad. Sci. U.S.A., 76:4511.

Geller, R., Levenson, R., and Housman, D., 1978, Significance of the cell cycle in commitment of murine erythroleukemia cells to erythroid differentiation, J. Cell Physiol., 95:213.

Groudine, M., and Weintraub, H., 1981, Activation of globin genes during chicken development, Cell, 24:393.

Gusella, J., Geller, R., Clarke, B., Weeks, V., and Housman, D., 1976, Commitment to erythroid differentiation by Friend erythroleukemia cells: a stochastic analysis, Cell, 9:221.

Harrison, P. R., 1977, The biology of the Friend cell, in: "Biochemistry of Cell Differentiation II", J. Paul, ed., University Park Press, Baltimore, Vol. 15, p.227.

Ho, C., and Armentrout, R. W., 1978, Replication of the genes coding for 5S RNA in synchronized HeLa cells, Biochim. et Biophys. Acta, 520:175.

Holtzer, H., Weintraub, H., Mayne, R., and Mochan, B., 1972, The cell cycle, cell lineages, and cell differentiation, Curr. Topics Dev. Biol., 7:229.

Kasupski, G. J., and Mukherjee, B. B., 1977, Effects of controlled exposure of L cells to bromodeoxyuridine (BudR), Expl. Cell Res., 106:327.

Levy, J., Terada, M., Rifkind, R. A., and Marks, P. A., 1975, Induction of erythroid differentiation by dimethylsulfoxide in cells infected with Friend virus: Relationship to the cell cycle, Proc. Natl. Acad. Sci. U.S.A., 72:28.

Lima de Faria, A., and Jaworska, H., 1968, Late DNA synthesis in heterochromatin, Nature, Lond., 217:138.

Lough, J., and Bischnoff, R., 1976, Differential sensitivity to 5-bromo-deoxyuridine during the S phase of synchronized myogenic cells, Develop. Biol., 50:457.

Marks, P. A., and Rifkind, R. A., 1978, Erythroleukemic differentiation, Ann. Rev. Biochem., 47:419.

McClintock, P. R., and Papaconstantinou, J., 1974, Regulation of hemoglobin synthesis in a murine erythroblastic leukemic cell: the requirement for replication to induce hemoglobin synthesis, Proc. Natl. Acad. Sci. U.S.A., 71:4551.

Mueller, G. C., and Kajiwara, K., 1964, Molecular events in the reproduction of animal cells. III. Fractional synthesis of deoxyribonucleic acid with 5-bromodeoxyuridine and its effect on cloning efficiency, Biochim. et Biophys. Acta, 91:486.

Ohta, Y., Tanaka, M., Terada, M., Miller, O. J., Bank. A., Marks, P. A., and Rifkind, R. A., 1976, Erythroid cell differentiation: Murine erythroleukemia cell variant with unique pattern of induction by polar compounds, Proc. Natl. Acad. Sci. U.S.A., 73:1232.

Osley, M. A., Sheffrey, M., and Newton, A., 1977, Regulation of flagellin synthesis in the cell cycle of caulobacter: Dependence on DNA replication, Cell, 12:393.

Ostertag, W., Melderis, H., Steinheider, G., Kluge, N., and Dube, S., 1972, Synthesis of mouse haemoglobin and globin mRNA in leukaemic cell cultures, Nature New Biol., 239:231.

Pardue, M. L., and Gall, J. G., 1970, Chromosomal localization of mouse satellite DNA, Science, 168:1356.

Rifkind, R. A., Marks, P. A., Bank, A., Terada, M., Reuben, R. C., Maniatis, G. M., Fibach, E., Nudel, U., Salmon, J. E., and Gazitt, Y., 1978, Regulation of differentiation in normal and transformed erythroid cells, In Vitro, 14:155.

Ross, J., Ikawa, Y., and Leder, P., 1972, Globin messenger-RNA induction during erythroid differentiation of cultured leukemia cells, Proc. Natl. Acad. Sci. U.S.A., 69:3620.

Rutter, W. J., Pictet, R. L., and Morris, P. W., 1973, Toward molecular mechanisms of developmental processes, Ann. Rev. Biochem., 42:601.

Stambrook, P. J., 1974, The temporal replication of ribosomal genes in synchronized Chinese hamster cells, J. Mol. Biol., 82:303.

Tobia, A. M., Schildkraut, C. L., and Maio, J. J., 1970, Deoxyribonucleic acid replication in synchronized cultured mammalian cells, J. Mol. Biol., 54:499.

Tobia, A. M., Schildkraut, C. L., and Maio, J. J., 1971, DNA replication in synchronized cultured mammalian cells, Biochim. et Biophys. Acta, 246:258.

REGULABILITY OF GENE EXPRESSION AND

DIFFERENTIATION DURING MYOGENESIS

Huw A. John

Department of Genetics
University of Edinburgh
Edinburgh EH9 3JN

Gene expression and differentiation during myogenesis may be regulated by various intrinsic and extrinsic control elements and influences. The role of some of these is discussed in this chapter.

INTRINSIC FACTORS IN INHERITANCE OF THE DIFFERENTIATED STATE

The myogenic cell can divide repeatedly in tissue culture and still "remember" its particular state of differentiation. Even though the culture conditions which promote division of myogenic cells are relatively nonspecific the myogenic cells do not convert to some other cell type. In some way the differentiated state is able to perpetuate itself.

Investigations using cell fusion techniques have aimed at an understanding of how the state of differentiation is remembered. Carlsson et al. (1974a) found that when dormant nuclei of chicken erythrocytes were introduced into rat myoblast cytoplasm, DNA and RNA synthesis were reactivated. Rat antigens characteristic of nucleoplasm and nucleoli migrated into the chick erythrocyte nuclei in the rat myoblast x chicken erythrocyte heterokaryons. It was proposed that these antigens might represent a class of molecules which could reprogramme the erythrocyte nuclei to express muscle genes. However, chick myosin could not be detected in the heterokaryons by immunological methods (Carlsson et al., 1974b) suggesting that signals controlling the muscle specific genes present in rat muscle cells were not received and acted upon by chicken erythrocyte nuclei. This lack of reponse might have been because the nuclei were from cells (erythrocytes) that had undergone terminal differentiation. Therefore Linder et al. (1979) tested the nuclei of multipotent stem cells to see if they were more susceptible to gene regulatory signals present in the foreign cytoplasm. The gene expression pattern of a mouse embryonal carcinoma cell nucleus was not detectably altered by exposure to myoblast cytoplasm.

When intact cells were reconstituted by fusion of nuclei (minicells) from enucleated rat myoblasts and cytoplasms from enucleated mouse fibroblasts, clones derived from the reconstituted cells formed myotubes which produced myosin and developed the cross-striated pattern typical of skeletal muscle. The myogenic programme of the rat myoblast persisted through the enucleation and reconstitution procedures, and was not obviously altered by a period of exposure to mouse fibroblast cytoplasm (Ringertz et al., 1978).

CRITICAL FINAL MITOSIS

After a period of cell division myogenic cells fuse to form multi-nucleated myotubes. It has been suggested that during the final cell cycle prior to fusion there is a crucial event that results in the repression of the genes involved in cell proliferation and the activation of genes needed for differentiation.

Gurdon and Woodland (1968) proposed that only during such a critical mitosis, when the nuclear membrane transiently disappears, can proteins easily leave the nucleus and new proteins responsible for reprogramming the genes move from the cytoplasm and bind to the decondensing chromosomes.

In the theory proposed by Holtzer (for review see Dienstman and Holtzer, 1977) the decision to differentiate is made during the phase of DNA synthesis (S) prior to the final or "quantal" mitosis. The rearrangement of chromatin that occurs during S may make it a particularly suitable time to make extensive changes in the genetic programme needed for the subsequent fusion and biochemical differentiation which start during the G1 phase after the "quantal" mitosis.

Holtzer stressed that a dividing myoblast does not and cannot transform into a postmitotic myoblast committed to differentiation without undergoing the critical "quantal" mitosis. The theory therefore predicts that myogenic cells at a given moment have only one alternative open to them - either to divide or to fuse - but not both. Tissue culture conditions can be manipulated so that cultured muscle cells will either multiply or differentiate. It is therefore possible to test the quantal mitosis theory by investigating whether cells which fuse into myotubes must first undergo a division in the presence of the culture conditions which cause them to differentiate or whether they can be switched from the mitotic cycle to the post mitotic state without going through a cell division. Several reports have indicated that cells can withdraw from the mitotic cycle and differentiate in the permissive medium without going through any part of a mitosis (reviewed by Nadal-Ginard, 1978).

Cell cloning and DNA labelling experiments showed a direct correlation between the time of culture in differentiating medium and a progressive loss of proliferative capacity of mononucleated rat myoblasts demonstrating that these cells become irreversibly committed to differentiation and withdraw from the cell cycle prior to and not as a consequence of cell fusion (Nadal-Ginard, 1978). The commitment step occurs during the G1 phase prior to fusion. This G1 phase has a latent period during which no irreversible step towards differentiation occurs and the cells remain ambivalent toward growth or differentiation. Under proper conditions, this period is followed by an irreversible commitment toward differentiation and a loss of proliferative capacity.

THE ROLE OF PROSTAGLANDIN E_1 AND CYCLIC AMP

Zalin and Montague (1974) observed a spontaneous but transient increase in intracellular cAMP 5-6 hours before the onset of fusion in primary chick myoblasts suggesting that cAMP is acting as a signal for muscle cell fusion. Cyclic AMP is known to exert control by activating protein kinase which transfers a phosphate group to other enzymes (phosphorylation) thus activating them. cAMP is generated from ATP by the enzyme adenylate cyclase which is triggered by prostaglandin E_1 binding to cell receptors. Physiological concentrations of prostaglandin E_1 added to muscle cultures produced similar intracellular increase in cAMP 4 hours earlier than that occurring normally and brought forward the fusion process by a corresponding 4 hours. Two inhibitors of prostaglandin synthesis produced a marked inhibition of cell fusion suggesting that synthesis of prostaglandin E_1 occurred normally in muscle cultures (Zalin, 1977).

When myoblast DNA synthesis and differentiation were prevented by FUdR there was no increase in cAMP in cultures suggesting that DNA synthesis is necessary and there is a link between the myoblast cell cycle and the transient cAMP increase (Zalin and Montague, 1974).

CELL FUSION

Yaffe and his colleagues (Shainberg et al., 1969; Yaffe and Dym, 1972) based their theory that fusion acts as a trigger for further differentiation on the observation that both cell fusion and creatine phosphokinase accumulation are prevented when Ca^{2+} is removed from culture medium. More recently there have been several reports that fusion can be blocked without inhibiting the synthesis of muscle specific proteins (for review, see John and Jones, 1981). Morris and

Cole (1979) found that external Ca^{2+} exerted rather different effects in two concentration ranges. They suggested as a possible explanation for the controversy in the literature, that workers who found inhibition of CPK and myosin synthesis in fusion-inhibited cultures may have been using higher effective external Ca^{2+} concentrations than those who found normal CPK and myosin synthesis.

It is now established that in addition to muscle characteristic forms of such proteins as myosin and actin there are isoenzymes and polymorphic forms present in the dividing myogenic cell populations and also in non-myogenic cells (Chi et al., 1975a, b; Whalen et al., 1976). Obviously, methods that distinguish the muscle specific forms of protein are therefore vital before interpreting results in terms of muscle-specific gene expression.

An alternative approach to detecting gene expression in relation to fusion is by direct measurement of specific RNA transcripts. John et al. (1977) prepared a cDNA copy of purified chick embryonic skeletal myosin heavy chain mRNA (MHCmRNA) which distinguished between myogenic and non myogenic cells compared by in vitro and in situ hybridization. The majority of cells in replicating mononucleate myogenic cell cultures contained no detectable message. A large amount of MHCmRNA was found in fusion blocked mononucleate cells suggesting that fusion was not necessary to trigger off transcription of this gene. Similar studies (Robbins and Heywood, 1978; Dym et al., 1979) showed the existence of MHCmRNA transcripts prior to cell fusion. Detection of specific transcripts can now be made far more reliably using cloned probes (see Yaffe and Buckingham in this book).

In summary most evidence suggests that fusion is not an essential signal for the expression of muscle specific genes. However, under normal circumstances fusion is accompanied by a massive increase in transcription and translation.

GENE REGULATION AT THREE DIFFERENT LEVELS OF CONTROL?

Experimental evidence has indicated that the genes expressed during differentiation of muscle may be regulated at three levels of control.

1. Transcriptional control in which qualitative or quantitative changes in protein synthesis are a direct reflection of changes in the transcription of genomic DNA into RNA.
2. Post-transcriptional control of the processing of nuclear RNA into cytoplasmic messenger.
3. Translational control in which gene expression is modulated by a mechanism that selects or activates particular mRNA sequences from a pre-existing pool of untranslated mRNA.

Transcriptional Control

The evidence for transcriptional control comes from numerous reports (reviewed by John and Jones, 1981) demonstrating that the amount of protein synthesized in a cell-free system by muscle specific mRNA was always directly proportional to the rate of synthesis of the same protein in the muscle cultures from which the RNA was extracted. There was no evidence for a high specific mRNA content and low synthesis of the corresponding protein or vice versa. It was suggested that RNA transcription immediately precedes translation into proteins.

Paterson and Bishop (1977) detected changes in the sequence complexity, distribution of abundance classes and coding capacity of the total mRNA population of primary chick embryo muscle cultures during myogenesis. The change from myoblast to myotube resulted in the appearance of a new abundant class consisting of six different mRNA sequences each present in 15 000 copies per cell and a complete change in mRNA abundance class distribution. The increased synthesis of the major muscle structural proteins in myotubes was correlated with increased synthesis of these proteins in a cell free system in response to total cellular RNA. The synthesis of these proteins was due to the new abundant class of mRNA found in the myotube.

Paterson and Bishop (1977) have pointed out that in experiments where the synthesis of muscle specific proteins in cell-free system is used as an assay of the level of translatable mRNA the following assumptions are made. First, that all mRNA present at all culture times are, in fact, translatable and that no mRNA exists in an inactive form that is not converted to an active form by the isolation procedure and/or by the translation machinery of the cell-free system. Second, that the RNA from one culture condition does not contain inhibitors for translation that are not present in other culture conditions.

Affara et al. (1980a, b) showed that new DNA sequences were transcribed in committed muscle cells and that a new set of mRNA entered the polysomes. This new mRNA class was absent from the nuclear RNA of undifferentiated cells implying that, in these cells, the corresponding genes were not transcribed. The new group of mRNAs which appear in the cytoplasm of myotubes were translated into muscle specific proteins. It thus appeared that the presence of new messages in the polysomes of myoblasts and myotubes is regulated at the level of transcription.

Post-transcriptional Control

Affara et al. (1980a) found that some RNA sequences common to myoblasts and myotubes and present in similar amounts in the nuclei of both showed increased abundance in myotube cytoplasm. The authors suggested that post-transcriptional mechanisms may selectively amplify some mRNAs which may have a role in determining differentiation, for example, by differential rates of turnover and exportation from the nucleus and/or differential mRNA stability.

Translational Control

Several lines of evidence have suggested that during myogenesis the onset of the intense synthesis of muscle specific proteins takes place by the activation of preformed mRNAs. Yaffe and Dym (1972) found that treatment of cultures just prior to cell fusion with actinomycin D, at concentrations that block RNA synthesis, did not prevent fusion or myosin synthesis indicating that muscle specific mRNA had already accumulated. Buckingham et al. (1974) observed a 26 S mRNA (presumptive myosin heavy chain mRNA) in calf myoblasts prior to fusion. The half-life of this 26 S mRNA increased from 10 hours in myoblasts to more than 50 hours in the myotubes. The authors therefore proposed that the transition to the differentiated myotube that synthesizes muscle specific proteins is effected by the stabilization of mRNA already being transcribed. In a later investigation, Buckingham et al. (1976) demonstrated that the short-lived 26 S RNA from dividing myoblasts was present in ribonucleoprotein (RNP) particles which do not enter polysomes. In contrast, the more stable 26 S RNA also initially present as RNP particles, just prior to and in the early stages of fusion, was shown to enter the large polysomes known to synthesize the myosin heavy chain later in fusion.

Heywood et al. (1975) and Bag and Sarkar (1976) independently isolated and characterized myosin mRNP particles from the muscle of chick embryos. Utilizing a myosin cDNA probe Dym et al. (1979) showed that just prior to cell fusion, when the burst of myosin synthesis had not yet occurred, the vast majority of cytoplasmic myosin mRNA transcripts were found in the stored mRNP particles with a minimal amount found in the large polysome fraction. However, in differentiated myotube cultures, when myosin synthesis was progressing at a high rate the reverse was found. Pulse-chase experiments indicated that during mid-fusion the myoblast-stored myosin heavy chain transcripts were precursors to the myotube polysomal myosin heavy chain mRNA, but later in differentiation the contribution of newly synthesized transcripts became predominant (Doetschman et al., 1980). Apparently, newly transcribed myosin mRNA does not necessarily enter polysomes directly but can be temporarily sequestered in a stored form as a cytoplasmic mRNP particle.

Several possible mechanisms have been suggested for maintaining mRNA in an inactive state prior to translation. A small oligo(U) containing translational control RNA may play a role (Bester et al., 1975; Kennedy et al., 1978). Another small RNA has been found to show differential inhibition of polysome translation and poly(A)+ mRNA translation (Pluskal and Sarkar, 1981). Bag and Sells (1979) have shown that in chick muscle there are three subpopulations of nonpolysomal cytoplasmic mRNA-protein complexes which differ with respect to the relative proportion and nature of their protein moieties. However, all three can be translated either as protein-free RNA or complexed with the proteins suggesting that the proteins do not have a role in control of translation.

Heywood and his colleagues have proposed another way in which muscle specific protein synthesis may be controlled at the post-transcriptional level. They suggest that muscle specific initiation factors are necessary so that mRNA may bind to ribosomes and translation of proteins be initiated. Such a discriminatory role of initiation factors would provide a mechanism by which the rates of translation of specific messages, or classes of messengers could be altered during differentiation. Heywood et al. reported that the translation of mRNA for globin, myosin and myoglobin by heterologous ribosomes required a specific factor - fraction 3 of an eukaryotic initiation factor (eIF-3) - isolated from a high concentration KCl wash of ribosomes prepared from reticulocytes, red muscle and white muscle respectively (Rourke and Heywood, 1972; Heywood et al., 1974). Specific components of muscle 15-18 S initiation factor preparations had a higher binding affinity for myosin mRNA than globin mRNA (Heywood and Kennedy, 1978). The muscle eIF-3 contained components which allowed it to recognise and stimulate specifically the translation of myosin mRNA in a muscle cell-free protein synthesis system (Gette and Heywood, 1979). A protein fraction from myosin mRNP particles was found to stimulate the synthesis of the myosin heavy chain when the isolated fraction was added to a wheat germ cell-free system containing myosin mRNA (Bester et al., 1980).

THE ROLE OF NEIGHBOURING CELLS

The differentiation of a myogenic cell is probably influenced by neighbouring myogenic cells and fibroblasts.

Neighbouring myoblasts

Konigsberg (1961) suggested that the progressive increase in the myogenic cell population may have two effects. At the higher cell densities cell contacts might occur more frequently, thus increasing the probability of encounters between cells competent to fuse. Alternatively, increased cell density would accelerate any alterations ("conditioning") of the medium generated by the metabolic activities of the cells. In this way high cell density might facilitate the establishment of a microenvironment in some way favourable for differentiation. Further investigations by Konigsberg (1971) indicated a diffusion-mediated, cell density-dependent control over the time of initiation of fusion. The fusion-promoting activity in the medium was larger than a spherical molecule with a molecular weight of 300 000.

Using a method for both collection and assay of "conditioning" factors in medium, Doering and Fischman (1977) demonstrated that conditioned medium activity was the result of both nutrient depletion, and the addition of macromolecules, probably protein in nature, with a molecular weight greater than 10 000 which was secreted by both pre- and post-fusion muscle cells.

In contrast, Yeoh and Holtzer (1977) found no evidence that increased cell density induced precocious fusion in myogenic cells. Conditioned medium prepared from advanced cultures by the method of Doering and Fischman did however induce earlier and more extensive myotube formation but these cultures degenerated beyond day 4 in culture.

In tissue culture, myoblasts divide several times and then some fuse and produce cell-specific proteins while others (possibly analogous to the satellite cells in vivo) remain as mononucleate cells. When the cells from advanced cultures are transferred to new culture dishes it is presumably these mononucleate cells that resume myogenic activity, since multinucleated myotubes will not normally stick down afresh. Some clue as to the mechanism which underlies this difference in myoblast behaviour has been provided by Nameroff and Holtzer (1969). Embryonic myogenic cells were induced to mimic the behaviour of satellite cells by seeding them on confluent, non-proliferating substrates such as 5-day cultures of muscle cells. On such substrates presumptive myoblasts (identified by ^{3}H-thymidine labelled nuclei) were quiescent. However, when removed from the inactivating influence of these cellular substrates and sub-cultured, they migrated, multiplied and fused to form typical myotubes. When the seeded cells were grown on Millipore filters above the inactivating substrate or in circular cuts where they did not touch the substrate cells they could divide and differentiate. Muscle differentiation could occur on substrates fixed with alcohol but not when fixed with glutaraldehyde. Nameroff and Holtzer suggested that glutaraldehyde was a superior fixative for cell surface components and intercellular matrix. They proposed that an extracellular polysaccharide matrix was responsible for the observed behaviour.

Neighbouring Fibroblasts

Neighbouring fibroblasts are responsible for laying down the muscle connective tissue system which involves synthesis of large amounts of collagen which may have an important influence acting as a substratum for muscle cells.

Hauschka and Konigsberg (1966) demonstrated that when the culture dish surface was coated with rat tail type I collagen, muscle cell growth and differentiation was promoted. Ketley et al. (1976) found that different polymorphic forms of collagen (Types I, II, III and IV) were equally effective as substrates for chick myogenesis. In contrast, John and Lawson (1980) found that type I collagen from chicken muscle caused parallel alignment of chick myotubes whereas on type V collagen substrates each myotube was rounded with nuclei bunched together. The alignment of cells suggested that the underlying cell recognition sites on collagen molecules were organized in such a way as to produce this effect. The way collagen might interact with muscle cells at the molecular level has been discussed previously (John and Lawson, 1980; John, 1981).

Bailey et al. (1979) showed that different collagen types were synthesized at different stages of chick myogenesis in culture. Type V collagen was synthesized during division of myoblasts while only after fusion was type I and III detected. The authors suggested that the presence of type V was probably involved in the sequence of events leading to myoblast differentiation.

THE ROLE OF FACTORS CIRCULATING IN THE BLOOD

Factors circulating in the blood including the serum proteins, nutrients and hormones almost certainly have a role in regulating muscle cell differentiation. Many of these factors are probably represented in tissue culture medium by the serum and embryo extract components which are essential to supplement the defined medium.

The type of serum used in tissue culture of muscle cells can affect both growth and differentiation. Foetal calf serum favours division of myogenic cells, while horse serum promotes differentiation (Yaffe, 1971; Morris and Cole, 1972). John (1981) used SDS-gel electrophoresis to analyse protein components of medium at different stages of myogenesis. Analysis of both horse serum- and foetal calf serum-containing medium showed no detectable change between the early stages of cultures and the period during and after fusion. However, increased amounts of serum albumin, α_2 macroglobulin, α_1 acid glycoprotein, β_2 glycoprotein, transferrin, haemopexin, ceruloplasmin and immunoglobulins were found in 10% horse serum-containing medium. The amount of these proteins in medium may affect myogenesis by promotion or inhibition of either cell division or cell differentiation.

Previous investigations (reviewed by Paul, 1970) indicated that α-acid glycoprotein and α_2 macroglobulin were the active components of serum and these disappeared during cultivation of cells. Both had to be present to stimulate growth and together they could replace serum. The macroglobulin could be replaced by non-protein molecules but the acid glyocoprotein was essential.

Kohama and Ozawa (1978) reported that low concentrations of chicken serum stimulated the proliferation of cultures of 12 day embryonic chick muscle cells. The activity of serum varied during the course of development. The activity of serum from the embryo was high from day 11 to day 13 of incubation and then decreased, reaching a minimum on day 16. This decrease correlated with the period of decreased cell division and increased fusion of myotubes in vivo (Hermann et al., 1970). Subsequently, the activity of the serum increased, reaching a peak around the day of hatching. After hatching, the activity of the serum began to decrease on day 4, reaching a minimum at day 18.

Chick embryo extract (CEE) is used to supplement medium for myogenic cultures. Yaffe (1971) suggested that a high CEE concentration was a non specific stimulator of cell proliferation. De la Haba et al. (1975) separated CEE into two fractions, a non-filterable fraction (including molecules of molecular weight greater than 10 000) which promotes both the fusion of myoblasts to form syncytia and the further differentiation of these syncytia into myotubes, and a filterable fraction which stimulated myoblast fusion to form syncytia only. The filterable fraction promoted myotube formation if the cells were grown on a collagen substratum. The nonfilterable fraction already contained a collagen-like molecule.

NEURAL INFLUENCE

Nerve cells exert a profound influence on muscle development and maintenance of the differentiated state.

It is now established that the type of innervation affects gene expression during the differentiation of muscle cells. Experiments in which fast and slow muscle were cross-reinnervated suggested that patterns of impulses may be important and this was confirmed by direct electrical stimulation of muscle via nerve in vivo (Salmons and Sreter, 1976).

Intermittent electrical stimulation of myotube cultures in the total absence of nerves leads to loss of sensitivity to iontophoretically applied acetylcholine and less binding of ^{125}I-labelled α-bungarotoxin in comparison with inactive fibres (Cohen and Fischbach, 1973), decreased synthesis of acetylcholine receptors (Shainberg and Burstein, 1976) and reduced acetylcholinesterase activity (Walker and Wilson, 1975). Repetitive stimulation increased protein synthesis and particularly myosin synthesis (Brevet et al., 1976).

However, there is also evidence that nerve cells can produce so-called neurotrophic factors which can influence differentiation of muscle cells. Oh et al. (1972) found that acetylcholinesterase activity in chick muscle cultures was increased by the presence of innervating spinal cord explants. The increased activity could still be obtained if the muscle cells were on one side of a coverslip and the nerve cells on the other indicating that a functional synapse was not necessary. Furthermore, soluble protein extracts of the spinal cord could induce the same effect, suggesting that nerve cells could produce a diffusible factor. There is evidence that proteins synthesized in the spinal cord can flow distally along the peripheral nerves in vivo (e.g. Komiya and Austin, 1974). A soluble extract of adult chicken sciatic nerve enhanced the rate and degree of mophological differentiation and maturation, DNA and protein synthesis and acetylcholinesterase activity of muscle cultures (Oh, 1976). The active factor had a native molecular weight of 25 000 (Markelonis and Oh, 1978, 1980).

Medium from spinal cord explant cultures was capable of producing the same reduction in transverse membrane resistance as the co-culturing of spinal cord explants with muscle (Engelhardt et al., 1977). Extracts of rat spinal cord added to L_6 cloned rat muscle cell line increased the number of ACh receptors and also increased the number of clusters of receptors (Podleski et al., 1978). The active component appears to be a protein with a molecular weight of about 100 000.

John and Jones (1975) found that radioactively labelled myosin from rat myotubes in culture contained only two light chains which co-electrophoresed with the LC_1 and LC_2 light chains of adult rat fast myosin. When the myotubes were co-cultured with spinal cord cells there was no effect on light chain composition of the Ca^{2+} ATPase level of myosin suggesting that if there was diffusion of a trophic influence from the spinal cord cells it had no effect on myosin structure and function. It was not shown whether the nerve cells in tissue culture sent impulses to the myotubes; regular contractions of the myotubes were observed in the presence and absence of nerve cells. The experiments did not rule out the possibility that a regular pattern of impulses from nerve to muscle in tissue culture could modify myosin either directly or by triggering off a trophic influence.

Recently Bonner (1980) used clonal anlysis of myoblast differentiation to assess the effect of denervation on developing skeletal muscle. The differentiation of skeletal muscle myoblasts isolated from 10-18 day embryos was affected only if functional denervation by d-tubocurarine was carried out on 5-6 day embryos (Hamburger and Hamilton stage 27-30) and not on earlier or later embryos.

CONCLUSION

During embryonic development certain cells of mesodermal origin are determined to become muscle cells. Further important influences determine the future relationship of muscle to other tissues resulting in organized patterns. Once the identity of cells as muscle cells is established they can exist in two forms. One form is as myoblasts which are rapidly dividing during embryonic development and growth and are quiescent as satellite cells in mature differentiated muscle. These are responsible for producing and perpetuating the myogenic population in the individual. The second form is the terminally differentiated myotubes which effectively carry out the muscle function. The foregoing review indicates that gene expression and differentiation in both these stages is regulated by a variety of control elements and influences operating at different levels.

REFERENCES

Affara, N. A., Robert, B., Jacquet, M., Buckingham, M. E., and Gros, F., 1980a, Changes in gene expression during myogenic differentiation. I. Regulation of messenger RNA sequences expressed during myotube formation, J. Molec. Biol., 140:441.

Affara, N. A., Daubas, P., Weydert, A., and Gros, F., 1980b, Changes in gene expression during myogenic differentiation. II. Identification of the proteins encoded by myotube-specific complementary DNA sequences, J. Molec. Biol., 140:459.

Bag, J., and Sarkar, S., 1976, Studies on a nonpolysomal ribonucleoprotein coding for myosin heavy chains from chick embryonic muscles, J. biol. Chem., 251:7600.

Bag, J., and Sells, B. H., 1979, Heterogeneity of the nonpolysomal cytoplasmic (free) mRNA protein complexes of embryonic chicken muscle, Eur. J. Biochem., 99:507.

Bailey, A. J., Shellswell, G. B., and Duance, V. C., 1979, Identification and change of collagen types in differentiating myoblasts and developing chick muscle, Nature, Lond., 278:67.

Bester, A. J., Kennedy, D. S., and Heywood, S. M., 1975, Two classes of translational control RNA: their role in the regulation of protein synthesis, Proc. Natl. Acad. Sci. U.S.A., 72:1523.

Bester, A. J., Durrheim, G., Kennedy, D. S., and Heywood, S. M., 1980, Isolation of myosin messenger ribonucleoprotein particles which contain a protein fraction affecting myosin synthesis, Biochem. Biophys. Res. Commun., 92:524.

Bonner, P. H., 1980, Differentiation of chick embryo myoblasts is transiently sensitive to functional denervation, Develop. Biol., 76:79.

Brevet, A., Pinto, E., Peacock, J., and Stockdale, F. E., 1976, Myosin synthesis increased by electrical stimulation of skeletal muscle cell cultures, Science, 193:1152.

Buckingham, M. E., Caput, D., Cohen, A., Whalen, R. G., and Gros, F., 1974, The synthesis and stability of cytoplasmic messenger RNA during myoblast differentiation in culture, Proc. Natl. Acad. Sci. U.S.A., 71:1466.

Buckingham, M. E., Cohen, A., and Gros, F., 1976, Cytoplasmic distribution of pulse-labelled poly(A)-containing RNA, particularly 26 S RNA, during myoblast growth and differentiation, J. Molec. Biol.

103:611.

Carlsson, S.-A., Luger, O., Ringertz, N. R., and Savage, R. E., 1974a, Phenotypic expression in chick erythrocyte x rat myoblast hybrids and in chick myoblast x rat myoblast hybrids, Expl. Cell. Res., 84:47.

Carlsson, S.-A., Ringertz, N. R., and Savage, R. E., 1974b, Intracellular antigen migration in interspecific myoblast heterokaryons, Expl. Cell Res., 84:255.

Chi J. C. H., Rubinstein, N., Stahs, K., and Holtzer, H., 1975a, Synthesis of myosin heavy and light chains in muscle cultures, J. Cell Biol., 67:523.

Chi, J. C., Fellini, S. A., and Holtzer, H., 1975b, Differences among myosins synthesized in non-myogenic cells, presumptive myoblasts and myoblasts, Proc. Natl. Acad. Sci. U.S.A., 72:4999.

Cohen, S. A., and Fischbach, G. D., 1973, Regulation of muscle acetylcholine sensitivity by muscle activity in cell culture, Science, 181:76.

De la Haba, G., Kamali, H. M., and Tiede, D. M., 1975, Myogenesis of avian striated muscle in vitro: role of collagen in myofiber formation, Proc. Natl. Acad. Sci., U.S.A., 72:2729.

Dienstman, S. R., and Holtzer, H., 1977, Skeletal myogenesis. Control of proliferation in a normal cell lineage, Expl. Cell Res., 107:355.

Doering, J. L., and Fischman, D. A., 1977, A fusion-promoting macromolecular factor in muscle conditioned medium, Expl. Cell Res., 105:437.

Doetschman, T. C., Dym, H. P., Siegel, E. J., and Heywood, S. M., 1980, Myoblast stored myosin heavy chain transcripts are precursors to the myotube polysomal myosin heavy chain mRNAs, Differentiation, 16:149.

Dym, H. P., Kennedy, D. S., and Heywood, S. M., 1979, Sub-cellular distribution of the cytoplasmic myosin heavy chain mRNA during myogenesis, Differentiation, 12:145.

Engelhardt, J. K., Ishikawa, K., Mori, J., and Shimabukuro, Y., 1977, Neurotrophic effects on the electrical properties of cultured muscle produced by conditioned medium from spinal cord explants, Brain Research, 128:243.

Gette, W. R., and Heywood, S. M., 1979, Translation of myosin heavy chain messenger ribonucleic acid in an eukaryotic initiation factor 3- and messenger-dependent muscle cell-free system, J. Biol. Chem., 254:9879.

Gurdon, J. B., and Woodland, H. R., 1968, The cytoplasmic control of nuclear activity in animal development, Biol. Rev., 43:233.

Hauschka, S. D., and Konigsberg, I. R., 1966, The influence of collagen on the development of muscle clones, Proc. Natl. Acad. Sci. U.S.A., 55:119.

Hermann, H., Heywood, S. M., and Marchok, A. C., 1970, Reconstruction of muscle development as a sequence of macromolecular synthesis Current Topics in Develop. Biol., 5:181.

Heywood, S. M., Kennedy, D. S., and Bester, A. J., 1974, Separation of specific initiation factors involved in the translation of myosin and myoglobin messenger RNAs and the isolation of a new RNA involved in translation, Proc. Natl. Acad. Sci. U.S.A., 71:2428.

Heywood, S. M., Kennedy, D. S., and Bester, A. J., 1975, Stored myosin messenger in embryonic chick muscle, FEBS Letters, 53:69.

Heywood, S. M., and Kennedy, D. S., 1978, Messenger RNA affinity column fractionation of eukaryotic initiation factor and the translation of myosin messenger RNA, Arch. Biochem. Biophys., 192:270.

John, H. A., 1982, Factors affecting gene expression and differentiation during myogenesis in tissue culture, in: "Differentiation in vitro," M. M. Yeoman and D. E. S. Truman, eds., Cambridge University Press, p.121.

John, H. A., and Jones, K. W., 1975, Tissue culture investigations of the effect of nerve on myosin structure and function, in: "Recent Advances in Myology," W. G. Bradley, D. Gardner-Medwin, and J. N. Walton, eds., Excerpta Medica, Amsterdam.

John, H. A., Patrinou-Georgoulas, M., and Jones, K. W., 1977, Detection of myosin heavy chain mRNA during myogenesis in tissue culture by in vitro and in situ hybridization, Cell, 12:501.

John, H. A., and Lawson, H., 1980, The effect of different collagen types used as substrata on myogenesis in tissue culture, Cell Biol. Int. Reports, 4:841.

John, H. A., and Jones, K. W., 1981, Tissue culture in muscle disease, in: "Disorders of Voluntary Muscle," J. N. Walton, ed., Churchill Livingstone, Edinburgh.

Kennedy, D. S., Siegel, E., and Heywood, S. M., 1978, Purification of myosin mRNP translational control RNA and its inhibition of myosin and globin messenger translation, FEBS Letters, 90:209.

Ketley, J. N., Orkin, R. W., and Martin, G. R., 1976, Collagen in developing chick muscle in vivo and in vitro, Expl. Cell Res., 99:261.

Kohama, K., and Ozawa, E., 1978, Muscle trophic factor: II. Ontogenic development of activity of a muscle trophic factor in chicken serum, Muscle and Nerve, 1:236.

Komiya, Y., and Austin, L., 1974, Axoplasmic flow of protein in the sciatic nerve of normal and dystrophic mice, Expl. Neurology, 43:1.

Konigsberg, I. R., 1961, Some aspects of myogenesis in vitro, Circulation, 24:447.

Konigsberg, I. R., 1971, Diffusion mediated control of myoblast fusion, Develop. Biol., 26:133.

Linder, S., Brzeski, H., and Ringertz, N. R., 1979, Phenotypic expression in cybrids derived from teratocarcinoma cells fused with myoblast cytoplasms, Expl. Cell Res., 120:1.

Markelonis, G. J., and Oh, T. H., 1978, A protein fraction from peripheral nerve having neurotrophic effects on skeletal muscle cells in culture, Expl. Neurology, 58:285.

Markelonis, G. J., and Oh, T. H., 1978, A sciatic nerve protein has a trophic effect on development and maintenance of skeletal muscle cells in culture, Proc. Natl. Acad. Sci. U.S.A., 76:2470.

Morris, G. E., and Cole, R. J., 1972, Cell fusion and differentiation in cultured chick muscle cells, Expl. Cell Res., 75:191.

Morris, G. E., and Cole, R. J., 1979, Calcium and the control of muscle-specific creatine kinase accumulation during skeletal muscle differentiation in vitro, Develop. Biol., 69:146.

Nadal-Ginard, B., 1978, Commitment, fusion and biochemical differentiation of a myogenic cell line in the absence of DNA synthesis, Cell, 15:855.

Nameroff, M., and Holtzer, H., 1969, Interference with myogenesis, Develop. Biol., 19:380.

Oh, T. H., 1976, Neurotrophic effects of sciatic nerve extracts on muscle development in culture, Expl. Neurology, 50:376.

Oh, T. H., Johnson, D. D., and Kim, S. V., 1972, Neurotrophic effect on isolated chick embryo muscle in culture, Science, 178:1298.

Paterson, B. M., and Bishop, J. O., 1977, Changes in the mRNA population of chick myoblasts during myogenesis in vitro, Cell, 12:751.

Paul, J., 1970, "Cell and Tissue Culture," Livingstone, Edinburgh.

Pluskal, M. G., and Sarkar, S., 1981, Cytoplasmic low molecular weight ribonucleic acid species of chick embryonic muscles, a potent inhibitor of messenger ribonucleic acid translation in vitro, Biochemistry, N.Y., 20:2048.

Podleski, T., Axelrod, D., Raudin, P., Greenberg, I., Johson, M. M., and Salpeter, M. M., 1978, Nerve extract induces increase and redistribution of acetylcholine receptors on cloned muscle cells, Proc. Natl. Acad. Sci. U.S.A., 75:2035.

Ringertz, N. R., Krondahl, U., and Coleman, J. R., 1978, Reconstitution of cells by fusion of cell fragments. I. Myogenic expression after fusion of minicells from rat myoblasts (L6) with mouse fibroblasts (A9) cytoplasm, Expl. Cell Res., 113:233.

Robbins, J., and Heywood, S. M., 1978, Quantification of myosin heavy-chain mRNA during myogenesis, Eur. J. Biochem., 82:601.

Rourke, A. W., and Heywood, S. M., 1972, Myosin synthesis and specificity of eukaryotic initiation factors, Biochemistry, 11:2061.

Salmons, S., and Sreter, F. A., 1976, Significance of impulse activity in the transformation of skeletal muscle type, Nature, Lond., 263:30.

Shainberg, A., Yagil, G., and Yaffe, D., 1969, Control of myogenesis in vitro by Ca^{2+} concentrations in nutritional medium, Expl. Cell Res., 58:163.

Shainberg, A., and Burstein, M., 1976, Decrease of acetylcholine receptor synthesis in muscle cultures by electrical stimulation, Nature, Lond., 264:368.

Walker, C. R., and Wilson, B. W., 1975, Control of acetylcholinesterase by contractile activity of cultured muscle cells, Nature, Lond., 256:215.

Whalen, R. G., Butler-Browne, G. S., and Gros, F., 1976, Protein synthesis and actin heterogeneity in calf muscle cells in culture,

Proc. Natl. Acad. Sci. U.S.A., 73:2018.

Yaffe, D., 1971, Developmental changes preceding cell fusion during muscle differentiation in vitro, Expl. Cell Res., 66:33.

Yaffe, D., and Dym, H., 1972, Gene expression during differentiation of contractile muscle fibres, Cold Spring Harbor Symp. quant. Biol., 37:543.

Yeoh, G. C. T., and Holtzer, H., 1977, The effect of cell density, conditioned medium and cytosine arabinoside on myogenesis in primary and secondary cultures, Expl. Cell Res., 104:63.

Zalin, R. J., 1977, Prostaglandins and myoblast fusion, Develop. Biol., 59:241.

Zalin, R. J., and Montague, W., 1974, Changes in adenylate cyclase, cyclic AMP and protein kinase levels in chick myoblasts and their relationship to differentiation, Cell, 2:103.

THE MURINE HAEMOPOIETIC STEM CELL:

PATTERNS OF PROLIFERATION AND DIFFERENTIATION

B. I. Lord

Paterson Laboratories
Christie Hospital & Holt Radium Institute
Manchester M20 9BX

The haemopoietic stem cell exists as the base for a large variety of cell types. As a result, in terms of cell numbers it is very much a minority population and consequently has escaped a definative morphological definition. The classical techniques for studying cell proliferation and differentiation are, therefore, not applicable. Its basic assay is the spleen colony-forming unit (CFU-S) assay (Till and McCulloch, 1961) which relies on the ability of these cells to form colonies of haemopoietic cells in the spleens of irradiated mice. About 3 bone marrow cells in 10^4 form macroscopic colonies. CFU-S are pluripotent and clonogenic (Becker et al., 1963).

Not all of the cells capable of forming spleen colonies, however, actually seed in the spleen - they are distributed around the various haemopoietic tissue sites in the body. A more realistic evaluation of the number of stem cells in the mouse, therefore, is obtained by determining the proportion of colony-forming cells which actually form colonies. This is done by a secondary transplantation experiment (Becker et al., 1965; Lord, 1971). Marrow cells are grafted into a primary irradiated recipient mouse whose spleen is removed 24 hrs later and assayed for CFU-S content in secondary irradiated recipient mice. The number of colony forming cells injected into the primary recipient is determined by a separate primary colony assay. Thus, the simple relationship between the number of colonies produced in the secondary recipients and the number of primary colony-forming cells injected gives the spleen seeding factor of the colony-forming cells. This figure is about 10 per cent which means that the total number of colony forming cells per mouse is about $1-2 \times 10^6$.

The CFU-S population is relatively quiescent, less than 10% normally being in DNA-synthesis, suggesting a population turnover time of about 5 days. At the same time it gives rise to a vast production of mature cells: about 5×10^9 cells per day in the red cell system alone.

Haemopoietic tissue, therefore, is a complex, heterogenous collection of cells (though it would be wrong to assume it to be a randomly distributed collection of free-living cells: it has a cellular organisation as exquisite as any epithelial tissue, for example (Lord, 1978; Lord and Hendry, 1972; Lord, et al., 1975; Xu and Hendry, 1981) which requires an equally complex set of regulatory functions to maintain proliferation and differentiation control over its stem and differentiated cell populations.

RELATIONSHIP OF THE COLONY-FORMING CELL TO THE STEM CELL

Since the CFU-S is a self-renewing population of cells giving rise to several lines of maturing cells, it is usually considered as the stem cell. With a population turnover time of 5 days, a CFU-S probably undergoes at least 200 population doublings in the lifetime of a mouse. Furthermore, CFU-S from an old mouse repopulate an irradiated mouse just as efficiently as those from a young animal. Since the population showed no evidence of ageing, CFU-S thus comply with an even more stringent definition of a stem cell: that its self-maintenance capacity must remain intact for at least the lifetime of the animal (Lajtha, 1979). However, when serially transplanted, the repopulating capacity of the CFU-S population does decline (Siminovitch et al., 1964) even when attempts are made always to transplant the same number of CFU-S (Lajtha and Schofield, 1971). This and the finding that some CFU-S are 'better' than others in that they generate more CFU-S in an individual colony (demonstrate a higher self-renewal capacity) led to the suggestion that the CFU-S population is 'age-structured' (Schofield and Lajtha, 1973; Rosendaal et al., 1979). Thus, although the population as a whole does not age, individual CFU-S apparently do. To overcome this problem Schofield proposed that the ageing CFU-S population is anchored and fed by a true stem cell which by virtue of its location relative to a specific micro-environmental cellular organisation (a 'niche') is retained as a purely self-maintaining cell: at each division, one daughter remains in the niche while the other proceeds into its ageing transit as a colony-forming cell (Schofield, 1978). Nevertheless, the haemopoietic system including immune-competence can be re-established in an irradiated mouse by an injection of normal bone marrow containing the full spectrum of ageing CFU-S. For most practical purposes, therefore, CFU-S can be considered as stem cells.

CONTROL OF STEM CELL PROLIFERATION

Studies of the kinetic behaviour of CFU-S following partial body irradiation (Croizat et al., 1970; Gidali and Lajtha, 1972) have shown that CFU-S proliferation is regulated locally rather than by circulating humoral factors. Haemopoietic tissue containing proliferatively quiescent CFU-S was subsequently shown to contain an inhibitor of CFU-S proliferation (Lord et al., 1976) and that containing proliferatively active CFU-S was shown to contain a stimulator (Lord et al., 1977). Both factors appear to act in an 'all-or-none' manner, switching proliferation respectively off and on. While the triggering action of the stimulator is extremely rapid (within about 30 min. and presumably mobilizing cells from the Go phase directly into DNA-synthesis) the effectiveness of the inhibitor is related to the phase of the cycle. Cells in DNA-synthesis are not directly affected. Interphase cells, however, are sensitive to inhibitor but their progress round the cycle continues through to the Go-phase where they leave the cycle. Both these factors are specific for the CFU-S population, having no effect on proliferation of the committed granulocyte/macrophage precursor cell (GM-CFC), an immediate descendant of the CFU-S.

The finding that all haemopoietic tissue, irrespective of the proliferative status of the CFU-S, contains both inhibitor- and stimulator-producing cells (Wright and Lord, 1979) and that the two can interact with each other, the one being able to override the effect of the other (Lord et al., 1977) suggested that the overall regulation of CFU-S proliferation depends on the balance of inhibitor and stimulator production. This principle was strengthened by the finding that the presence of inhibitor blocks the production of stimulator and vice-versa. The influence of inhibitor- and stimulator-producing cells on CFU-S proliferation and the fact that these regulators necessarily act over only a short-range, suggest environmental involvements and it is probable that the geometrical organisation of CFU-S with their regulator cells must form an integral part of the regulation processes. It would seem appropriate to consider the stem cell niche proposed by Schofield to constitute just such a microenvironmental array of cells. At the same time, differentiation must also be considered and it would be wrong to suggest that inhibitor- and stimulator-producing cells are necessarily the only cells making up that environment.

DIFFERENTIATION OF THE PLURIPOTENT STEM CELL

In its capacity as a CFU-S, the haemopoietic stem cell can immediately be seen to give rise to erythroid, granulocytic and megakaryocytic cells. In addition, a surprising variety of other cell types are now thought to originate from that same cell. These include macrophages and eosinophils (Metcalf and Moore, 1971), lymphocytes (Phillips and Jones, 1979), osteoclasts (Ash et al., 1980), tissue mast cells (Kitamura et al., 1981) and epidermal Langerhans cells (Katz et al., 1979). A vast amount of diversity is, therefore, generated from the haemopoietic stem cell and as yet there is little more than speculation by way of explanations.

Two main theories have been postulated and have resulted in a considerable degree of polarization between haematologists who favour HER (haemopoiesis engendered randomly) and those who favour HIM (haemopoietic inductive microenvironment). Based on the observation that in the early stages of spleen colony growth, the majority of cells are of one type, other types developing later, on the periphery of the initial development, Curry and Trentin (1967) came to the conclusion that microenvironmental factors determine the direction of differentiation. However, a variety of committed precursors develop in the early stages even of pure erythroid or granulocytic colonies suggesting that HIMs perhaps act at a later stage than CFU-S commitment.

The stochastic model, by contrast, requires that any CFU-S (and its daughter CFU-S) is always subject to the same chance of developing along any given differentiation pathway (Siminovitch et al., 1963). This random differentiation process was suggested by the observation that colonies differ from one another in their content of differentiated cell types and new CFU-S. David and MacWilliams (1978), however, showed that the variation in bone marrow clones is much larger than could be generated by a stochastic process alone but that the presence of microenvironmental regions with different self-renewal forces could be sufficient. The fact that the erythroid:myeloid ratio is 3:1 in the spleen but 1:3 in the marrow, however, does indicate at least some involvement of the microenvironment.

A third model for the generation of diversity in a cell system is that developed by Holtzer (1978) and has received little attention from experimental haematologists. It assumes that differentiation is a programmed sequence of changes tied to proliferation cycles. At each stage, a cell has two potential daughters only, each different from the other and its mother and each of which has its own and well defined, though different, bipotentiality. Symmetrical proliferation divisions for population amplification are permitted but diversification arises by asymmetrical divisions in what are termed 'quantal cell cycles'. In the classical sense, a pluripotent stem cell with three or more possible progeny may, at any time, turn into any one of these differentiated states. Holtzer points out, however, there is no experimental evidence supporting the concept that a cell exists with the capacity to give rise to three or more cell types without first passing through intermediate cell cycles. (Neither is there experimental evidence to the contrary.) But, there is now some evidence in support of the bipotentiality principle.

Humphries et al. (1979) showed that some of the burst forming units, the supposedly early erythroid committed precursor cells, lead to the development of both erythroid and megakaryocytic cells.

Table 1. Loss of differentiation potentials during successive diversification stages.

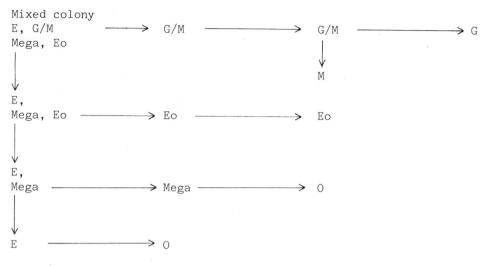

E - erythroid G/M - granulocyte/macrophage
Mega - megakaryocytic Eo - eosinophilic
(Schedule based on Johnson, 1981.)

Metcalf (1981) replated single cells from first and second generation daughters of GM-CFC and showed they would amplify as GM-CFC for several cell cycles after which the two individual lines of granulocytes and macrophages developed. Johnson (1981) used the mixed-colony culture technique - the near in vitro parallel of the CFU-S assay. He noted that the truly mixed colony was always characterised by the inclusion of erythroid cells and that on replating individual colonies, only the mixed ones were capable of generating further mixed colonies containing erythroid cells. Primary GM-CFC colonies could produce only GM-CFC secondary colonies. Similarly, primary eosinophilic colonies could give rise only to eosinophilic colonies while megakaryocytic and pure erythroid colonies were terminal in the sense that they could not support the growth of secondary colonies. These results are depicted schematically in Table 1, where striking similarities with the Holtzer bipotentiality principle are obvious.

This differs from Holtzer in two important aspects. (1) It carries specific differentiation potentials part or all the way through the baseline sequence e.g. erythroid. (2) Although diversification is illustrated in the table as resulting from a division, this is not essential. Incorporating the principle of an age-structured CFU-S population, the baseline sequence of mixed colony through to pure erythroid colony can readily be generated by allowing CFU-S of different 'ages' to express different differentiation

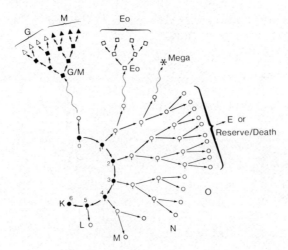

Figure 1. Generation of haemopoietic tissue starting from a single cell at position Ko. Movement from position 0 to position 1 etc. results in the deletion of cell Ko but this is a chronological movement rather than a spatial movement. G - granulocytic, M - macrophage, Eo - eosinophil, Mega - megakaryocyte, E - erythrocyte. K→0 - quantal cell stages. ○ - indicates a cell replaced by two cells in the next quantal stage. Reproduced, with permission, from Lord, 1981.

potentials. Figure 1 illustrates this principle more fully. A CFU-S in the Ko position is the primary stem cell - possibly the niche occupying cell. A mitotic division leaves one cell in the niche (K_1) and one daughter moves into the L-shell as an older CFU-S. In this shell it is available for differentiation to a GM-CFC. If not utilized in this way, the L_1 cell can divide to put two older daughters in the M-shell where they become available for eosinophil differentiation etc. Extra symmetrical amplification divisions may be inserted between any or all of these quantal shells.

This approach is largely speculative but two pieces of experimental evidence are in its favour. The identical response and recovery of CFU-S and GM-CFC to isopropyl methane sulphonate (Schofield, personal communication) was more indicative of a close temporal relationship between the cells than of differentiation arising at the end of the CFU-S ageing sequence. Secondly, extensive self-replication in vitro of GM-CFC has recently been reported (Dexter et al., 1980). If, indeed, the GM-CFC lies close to the CFU-S, one might expect it to retain many of the properties of CFU-S including a reasonable capacity for self-maintenance. With the correct stimulus of appropriate 'environmental factors' it is not unreasonable that GM-CFC may expand that capacity.

The means of regulating differentiation is equally little understood. Modification of the erythroid:myeloid ratio by diffusible factors from ARA-C or radiation-damaged marrow has been reported by Frindel (1979). Marrow fibroblasts may be involved in increasing differentiated cell output (Blackburn and Goldman, 1981) possibly enhancing differentiation by inducing GM-colony stimulating activity production (Wilson et al., 1974). These processes, however, remain a field for future evaluation.

REFERENCES

Ash, P., Loutit, J. F., and Townsend, K. M. S., 1980, Osteoclasts derived from haemopoietic stem cells, Nature, Lond., 283:669.

Becker A. J., McCulloch, E. A., Siminovitch, L., and Till, J. E., 1965, The effect of differing demands for blood cell production on DNA synthesis by haemopoietic colony forming cells of mice, Blood, 26:296.

Becker, A. J., McCulloch, E. A., and Till, J. E., 1963, Cytological demonstration of the clonal nature of spleen colonies derived from mouse marrow cells, Nature, Lond., 197:452.

Blackburn, M. J., and Goldman, J. M., 1981, Increased haemopoietic cell survival in vitro induced by a human marrow fibroblast factor, Br. J. Haemat., 48:117.

Croizat, H., Frindel, E., and Tubiana, M., 1970, Proliferative activity of the stem cells in the bone marrow of mice after single and multiple irradiations (total or partial body exposure), Int. J. Radiat. Biol., 18:347.

Curry, J. L., and Trentin, J. J., 1967, Hemopoietic spleen colony studies. I. Growth and differentiation, Develop. Biol., 5:395.

David, C. N., and MacWilliams, H., 1978, Regulation of the self-renewal probability in Hydra stem cell clones, Proc. Natl. Acad. Sci. U.S.A., 75:886.

Dexter, T. M., Garland, J. M., Scott, D., Scolnick, E., and Metcalf, D., 1980, Growth of factor dependent haemopoietic precursor cell lines, J. Exp. Med., 152:1036.

Frindel, E., 1979, Regulation of bone marrow stem cell kinetics, in: "Cell Lineage, Stem Cells and Cell Determination," INSERM Symp. No. 10, N. Le Douarin, ed., Elsevier (North Holland) Biomedical Press, Amsterdam, p.227.

Gidali, J., and Lajtha, L. G., 1972, Regulation of haemopoietic stem cell turnover in partially irradiated mice, Cell Tissue Kinet., 5:47.

Holtzer, H., 1978, Cell lineages, stem cells and the 'quantal' cell cycle concept, in: "Stem Cells and Tissue Homeostasis," B. I. Lord, C. S. Potten, and R. J. Cole, eds., Cambridge University Press, Cambridge, London, New York, Melbourne, p.1.

Humphries, R. K., Eaves, A. C., and Eaves, C. J., 1979, Characterisation of a primitive erythropoietic progenitor found in mouse marrow before and after several weeks in culture, Blood, 53:746.

Johnson, G. R., 1981, Is erythropoiesis an obligatory step in the commitment of multipotential haemopoietic stem cells? in: "Experimental Hematology Today," S. J. Baum, G. D. Ledney, and A. Khan, eds., S. Karger, Basel, p.13.

Katz, S. I., Tamaki, K., and Sachs, D. H., 1979, Epidermal Langerhans cells are derived from cells originating in the bone marrow, Nature, Lond., 282:324.

Kitamura, Y., Yokoyama, M., Matsuda, H., Ohno, T., and Mori, K. J., 1981, Spleen colony-forming cell as common precursor for tissue mast cells and granulocytes, Nature, Lond., 291:159.

Lajtha, L. G., 1979, Stem cell concepts, Differentiation, 14:23.

Lajtha, L. G., and Schofield, R., 1971, Regulation of stem cell renewal and differentiation: possible significance in ageing, in: "Advances in Gerontological Research," B. L. Strehler, ed., Academic Press, New York, London, 3:131.

Lord, B. I., 1971, The relationship between spleen colony production and spleen cellularity, Cell Tissue Kinet., 4:211.

Lord, B. I., 1978, Cellular and architectural factors influencing the proliferation of haemopoietic stem cells, in: "Differentiation of Normal and Neoplastic Hematopoietic Cells," B. Clarkson, P. A. Marks, J. E. Till, eds., Cold Spring Harbor Conferences on Cell Proliferation, Vol. 5, p.775.

Lord, B. I., 1982, Haemopoietic stem cells, in: "Identification and Characterisation of Stem Cell Populations," C. S. Potten, ed., Churchill Livingstone, Edinburgh, (in press).

Lord, B. I., and Hendry, J. H., 1972, The distribution of haemopoietic colony forming units in the mouse femur and its modification by X-rays, Br. J. Radiol., 45:110.

Lord, B. I., Mori, K. J., and Wright, E. G., 1977, A stimulator of stem cell proliferation in regenerating bone marrow, Biomed. Exp., 27:223.

Lord, B. I., Mori, K. J., Wright, E. G., and Lajtha, L. G., 1976, An inhibitor of stem cell proliferation in normal bone marrow, Br. J. Hemat., 34:441.

Lord, B. I., Testa, N. G., and Hendry, J. H., 1975, The relative spatial distribution of CFU-S and CFU-C in the normal mouse femur, Blood, 46:65.

Metcalf, D., 1981, Commitment of bipotential GM progenitor cells to granulocyte or macrophage formation, in: "Stem Cells," (in press), Meeting of the European Stem Cell Club, reported by B. I. Lord.

Metcalf, D., and Moore, M. A. S., 1971, "Haemopoietic Cells," North Holland Biomedical Press, Amsterdam, p.550.

Phillips, R. B., and Jones, E., 1979, Quoted by T. M. Dexter, Cell interactions in vitro. Cellular dynamics of haemopoiesis, in: "Clinics in Haematology," L. J. Lajtha, ed., W. B. Saunders Co., London, Philadelphia, Toronto, 8:453.

Rosendaal, M., Hodgson, G. S., and Bradley, T. R., 1979, Organization of haemopoietic stem cells: the generation-age hypothesis, Cell Tissue Kinet., 12:17.

Schofield, R., 1978, The relationship between the spleen colony-forming cell and the haemopoietic stem cell: A hypothesis, Blood Cells, 4:7.

Schofield, R., and Lajtha, L. G., 1973, Effect of isopropyl methane sulphonate (IMS) on haemopoietic colony-forming cells, Br. J. Hemat., 25:195.

Siminovitch, L., McCulloch, E. A., and Till, J. E., 1963, Distribution of colony-forming cells among spleen colonies, J. cell comp. Physiol., 62:327.

Siminovitch, L., Till, J. E., and McCulloch, E. A., 1964, Decline in colony-forming ability of marrow cells subjected to serial transplantation into irradiated mice, J. cell comp. Physiol., 64:23.

Till, J. E., and McCulloch, E. A., 1961, A direct measurement of the radiation sensitivity of normal mouse bone marrow cells, Radiat. Res., 14:213.

Wilson, F. D., O'Grady, L., McNeill, C. J., and Munns, S. L., 1974, The formation of bone marrow derived fibroblastic plaques in vitro: preliminary results contrasting these populations to CFU-C, Expl. Hemat., 2:343.

Wright, E. G., and Lord, B. I., 1979, Production of stem cell regulators by fractionated haemopoietic cell suspensions, Leuk. Res., 3:15.

Xu, C. X., and Hendry, J. H., 1981, The radial distribution of fibroblastic colony-forming cells in mouse femoral marrow, Biomedicine, (in press).

ABSTRACT

INHIBITOR AND STIMULATOR IN THE REGULATION OF HAEMOPOIETIC

STEM CELL PROLIFERATION

> B. I. Lord
>
> Paterson Laboratories
> Christie Hospital and Holt Radium Institute
> Manchester M20 9BX

It has been demonstrated that an extract of normal bone marrow, NBME-IV (Lord et al., 1976) and an extract of regenerating bone marrow, RBME-III (Lord et al., 1977) can be used respectively to inhibit and stimulate the proliferative activity of haemopoietic spleen colony forming cells. Although not mutually destructive, these two extracts can act in competition with one another and it was therefore suggested that stem cell proliferation is regulated by the prevailing balance of inhibitor and stimulator.

In this study, the interrelationships of the inhibitor and stimulator have been examined by investigating the effects of inhibitor on the synthesis of stimulator and <u>vice versa</u>. Endogenous factors were removed by washing a preparation of normal or regenerating marrow cells. Synthesis of further regulatory factors was then initiated by resuspending the cells in fresh medium and incubating them at 37°C. A normal complement of inhibitor and stimulator was obtained after 5 and 3 hours incubation of the appropriate cell populations respectively. However, if these experiments were carried out in the presence of added inhibitor or stimulator, the synthesis of the complementary factor was found to be blocked.

A model describing the interrelationships of NBME-IV and RBME-III and their roles in the regulation of haemopoietic stem cell proliferation will be introduced and discussed.

REFERENCES

Lord, B. I., Mori, K. J., and Wright, E. J., 1977, A stimulator of stem cell proliferation in regenerating bone marrow, Biomed. Exp., 27:223.
Lord, B. I., Mori, K. J., Wright, E. J., and Lajtha, L. G., 1976, An inhibitor of stem cell proliferation in normal bone marrow, B. J. Haemat., 34:441.

ABSTRACT

LYMPHOCYTE DIFFERENTIATION WITHIN THE THYMUS

J. J. T. Owen

Department of Anatomy
Medical School
Birmingham B15 2TJ

Studies on the differentiation of lymphocytes within the embryonic thymus are greatly aided by two factors. First, structural and functional maturation of thymic lymphocytes and stromal cells can be obtained in vitro. Second, a variety of monoclonal antibodies are available which detect cell membrane antigens on both cell types. Studies have been carried out on normal thymus ontogeny and on abnormal thymic development in nude mice using these approaches. For a recent review see Owen and Jenkinson (1981).

REFERENCES

Owen, J. J. T., and Jenkinson, E. J., 1981, Embryology of the lymphoid system, Prog. Allergy, 29:1.

THE ROLE OF A GROWTH FACTOR DERIVED FROM THE RETINA (EDGF)
IN CONTROLLING THE DIFFERENTIATED STAGES OF SEVERAL OCULAR AND
NON-OCULAR TISSUES

Y. Courtois, C. Arruti[1], D. Barritault[2], J. Courty,
J. Tassin, M. Olivie, J. Plouet, M. Laurent[3],
and M. Perry[4]

INSERM U 118 - CNRS ERA 842
29 Rue Wilhem
75016 Paris, France

INTRODUCTION

Over the last twenty years, a new family of molecules, collectively named growth factors, has been discovered, and some of its members purified (reviewed. by Gospodarowicz and Moran, 1976). Most of these factors have been isolated either from the blood, from tissues or produced *in vitro*. Although the physiological significance of these factors is not yet understood, we suspect that some of them have important regulatory functions during embryogenesis and development, and in the control of cellular proliferation, and differentiation, as well as in maintaining tissue homeostasis. *In vivo* experiments elucidating the biological significance of these factors and the relationships between the producing and the target tissues have not yet been achieved. (There is no known growth factor defective mutant, nor an associated pathology. Furthermore, some of these factors are found in essential organs that cannot therefore be removed.) One possible first approach is to study, *in vitro*, the action of a growth factor on specific target cells, before looking at the effect of this factor on the same cells *in vivo*.

[1] Laboratorio de Cultivo de Tejidos, Dpto de Histologia y Embriologia, Faculdad de Medicina, Montevideo, Uruguay.
[2] also, University PARIS XII, UER Sciences 94100, Creteil, France.
[3] INSERM U 235 27, rue du Faubourg Saint Jacques, 75014 Paris, France.
[4] Agricultural Research Council's Poultry Research Centre, Roslin, Midlothian, EH25 9PS, Scotland, U.K.

In the course of our study on lens cell differentiation and ageing, we were confronted with the task of finding in vitro conditions in which we could induce these cells to differentiate as well as in vivo. Taking as a basis the numerous in vivo experiments which had clearly shown that retina plays an important role in the control of lens induction and differentiation during embryogenesis, Arruti and Courtois (1978) added a soluble extract of adult bovine retina to cultures of adult bovine epithelial lens (BEL) cells. Two effects were readily detected. First, the cell population increased remarkably compared with non-treated cultures. Second, the cells changed in their morphology, they elongated and became fibroblast-like. We present some data on the investigations of aspects both of the effects of retinal extract on BEL cells, and other ocular and non-ocular target cells.

RETINA AND OTHER OCULAR TISSUES CONTAIN A GROWTH FACTOR OR FACTORS RESPONSIBLE FOR MITOGENIC ACTIVITY: THE EYE DERIVED GROWTH FACTOR(s) OR EDGF

Extracts of retina, vitreous body, iris, choroid with pigmented epithelium, aqueous humour and the lens cortex were tested for their ability to stimulate tritiated thymidine (^{3}HTdR) incorporation and cell multiplication of BEL cells (Table 1 and Barritault et al., 1981a). Mitogenic activity could be detected not only in crude extracts of retina but also in vitreous body, iris, and choroid with pigmented epithelium. No activity was found in aqueous humour and lens cortex. However acetic acid purified aqueous humour (see below, EDGF purification) was also active on BEL cells, suggesting the existence of an inhibitor in the crude preparation. Evidence that the mitogenic activity found in the various ocular tissues had all the characteristics of a growth factor has been presented in Barritault et al. (1981a) and the name of "Eye Derived Growth Factor(s)" or EDGF was chosen. Briefly, EDGF can induce several rounds of confluent or sparse quiescent target cells, and this effect was dose dependent. In the absence of serum, EDGF could also stimulate cell proliferation: however there may be a synergistic effect between the serum and the extract since a combination of both resulted in greater stimulation than the addition of either one separately. One possible explanation is that EDGF makes the cells more sensitive to the growth factors present in the serum. The addition of EDGF by itself is not sufficient to maintain a high proliferative rate in the complete absence of serum for long periods - cells become highly vacuolated and die. However in a low serum concentration, of 0.5%, the cells can be maintained, and proliferate at a very low rate, forming mutilayers, and can be stimulated to proliferate again by addition of the retinal extract.

In another set of experiments EDGF was added one day after plating to BEL cells seeded at cloning densities. After 13 days the

Table 1. Effects of different bovine ocular tissue extracts on H thymidine incorporation and cell number in BEL cultures.

Tissue of origin	µg of total protein per ml culture medium	Number of cells/cm^2	cpm H^3 thymidine (4h)
PBS alone	0	56.10	10 000
Retina	100	112.10	170 000
Iris	100	100.10	160 000
Choroid	100	110.10	130 000
Vitreous body	100	90.10	140 000
Aqueous humour	50	50.10	12 000
Lens cortex	100	55.10	10 000

Retina, iris, and choroid were crude extracts in PBS prepared after 100 000 g centrifugation and dialysis. Vitreous body was first concentrated by ammonium sulfate (AS) precipitation; the fraction used was the supernatant of 20% AS (precipitated by 60% AS). Aqueous humour was concentrated using a B 15 Amicon macrosolute concentration (mol. wt. cut-off above 15 000 Daltons). Undialyzed samples had a similar stimulatory effect. However aqueous humour was found mitogenic for BEL cells after the acetic acid purification step, suggesting that inhibitors were eliminated. Other non-ocular tissues were also tested such as rat liver or chick embryo extract and were inactive.

These results were obtained from experiments in which BEL cells were maintained at confluence, stimulated for 2 days and labelled for 4h. In each experiment, one control was done by adding PBS alone. Since the extracts are not purified, a dose-response curve has to be determined for every sample. Only the maximum stimulation is shown here. Similar results have been obtained with extracts from human, calf, and horse retina.

Table 2. Purification steps of EDGF (values for 1000 retinas).

Purification steps	Proteins (mg)	Protein yield	SU (µg per ml of culture medium)	% activity recovery	Specific activity (SU x mg^{-1})	Total units (x 10^{-3})	Purification fold
20-000 g supernatant	28000	100	15	100	67	1866	1
20-60 % (NH$_4$)$_2$SO$_4$ fraction	18000	64	10	96	100	1800	1,5
Acetic acid (0,1 N)	1000	3,6	0,4	134	2500	2500	38
Blue Cibacron chromatography	120	0,43	0,03	214	33300	4000	530
High performance liquid chromatography	38	0,16	0,014	123	71300	2300	1064

Protein concentration is measured by the method of Bradford (1976)
Detail on the purification steps is described in Barritault et al. (1981b)

dishes were stained and examination indicated that colonies had developed much faster in the presence of EDGF. The results demonstrated that EDGF is a cloning factor and can be used to clone single cells.

PURIFICATION AND BIOCHEMICAL CHARACTERIZATION OF EDGF

A routine test, based on stimulation of ^{3}HTdR incorporation in confluent BEL cells was developed to assay the purification of the growth factor. BEL cells were seeded in 24-well Limbro plates and when confluence was reached, fractions of retinal extract were added at various dose levels at 0 hours and again at 24 hours. Cells were labeled with ^{3}HTdR (1 µCi per well, 25 Ci/mM) at 44 hours of culture for 4 hours. The radioactivity in TCA precipitable material was counted. A standard (brain FGF, kindly given by D. Gospodarowicz, or previously tested preparations of partially purified EDGF) was used in each set of experiments to establish a dose response curve of ^{3}HTdR incorporation and to measure the maximal incorporation (defined as 100%), and a stimulation unit was defined as the amount of protein added to one milliliter of culture medium which could induce incorporation of ^{3}HTdR. Purification steps, values for stimulation units and yeilds obtained with retina are presented in Table 2. Further details on the purification procedure will be published later (Barritault et al., 1981b). Molecular weight measurements indicate that EDGF from retina is a protein of 17 500 d ± 3.500 d with an isoelectric point of pH 4.5 ± 0.5. The most purified EDGF fractions that we have obtained had a specific activity 1000 fold greater than the initial crude extract. At this stage of purification EDGF stimulates replicative DNA synthesis and cell proliferation of bovine epithelium lens cell at a concentration of 14 ng per milliliter of culture medium. No evidence of differences between EDGF from retina and vitreous body could be detected during the different purification steps suggesting that both retina and vitreous body contain the same growth factor.

EFFECT OF EDGF ON THE PROLIFERATION OF VARIOUS CELL TYPES

To compare EDGF with other known factors and to study the mechanism of action of EDGF as well as its physiological significance, we looked at the ability of EDGF to stimulate the proliferation of other ocular and non-ocular cells from other species and origins. Different types of cells, primary cultures or established cell lines were cultured in the presence or in the absence of EDGF. The results are summarized in Table 3 (see also Barritault et al., 1981a) and were obtained in our laboratory and in collaboration with other laboratories. The results obtained by others or by us in parallel experiments using EGF or FGF are also indicated. All three growth factors differ in their spectrum of action on various cell types. However, there is a strong similarity between EDGF and FGF in their mitogenic activity.

Table 3. Effects of EDGF from retina on the proliferation of various cell types. Comparison with EGF and FGF. (Modified from Barritault et al., 1981a.)

a) EDGF is used as a retinal extract (50 µg x ml^{-1} of culture medium) or as acetic acid purified fraction (1 µg x ml^{-1} of culture medium).

b) Experiments also performed in our laboratory with brain FGF (gift from D. Gospodarowicz), usually in parallel with EDGF, and also results previously reported by Gospodarowicz et al. (1978b).

c) The same as b but using EGF (gift from Gospodarowicz or bought from Kor biochemicals).

d) All the data with FGF and EGF are from Gospodarowicz et al. (1978b). For FGF and EGF the stimulation is obtaine with pg or fg ml^{-1} (+++), or ng ml^{-1} (++), no effect (-). Blanks correspond to non-tested.

For EDGF the stimulation is measured by an increase in cell number or compared with the response obtained in parallel experiments with FGF or EGF and non-treated cells. Cells usually counted after 8 days of stimulation.
+++ mean of increase by a factor of 4 or more in cell numbe
 ++ an increase by 2-4
  + a weak increase
  - absence of stimulation

1. Lens epithelial cells, as described here
2. Cells donated by H. Bloemendal (Nijmegen, Holland)
3,4,5. Unpublished experiments with R. Clayton (Edinburgh)
6. J. Tassin: this laboratory. Despite the lack of stimulation in long-term experiments, RE induced a morphologica change
7,8. In collaboration with I. Guedon and M. Prunieras' (Paris)
9,10. Experiments performed in D. Gospodarowicz's laborator (San Francisco)
11. C. Arruti (Montevideo and this laboratory)
12,13. In collaboration with M. Fizman, Paris
14. Gift from Tauber (Toulouse)
15,16. Articular and cartilage rabbit chondrocytes in collaboration with M.T. Corvol, Paris
17,21. The response of fibroblasts to RE is different from other cells. In several cases after 6 days or more, the addition of the growth factor was toxic to the culture.
22. Gift from J. Pouyssegur (Nice)
23. In collaboration with Dr. Lagault, Paris

Cell type	Species	EDGF	FGF[d]	EGF[d]
1. Lens epithelium	Bovine	+++	+++[b]	−[c]
2. Lens epithelium	Calf	+++		
3. Lens epithelium	Chick normal	++		
4. Lens epithelium	Chick strain Hy1	++		
5. Lens epithelium	Chick strain Hy2	+		
6. Lens epithelium	Human	−	−	
7. Epidermal	Human adult	++		−[c]
8. Epidermal	Human newborn	+++	−	++[c]
9. Aortic endothelial	Bovine	+++	+++[b]	−
10. Heart endothelial	Bovine (foetus)	+++	+++[b]	−
11. Corneal endothelial	Bovine	+++	+++	
12. Myoblasts	Bovine (foetus)	+++	++[b]	−[c]
13. Myoblasts	Chick (embryo)	+	++[b]	
14. Vascular smooth muscle	Bovine	+++	++	+[c]
15. Chondrocytes (cartilage)	Rabbit	++	++	+++
16. Chondrocytes (articular)	Rabbit	+	++	+++
17. Skin fibroblasts	Human	−	++	++
18. Kidney fibroblasts	Bovine	−	+	
19. Skin fibroblasts	Bovine	−		
20. Lung fibroblasts	Bovine	−		
21. 3T3 Swiss	Mouse	−	++	++
22. 3T3 NIH	Rat	+++	+++[b]	+++[c]
23. Neuroblastoma NIE115	Mouse	+++		

EDGF IS DIFFERENT FROM OTHER KNOWN GROWTH FACTORS

The properties of EDGF from retina have been compared with other known factors. The isoelectric point of EDGF is close to that reported of EGF (Taylor et al., 1972). However radioimmunoassay against mouse anti-EGF failed to show the presence of EGF-like molecules in our preparation. A radio receptor assay on human fibroblast using iodinated EGF indicated that EDGF has specific binding sites and did not compete for specific EGF receptors (Plouet et al., 1981). A comparison of the biological activities (Table 3) indicates that these factors do not stimulate the same cells, and BEL cells for example are not stimulated by EGF (Gospodarowicz et al., 1978b).

Comparison with brain or pituitary FGF indicates that although both molecules have similar molecular weights, they differ from their isoelectrical point (pH 9.3 to 9.6; Gospodarowicz, 1975; Gospodarowicz et al., 1978a) versus pH 4.5 for EDGF. Furthermore, biological properties indicate (see Table 3) that EDGF stimulates human keratinocytes (Guedon et al., 1981) while FGF has no effect on these cells, and conversely EDGF fails to stimulate fibroblast proliferation for a significant period.

Further purification and comparative studies need to be undertaken to see whether EDGF might be identical to the glial maturation factor, isolated as described by Pettmann et al. (1918), which was also found to be mitogenic for BEL cells in our system while EDGF was able to induce maturation of glial cells (Pettmann, personal communication). The endothelial cell growth factor (Maciag et al., 1981a) which is similar to EDGF in some biochemical properties (molecular weight and isoelectrical point; Maciag et al., 1981a) and which also stimulates the proliferation of adult human keratinocytes (Maciag et al., 1981b) would also have to be more thoroughly compared to EDGF. Moreover, EDGF possesses an angiogenic activity (Thompson et al., 1981a) and since an angiogenic factor has been partially purified from bovine retina (D'Amore et al., 1981), we need to investigate a possible identity between these two factors.

EFFECT OF EDGF ON BEL CELLS

a) Effect on Cell Morphology - Scheme for morphological interconversions

If BEL cells are cultured continuously in the presence of EDGF the cells maintain an epithelioid character which corresponds to the organization of lens epithelium on the lens capsule in vivo. This morphology is maintained at least up to passage 17 (about 60 generations). The epitheloid cells are referred to as "type I cells". However type I cells kept at confluence for extended periods of time do not remain strictly contact-inhibited and overlap. Monolayers in dense foci show no orientated organization. By these criteria they behave differently from bovine endothelial cells, which remain strictly contact-inhibited when cultured in the presence of EDGF, as described by Arruti and Courtois (1981) (see below), or in the presence of FGF (Vlodasky et al., 1979). If BEL cells are cultured in the absence of EDGF, after a few passages a considerable heterogeneity in size is obtained and most of the cells become enlarged (Hughes et al., 1975; Laurent et al., 1978). These morphologically varied cells are referred to as "type II cells". Depending on culture conditions (seeding and number of passages), type II cells keep their enlarged epitheloid morphology, with an accumulation of intracellular filaments, or undergo a spontaneous elongation. These spontaneously elongated cells are clearly distinguishable from elongated cells obtained from a culture in which type II cells have been stimulated by retinal extract for three days or more, as shown by Arruti and Courtois (1978), when the general appearance of the confluent culture is similar to a contact-inhibited fibroblast culture at high density. This fibroblast-like morphology is referred to as type III cells. It is a transient stage since these cells can reverse to type I cells if they are subcultured in the continuous presence of EDGF. From these observations on the modulation of cell morphology, which is described fully in Courtois et al. (1981) we have proposed a scheme (Fig. 1) which illustrates the relationships between the various cell configurations.

b) Effect of EDGF on the Cytoskeleton of BEL Cells in Culture

The morphological differences induced by EDGF were studied at the level of the cytoskeleton. Direct phase contrast microscopy showed the pronounced appearance of large intracellular fibrils in type II cells, while these seemed to be missing in type I cells. Immunofluorescence studies using DNase I, and fluorescent anti-DNase to visualize the actin filaments indicate (see Courtois et al. 1981) that in type I cells microfilaments appear as thick cables arrayed in parallels showing some interweaving, and in type II cells actin cables are arranged predominantly in criss-cross pattern with few strictly parallel arrays giving a polygonal network described as starlike structures (Lonchampt et al. 1976) or geodomes (Lazarides, 1976). This polygonal network does not seem to be confined to particular regions of the cell and appears to be a stable configuration of type II cells. In type III cells, actin filaments are predominently in parallel arrays stretching across the whole length of the cell.

The distribution of actin, α-actinin and tropomyosin has also been investigated in all three types of morphologies and correlates

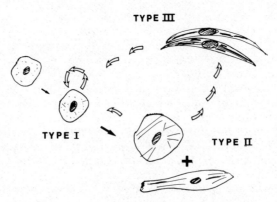

Figure 1. Modulation of the shape of BEL cells (from Courtois et al., 1981). The cell on the left corresponds to the initial cell morphology within the lens. The black arrows illustrate the spontaneous formation in the absence of EDGF of a type II culture with intermediate formation of a type I configuration. The open arrows correspond to the morphological conversions induced by or configurations maintained by EDGF addition.

with the distribution of actin. An illustration of the immunofluorescent distribution of actin, α-actinin and tropomyosin (using antiactinin and antitropomyosin kindly given by Dr. B. Jockush) is presented in Fig. 2 for type II cells, where the polygonal fibre networks can be seen. In presence of EDGF (in type I cells) this network is not so predominant and a diffuse distribution of actin, α-actinin and tropomyosin can be visualized.

c) Effect of EDGF on the Cellular Surface of BEL Cells

The various distributions of actin, α-actinin and tropomyosin containing filament systems observed in bovine lens cells cultured in the absence or presence of EDGF raised the question of whether these changes are mediated directly by EDGF or indirectly by other EDGF mediated effects. One possible answer is an effect of EDGF on the expression of the cell surface adhesive molecule fibronectin and/or on the expression of extracellular components such as the different types of collagens. Immunofluorescent studies shown in Courtois et al. (1981) on the distribution of fibronectin indicates that in type I cell fibronectin is hardly detected as opposed to type II cells which are brightly stained by fluorescent antifibronectin. In these cells, fibronectin is concentrated at areas of contact between cells and also accumulates in high density cultures to cover completely the upper surface of some cells. Labelling experiments indicated that non EDGF treated cells synthetize and possess about 100 fold more fibronectin on their cellular surface than type I BEL cells.

Collagen synthesis was also studied. Laurent et al. (1981) reported that type II BEL cells synthetize type I, III and IV collagen. As the length of BEL cell culture increased, type IV collagen, which is the only collagen present in vivo and found at the early passages, is progressively replaced by a mixture of other collagens. Preliminary results indicate that EDGF reduces the total amount of synthetized collagen (by a factor of 2 to 4) and changes the repartition in the culture of collagens synthetized on long term culture, so that more type IV collagen is synthetized (Laurent et al., unpublished results).

d) Fine Structure Modifications Induced by EDGF

Changes in morphology of BEL cells induced by EDGF were studied by electron microscopy by Perry et al. (1981). Sections of type I and type II cultures are presented in Fig. 3. A vertical section through a type I mutilayer culture (Fig. 3a) and type II (Fig. 3b) illustrates the differences between the two cultures. In type I, cells in multiple layers are closely opposed, with a few exceptions in some regions where plasma membranes diverge to accomodate a layer of intercellular material. These cells are closely associated with the plastic substratum. Type II cells on the other hand tend to lose their attachment to the plastic substratum over progressively wider areas as a substratum matrix is deposited at confluence. Cells in multiple layers are frequently separated by broad layers of intercellular material. The intercellular material deposited by type I cells (Fig. 3c) has a fairly homogeneous structure and no particular network can be seen. Extracellular material in type II cultures is presented in Fig. 3d and consisted of a loose meshwork of fibrils and granules.

The results of this survey on the effect of EDGF on BEL cells is summarized in Table 3 (from Perry et al., 1981).

EFFECT OF EDGF ON BOVINE CORNEAL ENDOTHELIAL (BEC) CELLS

Another type of cell, the corneal endothelial cell, which might well be another target cell for EDGF in vivo, was studied in tissue culture. The natural organization of these cells is to form in vivo a monolayer and therefore they are particularly well adapted to tissue culture conditions. However under standard conditions these cells grow poorly and during subculture rapidly lose their differentiated traits, such as contact inhibition, regular deposition of extracellular material, normal morphology, etc., and die after a few passages.

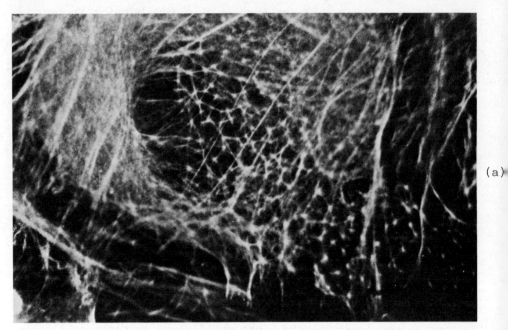

Figure 2. Visualization by indirect immunofluorescence of cytoskeleton elements in type II BEL cells.
2a. Actin cables (using anti-actin).
2b. α-actinin (using anti-α-actinin given by B. Jockush).
2c. Tropomyosin (using anti-tropomyosin given by B. Jockush).

Arruti and Courtois (1981) showed that EDGF from retina (mainly used as a crude extract) strongly stimulates cell proliferation and the increased growth rate is kept constant during successive passages. In the continuous presence of EDGF, BEC cells organize a confluent monolayer composed of closely opposed polygonal cells lying on a basement membrane. Thus over a large number of generations these cells contrive to express most of their differentiated traits: contact inhibition of movement, proliferation arrest in monolayers and the extracellular deposition of material organized as a typical basement membrane. The presence of EDGF enormously increases the number of cellular generations expressing these differentiated traits.

A study on the effect of the addition or withdrawal of EDGF on the shape and monolayer morphology of BEC cells has led to conclusions similar to those for BEL cells. Three morphologically different types of cells could also be obtained and the interconversion of one type to another could be modulated in the same manner as with BEL cells.

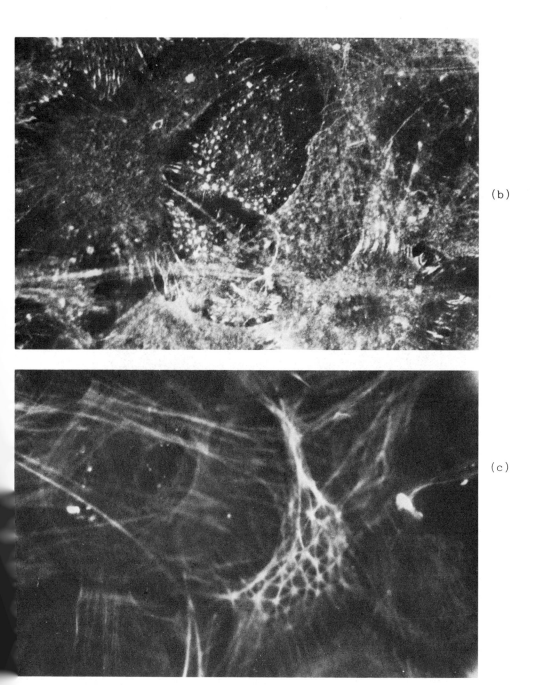

(b)

(c)

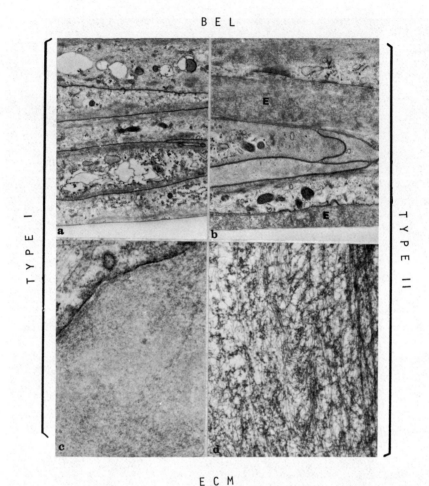

Figure 3. Electron microscopy studies of BEL cells (from Perry et al., 1981).
3a. The cells in a vertical section through type I multilayer culture are closely apposed (x 7 000).
3b. A layer of extracellular material (E) separates cells in a type II multilayer culture. A similar layer is evident beneath the culture (x 11 900).
3c. The intercellular material formed in type I cultures after 14 days is fairly homogeneous in structure (x 33 600).
3d. Extracellular material in type II cultures consists of a loose meshwork of fibrils and granules (x 16 800)

DO EDGF AND GROWTH FACTORS IN GENERAL MAINTAIN HOMEOSTASIS IN VIVO?

The effects of EDGF on the various cells present in the eye such as epithelial and endothelial corneal cells, vascular endothelial cells and lens epithelial cells led us to investigate its action on wound healing. Recent data (Thompson et al., 1981b) show that EDGF applied topically on wounded rabbit cornea increases significantly the rate of epithelium recovery and also improves the quality of the wound recovery. Thus EDGF has a similar action to EGF (Daniele et al., 1979) or MGF (Smith et al., 1981) which also seem effective in wound healing. The interesting feature with EDGF is that it comes from the eye itself, while the other growth factors are extracted from mouse male submaxillary gland. This raises the question of its physiological role in vivo during development and ageing.

CONCLUSION

We have isolated a growth factor from adult eye tissue which seems to have a broad spectrum of activity on different embryonic, adult or transformed cells from different species. From our study, and by comparing its mechanism of action in vitro and in vivo with other similar growth factors such as EGF and FGF, or other growth factors less well purified, one may conclude that these molecules might be implicated in the exchange of information between tissues not only in embryogenesis but also in adult homeostasis. With the exception of EGF, very little information on the concentration of the various growth factors mentioned above in the different body fluids is known. One problem is the lack of specificity observed in vitro of these factors towards the target cells. (Furthermore all the cells tested do not respond to the same extent to all the factors.) For instance, the Platelet Derived Growth Factor (PDGF) does not stimulate endothelial vascular cells, and EDGF has no action on fibroblasts (see Table 3). A given cell type is not necessarily a target at all stages of its growth. Gospodarowicz et al. (1978c) have clearly demonstrated that the same cells adhering to different supports may change their receptivity for EGF. In addition, as shown for bovine corneal or vascular endothelial cells, a growth factor can stimulate cells to divide while they are in exponential phase, but at the same time it can drive them to form a perfectly contact-inhibited monolayer at confluency. In this case as well as in many others the mechanism of action of the growth factor is to control the extracellular deposition of the basement membrane.

Thus it appears from in vitro studies as well as from in vivo observations that the determination and the stability of the differentiated states of a cell can be modified by a class of molecules described (perhaps mistakenly) as growth factors, whose mechanism of action is not yet clearly understood, but is readily accessible to analysis by cultivating the cells in a defined medium.

REFERENCES

Arruti, C., and Courois, Y., 1978, Morphological changes and growth stimulation of bovine epithelial lens cells by a retinal extract in vitro, Expl. Cell Res., 117:283.

Arruti, C., and Courtois, Y., 1981, Monolayer organization by serially cultured bovine corneal endothelial cells: effects of a retina derived growth promoting activity, Expl. Cell Res., in press.

Barritault, D., Arruti, C., and Courtois, Y., 1981a, Is there an ubiquitous growth factor in the eye? Differentiation, 18:29.

Barritault, D., Plouet, J., Courty, J., and Courtois, Y., 1981b, Partial purification of the Eye Derived Growth Factor from retina, submitted for publication.

Bradford, M. M., 1976, A rapid and sensitive method for the quantitation of microgram quantities of protein utilizing the principle of protein-dye binding, Anal. Biochem., 72:248.

Courtois, Y., Arruti, C., Barritault, D., Tassin, J., Olivie, M., and Hughes, R. C., 1981, Modulation of the shape of epithelial lens cells in vitro directed by a retinal extract factor: a model of interconversion and the role of actin filaments and fibronectin, Differentiation, 18:11.

D'Amore, P., Glaser, B., Brunson, S., and Fenselan, A., 1981, Angiogenic activity from bovine retina: Partial purification and characterization, Proc. Natl. Acad. Sci. U.S.A., 78:3068.

Daniele, S., Frati, L., Fiore, C., and Santoni, G., 1979, The effect of epidermal growth factor (EGF) on the corneal epithelium in humans, Albrecht. V. Graefes Arch Klin exp. ophthal., 210:159.

Gospodarowicz, D., 1975, Purification of a Fibroblast Growth Factor from bovine pituitary, J. Biochem., 250:2515.

Gospodarowicz, D., and Moran, J., 1976, Growth factors in mammalian cell culture, Annual Rev. Biochem., 531:558.

Gospodarowicz, D., Bialecki, H., and Greenburg, G., 1978a, Purification of the fibroblast growth factor activity from bovine brain, J. Biol. Chem., 252:3743.

Gospodarowicz, D., Greenburg, G., Bialecki, H., and Zetter, B., 1978b, Factors involved in the modulation of cell proliferation in vivo and in vitro: the role of fibroblast and epidermal growth factors in the proliferative response of mammalian cells, In Vitro, 14:118.

Gospodarowicz, D., Greenburg, G., and Birdwell, C., 1978c, Determination of cellular shape by the extracellular matrix and its correlation with the control of cellular growth, Cancer Res., 38:4155.

Guedon, I., Barritault, D., Courtois, Y., and Prunieras, M., 1981, Culture and cytogenetic studies of adult human keratinocytes using a new growth factor, Differentiation, 19:109.

Hughes, R. C., Laurent, M., Laurent, M., Longchampt, M. O., and Courtois, Y., 1975, Biosynthesis in cultured epithelial cells of bovine lens, Eur. J. Biochem., 52:143.

Laurent, M. V., Lonchampt, M. O., Regnault, F., Tassin, J.,

Courtois, Y., 1978, Biochemical, ultrastructural and immunological study of in vitro production of collagen by bovine lens epithelium cells in culture, Expl. Cell Res., 115:127.

Laurent, M., Kern, P., Courtois, Y., and Regnault, F., 1981, Synthesis of type I, II and IV collagen by bovine lens epithelium cells in long-term culture, Expl. Cell Res., 134:23.

Lazarides, E., 1976, Actin, α-actinin and tropomyosin interaction in the structural organization of actin filaments in non-muscle cells, J. Cell Biol., 68:202.

Longchampt, M. O., Laurent, M., Courtois, Y., Trenchev, P., and Hughes, R. C., 1976, Microtubules and microfilaments of bovine lens epithelial cells: Electron microscopy and immunofluorescence staining with specific antibodies, Exp. Eye Res., 23:1.

Maciag, T., Cerundolo, J., Ilsley, S., Kelley, P. R., and Forand, R., 1979, An endothelial growth factor from bovine hypothalamus: identification and partial characterization, Proc. Natl. Acad. Sci. U.S.A., 76:5674.

Maciag, T., Hoover, G. A., Stemerman, H. B., and Weinstein, R., 1981a, in:"Growth of Cells in Hormonally Defined Medium," D. A. Sirbasku, G. M. Sato and A. B. Purdee, eds., Ninth Cold Spring Harbor Conference on Cell Proliferation, in press.

Maciag, T., Nemore, R., Weinstein, R., and Gilchrest, B., 1981b, An endocrine approach to the control of epidermal growth: Serum-free cultivation of human keratinocytes, Science, 211:1452.

Perry, M., Tassin, J., and Courtois, Y., 1981, Fine structure of bovine lens epithelial cells in vitro in relation to modifications induced by a retinal extract (EDGF), Expl. Cell Res., 136:379.

Pettmann, B., Sensenbrenner, M., and Labourdette, G., 1981, Isolation of a glial maturation factor from beef brain, FEBS Letters, 118:195.

Plouet, J., Barritault, D., Courtois, Y., and Ladda, R., 1981, Epidermal growth factor and eye derived growth factor from retina are immunologically distinct and bound to different receptors on human skin fibroblasts, submitted for publication.

Smith, R., Smith, L., Rich, L., and Weimar, V., 1981, Effects of growth factors on corneal wound healing, Assoc. for Res. in Vis. and Ophthal. Inc., 20:222.

Taylor, J. M., Mitchell, W. M., and Cohen, S., 1972, Epidermal growth factor: physical and chemical properties, J. Biol. Chem., 247:5928.

Thompson, P., Arruti, C., Maurice, D., Plouet, J., Barritault, D., and Courtois, Y., 1981a, in:"International Workshop on Problems of Normal and Genetically Abnormal Retina," R. Clayton, ed., Academic Press, New York and London, in press.

Thompson, P., Desbordes, M. J., Giraud, J., Pouliquen, Y., Barritault, D., and Courtois, Y., 1981b, The effect of an eye derived growth factor (EDGF) on corneal epithelial regeneration, Exp. Eye Res., in press.

Vlodasky, I., Johnson, L., Greenburg, G., and Gospodarowicz, D., 1979,

Vascular endothelial cells maintained in the absence of fibroblast growth factor undergo structural and functional alterations that are incompatible with their in vivo differentiated properties, J. Cell Biol., 83:468.

TISSUE CULTURE OF CHICK EMBRYONIC CHOROIDAL CELLS:

CELL AGGREGATION AND PIGMENT ACCUMULATION

Gerald J. Chader, Eileen Masterson and Arnold Goldman*

Laboratory of Vision Research
National Institutes of Health
Bethesda MD 20205

*Departments of Ophthalmology and Anatomy
Medical College of Wisconsin
Milwaukee WI

SUMMARY

Cells originating from the chick embryo choroid were grown in culture for up to 1 month. The cells did not form coherent monolayer colonies but rather formed large transparent cell aggregates surrounded by a random, loose network of cells. The cell aggregates avidly took up and retained pigment granules while the individual cells did not. Choroidal cell aggregates were not similar to the "lentoid" bodies described by Okada and coworkers in cultures from other ocular tissues; they did not develop fiber cells and did not synthesize δ-crystallin. Thus aggregation of ocular cells appears to be a necessary but insufficient step for the formation of a "lentoid" body.

INTRODUCTION

Tissue culture provides an opportunity to investigate cellular differentiation and metabolism under controlled conditions. Cells from several tissues of the chick embryo eye have been studied in this regard since chick ocular tissues at this stage of development are particularly easy to cleanly dissect and grow in culture. Studies on chick retinal pigment epithelial (PE) cells (Newsome and Kenyon, 1973; Eguchi and Okada, 1973; Rodesch, 1973; Newsome et al., 1974; Redfern et al., 1976), lens (Okada et al., 1971; de Pomerai et al., 1977) and neural retina (Redfern et al., 1976; de Pomerai et al, 1977; Clayton et al., 1977; Araki and Okada, 1977) have yielded much information about the basic morphological and biochemical development of these tissues. No similar reports of choroidal cell culture have appeared in the literature. In the present communication, we demonstrate that cells derived from the chick choroid grow well in culture and that they exhibit interesting properties of cell aggregation.

MATERIALS AND METHODS

Embryos of Gallus domesticus (Truslow Farms, Chestertown, MD) at day 12 of incubation were used in this study. Choroidal tissue was obtained by dissecting out the pigment epethelial-choroid unit, and placing the unit in sterile 10% EDTA solution (in $Ca^{++}-Mg^{++}$ free Dulbecco's PBS) for ten minutes. At this point, the choroid is easily separated from the black pigmented epithelium as a white, fibrous material. Choroid tissue was placed in Coon's CTC-enzyme solution (6 units/ml collagenase, 0.1% trypsin, 2% chick serum, 4 mM EDTA) for ten minutes, and vortexed gently. The resultant individual choroidal cells were seeded on plastic tissue culture dishes in 3 ml of Eagles' Minimum Essential Medium (MEM) which was supplemented with 5% fetal calf serum, (heat inactivated; GIBCO, Grand Island, NY). In experiments designed to investigate the uptake of pigment granules into cells of choroidal origin, a slightly different technical approach was taken. Pigment epithelial cells are fragile and do not withstand prolonged treatment in EDTA solution if no serum is present to protect the cells. We thus placed the entire PE-choroid unit in 10% EDTA without serum at 37°C for 15 minutes, vortexed vigorously for one minute, placed it in CTC-enzyme solution for ten minutes, and vortexed again. Separate experiments (unpublished) demonstrate that choroid cells are able to withstand this treatment, but the PE cells are completely disrupted and neither attach nor grow in culture after such treatment. Choroidal cells, pigment granules and pigment epithelial cell debris were then seeded together as described above.

Pigment epithelial cells from 12 day old embryos were obtained by placing the pigment epithelium-choroid unit in CTC-enzyme solution for ten minutes, and selecting the sheets of pigment epithelial cells which cleanly detached from the unit with a sterile pasteur pipet. These pieces were incubated another five minutes in CTC to dissociate the sheets into individual cells, and seeded as outlined above. Culture medium was replaced every three days, and all cells were maintained in a 5% CO_2, 95% air atmosphere at 37.5°C.

Phase microscopy was performed with a Zeiss inverted tissue culture microscope and camera (Polaroid film Type 665). For electron microscopy, the culture dishes were emptied of medium and filled with 4% glutaraldehyde in 0.12 M cacodylate buffer pH 7.0 for 1 hour. Cells in the dishes were subsequently postfixed in 1% OsO_4 in cacodylate buffer for 1 hour at 0°C, dehydrated in an ascending series of ethanol concentrations and infiltrated with Araldite in situ. Polymerization was for 3 days at 60°C. Selected sites were then cut from the dish and re-embedded in Araldite in BEEM capsules. Thin sections were cut on an LKB Model III ultrmicrotome, collected on 200 mesh uncoated copper grids, stained with uranium and lead and viewed on and RCA Em U 46 microscope.

The possible synthesis of δ-crystallin was investigated in cultures of pure choroidal cells and in cultures of choroidal cells containing pigment granules. For this purpose, culture dishes were incubated overnight with 500 µCi of L-^{35}S-methionine (Amersham, specific activity-755 Ci/mmol). Cells were then rinsed four times with Dulbecco's PBS, scraped off the plate and homogenized in 0.3 ml tris buffer (10 mM, pH 7.0). The homogenate was then split into two samples. One sample was centrifuged five minutes at 10 000 r.p.m. and the pellet washed and recentrifuged four times, saving the supernatant each time. The supernatants were pooled, and 3 µl of cold δ-crystallin (1 mg/ml) added plus 15 µl of δ-crystallin antibody. A precipitate was allowed to form overnight at 4°, and was centrifuged the next day for ten minutes at 10 000 r.p.m. The pellet was washed four times with 10 mM Tris, and both the pellet obtained from this procedure, and a sample of the original homogenate were dissolved in electrophoretic buffer. Aliquots corresponding to approximately 100 000 c.p.m. of the pellet and homogenate solutions were applied to an SDS-polyacrylamide gel, as was an authentic ^{35}S-crystallin standard. The gel ran for five hours at 125 volts, and was subsequently stained with Coomassie Brilliant Blue and destained thirty minutes for protein. Autoradiography was also carried out on the gel with a development time of 24 hours.

RESULTS

Cells of choroidal origin in culture initially exhibit a typically fibroblastic appearance (Fig. 1A). The cells do not form discrete colonies as do cultured PE cells where individual cells abut and adjoin each other in a tight regular repeating mosaic pattern. The cells are slow-growing and, after about 3-4 weeks in culutre, cell aggregates can be seen (Fig 1B-D). The cells are not pigmented, and, under the inverted phase microscope, the aggregates appear to be quite similar to the "lentoid" bodies described by Okada (1977). Only in the aggregates do cells exhibit extensive cell-cell contact with each other.

Choroidal culture cells exhibit distinct characteristics when examined by electron microscopy. Unlike PE cells, which generally form a monolayer, these cells tend to form aggregates. Those cells at the apical portion of the aggregate have good access to the culture medium, and thus appear healthy (Fig. 2A). These cells do not contain melanin, but appear to have a well developed rough endoplasmic reticulum, and some demonstrate areas of dense Golgi. Other cells near

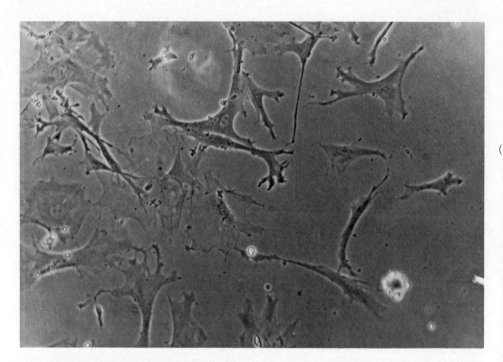

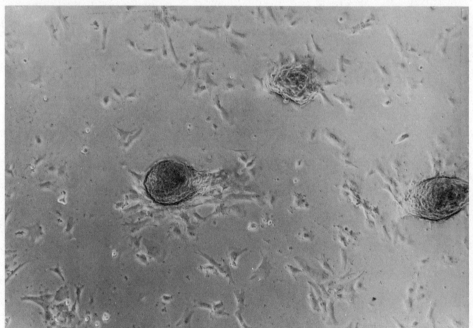

Figure 1. Typical cells of choroidal origin growing in culture
A) 14 days (369x) B) 27 days (232x)

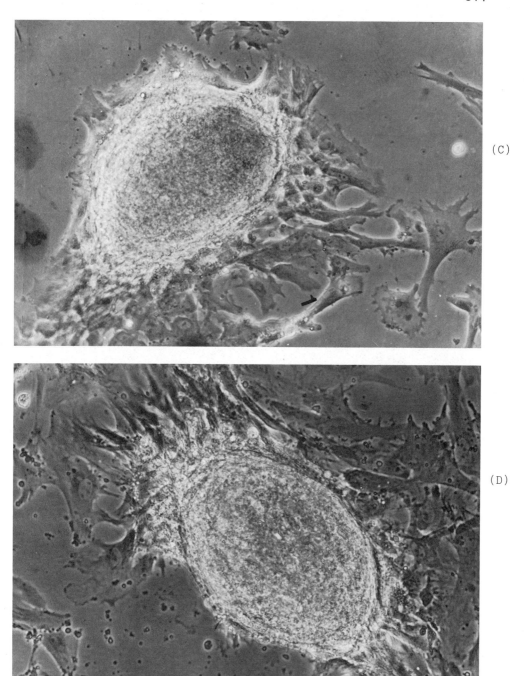

C) 23 days (369x)
D) 27 days (369x)

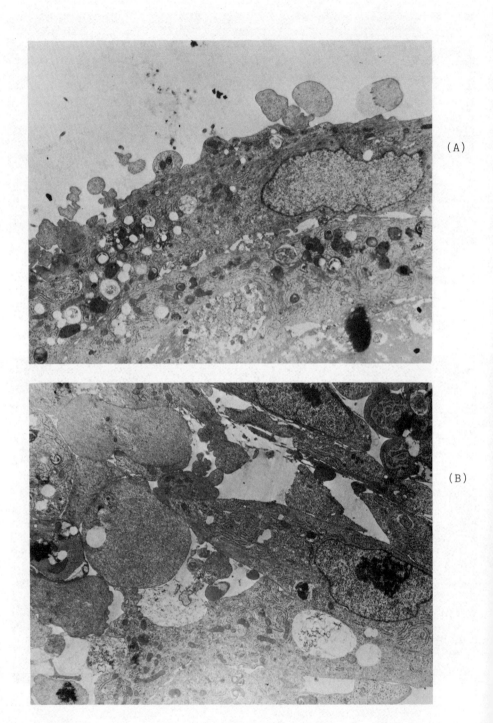

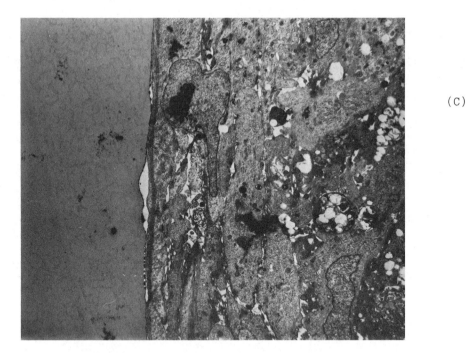

(C)

Figure 2. Typical electron micrographs of choroidal cells in culture
A) Choroidal cells near apex of aggregate: Cells are fibroblastic in appearance, with abundant rough endoplasmic epithelium. The extracellular space contains relatively little extracellular debris. 4960x B) Choroidal cells near apex of aggregate: These cells, from another location than the cells in Fig. 2A form a bilayered capsule surrounding a large quantity of amorphous debris. Portions of the apex of the outer cells appear to be in the process of budding away from the rest of the cell. 4960x
C) Choroidal cells near base of aggregate: These cells are denser and show less RER than did the cells near the apex. Cells abutting the culture dish contain few inclusion bodies, while those more central to the aggregate contain increased inclusion bodies. 3620x

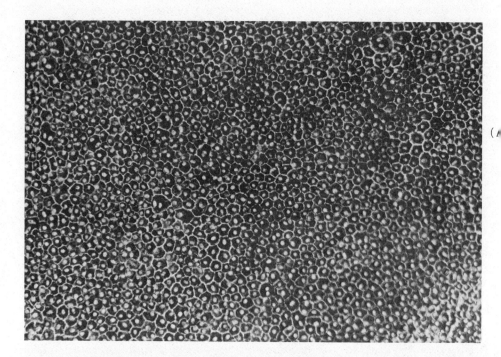

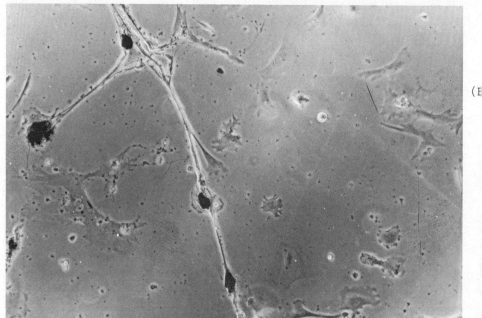

Figure 3. A) Typical pigmented epithelial cells growing in culture for 20 days (229x). Typical aggregates of choroidal cells

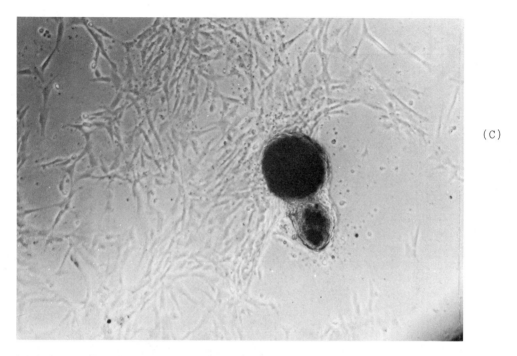

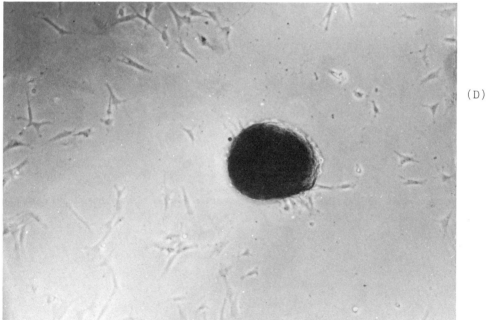

grown in the presence of pigment granules for B) 54 days (360x) C) 43 days (229x) D) 9 days (229x).

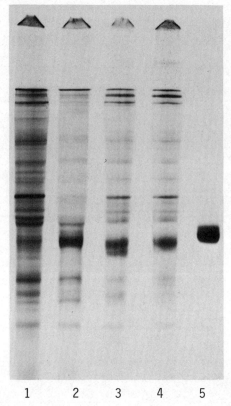

Figure 4. Autoradiographic pattern of SDS gel electrophoresis of cellular proteins labeled with L-^{35}S-methionine. Slot 1: membrane fraction of choroidal cells grown with pigment granules; Slot 2: membrane fraction of choroidal cells alone; Slot 3: homogenate of choroidal cells grown with pigment granules; Slot 4: homogenate of choroidal cells alone; Slot 5: sample of purified chick lens ^{35}S-δ-crystallin.

the surface (Fig. 2B) are surrounded by large amounts of extracellular debris. These cells have small, broad extensions out of the aggregate which resemble buds. Such cells are heavily laden with inclusion bodies, and tend to form a capsule 2-4 cells thick which surrounds a large mass of amorphous debris. At the base of the aggregate, adhering to the culture dish, are multilayered cells (Fig. 2C) which exhibit a gradient in the concentration of inclusion bodies. Very few such bodies are found in the most basal cells, although the number of inclusions increases as the center of the aggregate is approached.

Cultured PE cells from 12 day old embryos form pigmented colonies of cuboidal-like cells (Fig. 3A), very similar to colonies previously reported in PE cultures of 6-7 day old embryos (Redfern et al., 1976). If pigment granules from disrupted PE cells are in the medium with the choroidal cells, the choroidal cells will take up the pigment granules primarily at the site of the aggregates (Fig. 3B-D). Although some pigment granules could also be seen in non-aggregated cells, pigment was concentrated in the aggregates and appeared as dark, inky spots.

Figure 4 shows the autoradiographic pattern of newly synthesized ^{35}S-labeled protein from choroid cultures after treatment with δ-crystallin antibody. The labeling pattern in choroidal cell homogenate (gel 4) is generally similar to that in the homogenates of choroidal cells containing ingested melanin granules (gel 3). The choroid pellet fraction consists mainly of membranes (gel 2). It shows some marked differences from the choroid pellet fraction containing ingested pigment granules (gel 1). This is most obvious with the band (see arrow) that is weak in the choroid pellet but strong and distinct in the other three cases. The reason for such differences is yet unknown. We cannot rule out the possibility that differences in protein patterns between pure choroidal cultures and choroidal cultures containing pigment granules may in some measure be due to PE cell contamination. This problem is thought to be minimal however because of the fragile nature of the PE cells at the time of isolation and initial seeding (as also described by Saari et al., 1977) leading to very few if any viable PE cells. Moreover, no cells of typical PE cell morphology and growth characteristics are observed during the culture period under consideration (i.e. 3-6 weeks).

Most important however, no band was observed in any of the preparations that co-migrated with the δ-crystallin standard (gel 5). This is particularly evident in the pellet fractions precipitated with δ-crystallin antibody where one would expect the presence of crystallins to be accentuated.

DISCUSSION

There is presently no information in the literature with regard to choroidal cell culture. In this report, we present the observation that fibroblastic-like cells grow easily out of the choroidal vessels. The type of cell or cells in our cultures is yet unknown. Somewhat parallel studies on outgrowth from retinal capillary explants indicate that it is the mural cells (intramural pericytes) which proliferate in such cultures (Buzney et al., 1975). Cells in our cultures easily form aggregates after several weeks in culture and are also able to phagocytize pigment granules. This latter finding parallels the well-known transfer of pigment in epithelial cells of the skin and is similar to the phagocytic nature of the choroidal endothelium in the aschaemic rabbit eye (Johnson, 1975). Alternatively, it appears that cultured cells in general possess the ability to phagocytize non-specifically. Retinal PE cells can ingest a wide range of substances including polystyrene spheres, carbon particles and bacteria (Hollyfield and Ward, 1974; Custer and Bok, 1975; Funahashi et al., 1976). Also, studies by our group (Masterson et al., 1980) have demonstrated that many different types of epithelial cells are able to phagocytize in culture. Further studies thus will be necessary to determine the specificity of pigment granule uptake under our culture conditions.

The formation of "lentoid bodies" and the process of "transdifferentiation" as described by Eguchi and Okada (1973) is now well established in neural retinal and pigment epithelia cell culture (Okada et al., 1971; Redfern et al., 1976; Araki and Okada, 1977; de Pomerai et al., 1977; Okada, 1977). A prerequisite or at least a contributing factor to this process appears to be the disruption of normal cell-cell interactions prior to seeding of the cells in monolayer culture. Subsequently, in culture, cell aggregates form that have many of the hallmarks of lens cells including a non-pigmented, lens-fiber appearance and the presence of lens crystallins. Over several weeks in choroidal cultures cell aggregates form that resemble the lentoid bodies of the retina or pigment epithelium when viewed under the tissue culture microscope. This resemblance is only superficial however; the internal structure of the aggregates are not lenslike as determined by electron microscopy. Perhaps more diagnostic is the lack of synthesis of δ-crystallin by the cells as assessed by autoradiography of SDS gels after precipitation of the cultured choroidal preparations with authentic chick δ-crystallin and δ-crystallin antibody. In contrast, de Pomerai et al. (1977) have previously demonstrated the presence of α, β and δ-crystallin in lentoid bodies of neural retinal cultures. In any event, the nature of the cellular development of the cultured choroidal cells and formation of "lentoid bodies" in the cultures appears to be unique and not merely redifferentiation along the lines seen in cultured retina and pigment epithelium.

ACKNOWLEGEMENTS

We are deeply indebted to Dr. Joram Piatigorsky for supplying the δ-crystallin antibody and for a most helpful critique of the manuscript.

REFERENCES

Araki, M., and Okada, T., 1977, Differentiation of lens and pigment cells in culture of neural retinal cells of early chick embryos, Develop. Biol., 60:278.
Buzney, S., Frank, R., and Robison, W., 1975, Retinal capillaries:

proliferation of mural cells in vitro, Science, 190:985.
Clayton, R., de Pomerai, D., and Pritchard, D., 1977, Experimental manipulation of alternative pathways of differentiation in cultures of embryonic chick neural retina, Develop. Growth and Differ., 19:319.
Custer, N., and Bok, D., 1975, Pigment epithelium-photoreceptor interactions in the normal and dystrophic rat retina, Expl. Eye Res., 21:153.
Eguchi, G., and Okada, T., 1973, Differentiation of lens tissue from the progeny of chick retinal pigment cell culture in vitro. A demonstration of a switch of cell types in clonal culture, Proc. Natl. Acad. Sci. U.S.A., 70:1495.
Funahashi, M., Okisaka, S., and Kuwabara, T., 1976, Phagocytosis by the monkey pigment epithelium, Expl. Eye Res., 23:217.
Hollyfield, J., and Ward, A., 1974, Phagocytic activity in the retinal pigment epithelium of the frog R. pipiens I. Uptake of polystyrene spheres, J. Ultrastruct. Res., 46:327.
Israel, P., Masterson, E., Goldman, A., Wiggert, B., and Chader, G., 1980, Retinal pigment epithelial cell differentiation in vitro. Influence of culture medium. Invest. Ophthalmol. Vis. Sci., 19:720.
Johnson, N. F., 1975, Phagocytosis in the choroidal endothelium of the ischaemic rabbit eye, Acta Ophthal., 53:321.
Masterson, E., Goldman, A., and Chader, G., 1980, Phagocytosis of rod outer segments by cultured epithelial cells, Vis. Res., 21:143.
Newsome, D., Fletcher, R., Robison, W., Kenyon, K., and Chader, G., 1974, Effects of cyclic AMP and sephadex fractions of chick embryo extract on cloned retinal pigmented epithelium in tissue culture, J. Cell Biol., 61:369.
Newsome, D., and Kenyon, K., 1973, Collagen production in vitro by the retinal pigmented epithelium of the chick embryo, Develop. Biol., 32:387.
Okada, T., Eguchi, G., and Takeichi, M., 1971, The expression of differentiation by chicken lens epithelium in in vitro cell culture, Develop. Growth and Differ., 13:323.
de Pomerai, D., Pritchard, D., and Clayton, R., 1977, Biochemical and immunological studies of lentoid formation in cultures of embryonic chick neural retina and day old chick lens epithelium, Develop. Biol., 60:416.
Okada, T., 1977, A demonstration of lens-forming cells in neural retina in clonal cell culture, Develop. Growth and Differ., 19:323.
Redfern, N., Israel P., Bergsma, D., Robison, W., Whikehart, D., and Chader, G., 1976, Neural retina and pigment epithelial cells in culture: patterns of differentiation and effects of prostaglandins and cyclic AMP on pigmentation, Expl. Eye Res., 22:559.
F. Rodesch, 1973, Differentiation, contact inhibition and intercellular communication in retinal pigment cells, Expl. Cell Res., 76:55.
Saari, J., Bernt, A., Futterman, S., and Berman, E., 1977, Localiz-

ation of cellular retinol-binding protein in bovine retina and retinal pigment epithelium with a consideration of the pigment epithelium isolation technique, Invest. Ophthalmol. Vis. Sci. 16:797.

ABSTRACT

DEVELOPMENT OF MEMBRANE PROPERTIES OF NERVE CELLS IN CULTURE

R. Balazs

Medical Research Council
Developmental Neurobiology Unit
33 Johns Mews
LONDON WC1N 2NS

Neuronal monolayer cultures derived from 8-day old rat cerebellum are mainly composed of excitatory and inhibitory interneurones. The development of the cells has been followed by studying the changes during cultivation (a) in the protein profiles after lactoperoxidase catalysed iodination and (b) in the expression of neurotransmitter receptor binding.

(a) The amount of ^{125}I combined in surface constituents is relatively high in the first few days in culture, when the major ^{125}I-iodinated protein has an apparent M.W. of 140 kd. This comprises the D_2 protein. The decrease with cultivation time in the ^{125}I-iodination of 140P parallels changes in the molecular form of D_2, involving desialidation. D_2 protein is phosphorylated in the plasma membrane, and it is also released into the medium, but in a non-phosphorylated form. Thus it seems that a significant part of the polypeptide chain of D_2 is exposed on the cell surface, though with an anchorage within the membrane that can be phosphorylated, but is not released. An antiserum obtained by immunization with plasma membrane preparations from immature cerebellum stains selectively the surface of nerve cells in the cultures. The dominant antigen recognized by the antiserum has been identified as the D_2 protein. The antiserum, and derived IgG-s and $F(ab')_2$ fragments, though not the Fab' fragments, do not affect the initial differentiation of cells when included in the culture medium, but result selectively in the death of nerve cells after 3-5 days _in vitro_.

(b) Both muscarinic cholinergic and GABA receptors are expressed in the interneurone cultures. Receptor binding increases during 14 days of cultivation to levels which, in terms of unit protein, exceed the adult in vivo values. The affinity of the receptors for the ligands is of the same magnitude as in vivo.

SUMMARY OF DISCUSSION ON STRATEGIES OF REGULATION:

EXTERNAL SIGNALS, RECEPTORS AND EFFECTOR SYSTEMS

Much of the discussion centred on growth factors, their origins, the specificity of their effects and their receptors.

Lord pointed to the diverse origins of growth factors which acted on tissues remote from them; however they did not resemble hormones since they were the products of non-endocrine tissues with specific functions, for example platelets. All were identified by effects on cell lines in vitro. Were they generalised factors affecting growth and viability in a non-essential way, or were they necessary but not necessarily specific - could BPA be replaced by EGF for proliferation of early committed erythroid cells for example? Courtois said that the effect of EDGF on corneal wound healing was to accelerate its rate by 10-20%: it was not necessary but had an effect. Iscove thought that if there were a receptor for EGF then it could probably replace BPA but Clayton thought that a receptor was insufficient since an affected system must also be available. Courtois said there was evidence that cells without receptors did not respond, and confluent cell cultures did not respond. McDevitt asked about receptors on extra-ocular tissues and Courtois referred to the response of vascular endothelial cells to EDGF, a response which might be important in the pathogenesis of ocular vascular disease in diabetes. The response of fibroblasts was due to a different receptor.

Iscove was asked about the specificity of the response of CFUs in early and late stages of committment to erythropoietin and BPA. He replied that stem cells could go through the stages on high levels of either one and low levels of the other. The effects of these two factors may be additive, a low dose of one needed to be balanced by a high dose of the other to get cells through the mid-point of the sequence, which was the most vulnerable. Radbruch asked if BPA represented more than one molecular species but Iscove said that evidence at present was compatible with there being one.

As to the origins of growth factors, Courtois pointed out the EGF was found in rat submaxillary gland, yet if this were removed, circulating EGF was still present. Chader wondered whether EDGF was synthesised in all of the diverse tissues of the eye, or whether some

of them were merely storing it. Courtois said there was a possible
similarity between various growth factors in the eye studied by different investigators using different test systems.

Iscove raised a problem in the interpretation of growth factor
data. A factor might be required in vitro but did not necessarily
act as a regulator in vivo in the steady state, but other factors
might not be. He suggested that in his system the differentiation
sequence of the cells appears to be programmed. Death of cells in
the sequence might also be a regulatory factor, if cells had a programme of self-destruction in the absence of specific growth factors.
Malignancies might be a breakdown of this mechanism.

There was further discussion on receptors as differentiation
markers. Balazs suggested that β-adrenergic receptors defined astrocytes so that their detection in pigment cell cultures might be due
to the appearance of astrocytes or to cells of mixed phenotype.
Chader said that β-adrenergic receptors were not detectable in freshly
isolated pigment cells, but were detectable within 1-2 weeks, long
before any transdifferentiation changes were seen. Pessac asked
Rifkind about partly committed clonally derived colonies and Rifkind
replied that such mixed colonies suggest that committment was metastable in some progeny, and memory for exposure to an inducer had a
finite half life. Experiments with phorbol esters and dexamethasone
showed that committment was multistep. Such metastable cells were
most readily obtained after suboptimal exposure to inducer. He asked
Lord for evidence that CFUs were drawn on to provide haemopoietic precursors in normal non-stressed situations. Lord said that suicide
experiments showed that there was some turnover of the population of
CFUs, and referred to Dexter's work on a factor which permitted CMGFC
cells to replace themselves in culture and give rise to differentiated
macrophages. This population did not need to draw on stem cells.
Okada asked Owen about the origin of striated muscle cells from mammalian thymus in culture - might they be of neural crest origin? Owen
replied that a variety of tissue types originating from thymus had
been recorded. There was a mesenchymal component and there were probably cells of neural crest origin.

There was some discussion springing from John's review which mentioned Haywoods's translational control factor: Bloemendal thought
it unlikely that such control should be special to muscle and John
agreed that it was not generally accepted but had not been specifically retested. Buckingham suggested that this was part of a wider
problem. Nuclear RNA was well established but small RNA molecules
in cytoplasm with effects on several cell properties and translation
had been described in a number of systems. Although there was some
scepticism regarding specific chains, the experiment had not been repeated in exactly the same conditions and the possible roles of small
RNA species required further investigations. In reply to a question

from de Pomerai, she said that the specificity of muscle IF3S was quantitative not qualitative.

INTRODUCTORY REVIEW:

QUANTITATIVE REGULATION

        R. M. Clayton, D. E. S. Truman and A. P. Bird*

        Department of Genetics
        University of Edinburgh
        Edinburgh   EH9 3JN

        *MRC Mammalian Genome Unit
        Edinburgh   EH9 3JT

The idea that a pattern of specificity may be generated by controlling the relative proportions of components in a system rather than by exclusively non-quantitative, on-off control, is illustrated by the tissue specificity of the patterns of lactate dehydrogenase isozymes. The relative amounts of two gene products, the A and B subunits of LDH, can generate a wide range of isoenzyme patterns, each characteristic of one tissue (Markert and Ursprung, 1962; Markert, 1965). This model of tissue specificity based on quantitative control as opposed to qualitative regulation of tissue specific substances has been supported by a wide range of examples and the papers by Paul and by Clayton in these proceedings argue the case in detail.

More complex examples of quantitative control are provided by the contribution of such genetically specified molecules as different histocompatibility antigens to the surface of a range of cell types. The use of labelled antisera has shown that one product may make widely different contributions to different cell types (see for example Snell, 1980).

Indeed the overlapping combinatorial and quantitative nature of the contribution of different macromolecules to the surface of different cells was first shown for blood group antigens. These were originally thought to be confined to the red cell, but have since been detected on platelets and other cell types, as the technology used for their detection became sufficiently sensitive (Race and Sanger, 1968). Similarly knowledge of the quantitative variation in isoenzymes attended upon the application of electrophoretic techniques of sufficiently high resolution to enzyme analyses. Nucleotide hybridisation and more sensitive immunological techniques have shown, by detecting macromolecules at lower levels than hitherto, that many substances have a wider distribution than previously thought.

It is possible to produce models of the regulation of transcription of DNA which permit control of the quantity of the trancript, and Schibler, in this volume, has emphasised the role of the promotor site as the basis for tissue-specific variation of levels of α-amylase. However it should also be remembered that in eukaryotes there are many stages in the flow of information from gene to protein, and quantitative regulation may be applied at many or all of them. Following the formation of the primary trancript, capping, splicing, methylation and adenylation modify the RNA. Transport across the nuclear membrane may be regulated. mRNA may be sequestered before translation, and it may be broken down at varying rates. The rate of translation may be controlled by initiation factors, tRNA availability or the state of the ribosomes. Some of these potential points of regulation have been shown to be of developmental significance in certain systems (Revel and Groner, 1978; Ochoa and de Haro, 1979).

That quantitative regulation may be effected at transcriptional and different post transcriptional levels is evidenced from analyses of the defects which can lead to the β-thalassemias (see for example Maniatis et al., 1980).

While studies of gene arrangement, transcriptional regulation, selective processing and mRNA characteristics including sequence variations, sequestration, and stability, the availability of isoaccepting tRNAs initiation factors and so on will illuminate the mechanisms of quantitative regulation, the other aspect must be understood: the use made by the cell of the products at low as well as high levels.

It may be important to realise that different organisms may resolve problems of quantitative regulation in different ways. During early cleavage it is necessary to provide histones for newly formed chromatin. In the sea urchin this is facilitated by a large number of histone genes and by the accumulation of histone mRNA at earlier stages; in Xenopus, on the other hand, there are fewer histone genes and it is histone itself, rather than histone message, which is accumulated before the cleavage stages (Bird, 1980).

Some cells contain proteins, such as globins or immunoglobulins which operate optimally as obligate double dimers: in such cases a quantitative change in regulation of one gene product is disadvantageous for the cell. In other cases, non-coordinate regulation appears to be required, for example in the crystallins or isozymes such as LDH. The evolutionary pressure of regulatory mechanisms may not be similar between such cells.

A selective value in such regulation is evident for isoenzymes, but where a major tissue 'specific' product is detectable at particularly low levels in certain other tissues, it is not clear at present whether this was significant for the cell, or whether we are now able to detect noise in the system. Nor do we know whether, if the latter is the case, noise per se is an inexcapable corolary of synthetic activity or has a selective advantage of its own.

The possibility of investigating the role of post-transcription control depends to some extent on the techniques available and it should always be remembered that our present knowledge of the control of gene expression during development is heavily dependent upon systems in which cells make a very limited number of products to a very high abundance, such as globin, ovalbumin, immunoglobulin or crystallin. It has been suggested (Davidson and Britten, 1979) that such systems, which are not typical of cell differentiation may have atypical system of regulation.

REFERENCES

Bird, A. P., 1980, Gene reiteration and gene amplification, in:"Cell Biology," L. Goldstein and D. M. Prescott, eds., Academic Press, New York, Vol. 3, p.61.

Davidson, E. H., and Britten, R. J., 1979, Regulation of gene expression: possible role of repetitive sequences, Science, 204:1052.

Maniatis, T., Fritsch, E. F., Laver, J., and Lawn, R. M., 1980, The molecular genetics of human haemoglobins, Ann. Rev. Genet., 14:145.

Markert, C. L., 1965, Developmental genetics, Harvey Lect., 59:187.

Markert, C. L., and Ursprung, H., 1962, The ontogeny of isozyme patterns of lactate dehydrogenase in the mouse, Develop. Biol., 5:363.

Ochoa, S., and de Haro, C., 1979, Regulation of protein synthesis in eukaryotes, Ann. Rev. Biochem., 48:549.

Race, R. R., and Sanger, R., 1968, "Blood Groups in Man," 5th ed., Blackwell Scientific Publications, Oxford, Chap. III.

Revel, M., and Groner, Y., 1978, Post-transcriptional and translational controls of gene expression in eukaryotes, Ann. Rev. Biochem., 47:1079.

Snell, G. D., 1980, Recent Advances in Histocompatibility Genetics, Adv. Genet., 20:291.

SKELETAL MUSCLE MYOGENESIS:

THE EXPRESSION OF ACTIN AND MYOSIN mRNAs

M. E. Buckingham, M. Carvatti, A. Minty, B. Robert,
S. Alonso, A. Cohen, P. Daubas, and A. Weydert

Department of Molecular Biology
Pasteur Institute
25 rue du Dr. Roux
Paris, France

Friedrich Miescher Institute
Postfach 273
Basel CH-4002, Switzerland

The formation of skeletal muscle fibres is one of the more spectacular examples of terminal differentiation, which can take place spontaneously in tissue culture (for reviews see Yaffe, 1968; Buckingham, 1977). Myoblast cell fusion normally accompanies the characteristic biochemical changes which result in differentiated muscle; namely increased synthesis of contractile proteins (e.g. Devlin and Emerson, 1978; Garrels, 1979) and their organization into sarcomeric structures (Fischman, 1970), the accumulation of enzymes important in muscle metabolism (e.g. Caravatti et al., 1979), and the appearance of membrane components such as the acetyl choline receptor (e.g. Merlie et al., 1975) essential for nerve-muscle interaction. It has become evident in the last few years that the contractile proteins present in differentiated muscles are different from those of non-muscle tissues (e.g. Vandekerckhove and Weber, 1978) and furthermore that additional isoforms may be expressed during development. These results for mammalian muscles are summarized in Table 1. For the myosin heavy chain (MHC), embryonic (MHC_{emb}) and new born (MHC_{NB}), peptides can be distinguished (Whalen et al., 1979; 1981), in addition to the proteins of adult muscle tissues (see Buckingham, 1977). In the case of the light chains, a form of LC_1, LC_{1emb} also found in adult heart atria (Whalen et al., 1980) and embryonic heart (Whalen and Sell, 1980), is expressed during skeletal muscle development. Our results would suggest that this may also be the case for the actins, since a cardiac-like actin mRNA is accumulated in differentiating skeletal

Table 1. Numbers of myosin and actin peptides.

	Myosin heavy chain	Myosin light chains			Actin
		LC_1	LC_2	LC_3	
Skeletal muscle					α_{SM}
Fast	MHC_F (2)	LC_{1F}	LC_{2F}	LC_{3F}	
Extra-fast	MHC_{SF}				
Slow	MHC_S	$LC_{13}(LC_{1F})$	LC_{2F}	(LC_{3F})	
New born	MHC_{NB}				
Embryonic	MHC_{emb}	$LC_{1emb}(LC_{1F})$	LC_{2F}	(LC_{3F})	
Heart muscle					α_C
Atria	$MHC_{CA}(?=MHC_{CIV})$	LC_{1emb}	LC_{2SA}		
Ventricle	MHC_{CIV}, MHC_{CV3}	LC_{1S}	LC_{2S}		
Embryonic	MHC_{CV3}	LC_{1emb}	LC_{2S}		
Smooth muscle					α_S, γ_S
e.g. Stomach	MHC_{SM}	LC_{1NM}	LC_{2NM}		γ_S
Aorta					α_S
Non-muscle	MHC_{NM} (2)	LC_{1NM}	LC_{2NM}		β, γ
Minimum no. genes	11-12	5	4	1	6

muscle cells. The embryonic isoforms of myosin MHC_{emb} and LC_{1emb} do not appear to be regulated together, the LC_{1emb} to LC_{1E} transition apparently taking place earlier (see Whalen, 1980).

The way in which the expression of different tissue and developmental programmes is regulated is thus of major interest in the muscle system. In addition the co-ordination between the components of any one muscle programme is also an important consideration since a func-

Table 2. Actin and myosin recombinant plasmids isolated from new born mouse skeletal muscle.

Clone No.	Identification	Coding/Non-coding		Size of insertion	Size of homologous RNA
p32	$MHC_{(F\ or\ NB)}$	+	(+)	1150	6900
p161	LC_{1F}, LC_{3F}	–	+	400	1050, 900
p91	α_{SM}-Actin	+	+	1350	1600
p91 (200)	α_{SM}-Actin	–	+	200	1600
p81	α_{CM}-Actin	+	(+)	1080	1600
p41	Non-muscle actin (cloned from lymphoma)	+	(+)	1150	2000

Abbreviations: MHC – myosin heavy chain; F – adult fast skeletal muscle; NB – new born; SM – skeletal muscle; CM – cardiac muscle.

tional sarcomere requires a strict stoichiometry of the accumulated contractile proteins. In order to understand how the expression of a muscle phenotype is regulated, information is necessary both on the structure and chromosomal organization of the muscle genes, on their activation and transcription, and on the processing, accumulation and translationial efficiencies of the resultant transcripts. The advent of DNA recombinant technology has made these questions directly accessible to experimental study.

RECOMBINANT PLASMIDS: THE SIZE AND HOMOLOGIES OF ACTIN AND MYOSIN mRNAs

We have used this approach to obtain purified probes to muscle coding sequences. Recombinant plasmids containing DNA complementary to different size fractions of poly(A)+ RNA from new-born (about 10 day) mouse skeletal muscle were constructed by GC tailing and integration at the PstI site of pBR322 (for details, see Minty et al., 1981). The plasmids are listed in Table 2. Analysis of the size of messenger RNAs (mRNAs) and their cross-hybridization with the inserted sequence in the plasmid is based on the results of Northern blots. RNA is fractionated on agarose gels, transferred to activated paper and the "blot" hybridized with the appropriate radioactive plasmid sequence. The hybrid band is revealed by autoradiography (see Alwine et al., 1979).

(a) Myosin Heavy Chain

Initial identification of the myosin heavy chain plasmid (p32) was based on its hybridization to a muscle mRNA of 6 900 nucleotides, and on the pattern of the partial translation products of this RNA after filter-hybridization to the plasmid. This was confirmed by its cross-hybridization with the rat heavy chain probe described by Nudel et al. (1980). DNA sequence data on part of the insertion shows that it lies near the -COOH terminus of the molecule and codes for a protein sequence very similar to that of the adult rabbit skeletal muscle MHC (see Umeda et al., 1981). We conclude that it corresponds either to the adult or the newborn isoform of the myosin heavy chain expressed in mouse skeletal muscle, since there is also strong hybridization with RNA from these sources. It cross-hybridizes less well with the embryonic form, suggesting that the latter has diverged further from the adult isoform. As might be expected for the coding sequences of this gene family, there is also cross-tissue (e.g. with heart RNA) and cross-species (e.g. human, rat, chick) hybridization illustrating the conservation of myosin heavy chain sequences. This kind of information derived mainly from the analysis of Northern blots is of interest in an evolutionary context indicating the tissue and species divergence within a multigene family.

(b) The Actins

The actin multigene family is one in which the proteins are very highly conserved (see e.g. Vandekerckhove and Weber, 1978). We have isolated a number of recombinants with sequences complementary to different mouse actins (Table 2). These have been characterized by DNA sequence analysis. Plasmid 91 contains coding and non-coding sequences for the α-skeletal muscle actin mRNA (Minty et al., 1981). Not surprisingly, given the conservation of the protein, the plasmid cross-hybridizes with actins from different tissues and species. Messengers coding for the non-muscle β and γ actins are clearly distinguishable from the muscle actins by their larger size of 2 000 nucleotides compared with 1 600 for the muscle actins. This difference is due to additional non-coding sequences. On some Northern blots when hybridization with the non-muscle actin plasmid p41 is compared with that given by p91, two bands can be distinguished of 2 100 and 1 950 nucleotides, probably corresponding to the two non-muscle messengers. Plasmid 81, also cloned from new-born skeletal muscle, contains a "cardiac" like actin coding sequence (Minty et al., 1981, in preparation). The expression of this isoform will be discussed in a later section. The actin recombinants p91, p81 and p41

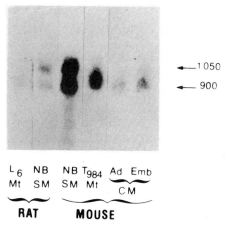

Figure 1. A Northern blot showing mRNAs cross-hybridizing with the myosin light chain plasmid, p161. From left to right: Total RNA prepared from myotubes (Mt) of the rat muscle cell line L_6 (44 h) poly(A)+ RNA from new-born rat skeletal muscle (NB, SM) (14 h), from myotubes (Mt) of the mouse muscle cell line T984 (44 h), from adult mouse heart (Ad, CM) (44 h) and from embryonic mouse heart (Emb, CM) (44 h). 1 μg poly(A)+ RNA of each sample was denatured and electrophoresed on the same 1.5 % agarose gel (McMaster and Carmichael, 1977). After transfer to DBM paper, the blots were hybridized with 5×10^6 cpm of p161 DNA (^{32}P) labelled by nick translation (Minty et al., 1981) and treated as described by Alwine et al. (1979). The exposure times for autoradiography of the different samples on the gel are shown in brackets above. Size markers (pBR322 and phage lambda DNA cut by various restriction enzymes and denatured as for the RNA) were used as standards to estimate mRNA size.

increases rapidly thereafter. The major alkali myosin light chain form synthesized initially in these cultures is LC_{1emb}. Synthesis of LC_{3F} is detectable on two dimensional gels from 144 h and increases slowly thereafter. The adult form LC_{1F} is also synthesized later during muscle fibre formation in these cultures.

RNA extracted from cultures of different ages has been analysed on Northern blots with the actin and myosin probes as shown in Fig. 3. No mRNA is detectable with p32, p91(200), or p161 in dividing myoblasts. The messengers for α-actin and the myosin heavy chain are first detectable at 96 h and accumulate rapidly thereafter reaching a maximum at 144 h. The messengers for the myosin light chains are first detectable at 120 h and continue to accumulate during the next two days. The messenger for LC_{1emb}, the alkali light chain initially

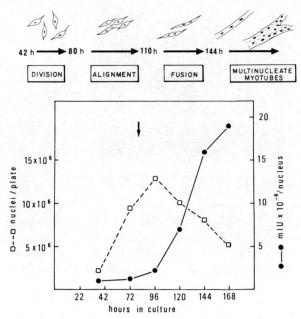

Figure 2. Proliferation and differentiation in the mouse muscle cell line, T984. Different stages of muscle cell fusion are presented schematically. Myotubes begin to appear after the cultures have become confluent. Large cross-striated fibres are present from about 144 h. The number of nuclei per plate (☐---☐) was calculated from the DNA content of culture dishes. The T984 cell line is aneuploid (Jakob et al., 1978) and an average value of 8.5 pg DNA per cell nucleus was used. Total creatine kinase activity per nucleus (●—●) is expressed in international units (IU) where 1 unit of creatine kinase is the amount required to catalyse the formation of 1 μmole ADP per min at 25°C. The enzyme was measured in the supernatant of total cellular homogenates by using the "CK NAC-activated UV-system" (Boehringer, Germany). The arrow indicates when the medium was changed from Dulbecco's modified Eagle's medium containing 15 % foetal calf serum to the same medium with 2 % foetal calf serum. This step down reduces overgrowth of the culture.

present, cross-hybridizes weakly with the probe which principally detects LC_{3F} (and LC_{1F}) (Fig. 1). At 120 h LC_{1emb} is still the major mRNA. Later accumulation of the mRNA for LC_{3F} (and LC_{1F}) reflects later synthesis of the adult form in vivo. These experiments therefore indicate a close correlation between mRNA accumulation and synthesis of the corresponding muscle protein during myogenesis.

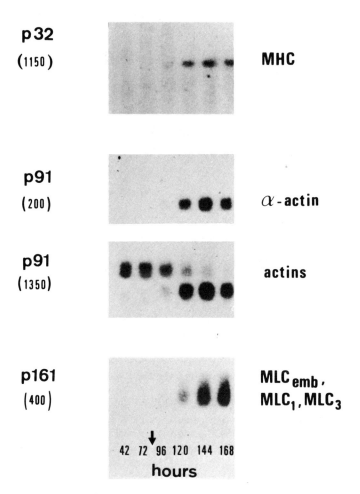

Figure 3. Northern blot analysis of accumulated actin and myosin mRNAs. Poly(A)+ RNA (2 μg per preparation) from T984 muscle cells at different days after plating was fractionated on 1.25 % agarose gels and transferred to DBM paper, as in Fig. 1. The paper was cut into 3 pieces containing different RNA size classes and hybridized with nick translated plasmid probe as indicated in Fig. 1. The exposure times for the different parts of the blot were as follows: MHC 82 h, α-actin 22 h, the actins 72 h, and the MLC 60 h.

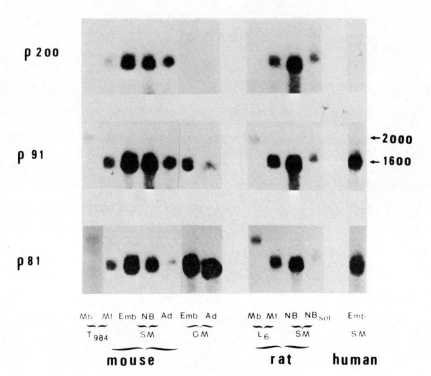

Figure 4. A Northern blot showing the relative homologies of actin mRNAs from different tissues, with the actin probes p81 and p91. Samples of poly(A)+ RNA were separated on 1.25 % agarose gels, transferred to DBM paper and hybridized with the nick-translated probes essentially as described for the actins in Fig. 3. The same region of the blot was re-used after loss of counts from the previous probe. The degree of hybridization with the different probes p81, p91 and p91-200 (p200) is therefore directly comparable. The samples shown are from left to right RNA from myoblasts (Mb) and myotubes (Mt) of the mouse muscle cell line T984, from embryonic (Emb), new-born (NB) and adult (Ad) skeletal muscle (SM) and embryonic (Emb) and adult (Ad) cardiac muscle (CM) of the mouse, from myoblasts (Mb) and myotubes (Mt) of the rat muscle cell line L_6, from new-born fast (NB) and new-born slow (soleus) (NB_{sol}) skeletal muscle of the rat and from human foetal (3 months) skeletal muscle (Emb, SM).

The question of whether a few transcripts of the muscle genes are present in dividing myoblasts is not resolved by these experiments. With the p91 (200) probe which is homologous to the mRNA for skeletal muscle α-actin, assuming that synthesis of α-actin is 2.5-5 % of total protein synthesis in myotubes, 40-80 molecules per cell of the mRNA can be detected by the technique used here. A minimum estimate for the accumulation of this mRNA is 130 fold during myogenesis. In the case of the myosins, neither the heavy chain nor the light chain probes are homologous to the embryonic forms and the sensitivity of detection is therefore less. (These points are discussed in detail in Caravatti et al. where the Northern blot results are presented (1981)). It has been reported that part of the muscle mRNA for the myosin heavy chain which accumulates during differentiation is not retained on oligo-dT cellulose and contains a short poly(A) sequence (Benoff and Nadal-Ginard, 1979). With RNA preparations from fused cultures we obtained similar results when poly(A)$^+$ RNA were analysed on Northern blots. This experiment, therefore, gives no indication of the presence of significant quantities of poly(A)$^-$ muscle mRNAs.

In dividing myoblasts, hybridization with the p91 probe reveals messengers for other isoforms of actin. The non-muscle (β, γ) actin mRNAs are larger (about 2 000 nucleotides) than the muscle forms (about 1 600 nucleotides) and can be seen on the blot migrating above the α-actin band. These mRNAs cease to accumulate at fusion, as the amount of α-actin mRNA increases. In dividing myoblasts a small amount of an α-actin type of mRNA, migrating rather faster than the skeletal muscle α-actin mRNA, is detected with the p91 plasmid, but not with the α-actin specific p91(200) probe. It probably corresponds to a smooth muscle actin mRNA present in myoblasts (see Minty et al., 1981).

EXPRESSION OF A CARDIAC ISOFORM OF ACTIN IN SKELETAL MUSCLE DURING DEVELOPMENT

Experiments using the actin probe p81 would suggest that skeletal muscle α-actin is not the only muscle actin sequence to accumulate during myotube formation. The p81 sequence was also cloned from RNA prepared from new-born mouse skeletal muscle. Northern blot analysis, however, shows that p81 has more homology with cardiac actin mRNA than with that from adult skeletal muscle. This is in direct contrast to the situation with p91 (Fig. 4). With embryonic skeletal muscle RNA and with RNA from the differentiated cells of the mouse muscle cell line T984 the difference between the plasmids is much less striking. We therefore conclude that in RNA from these tissues part of the muscle actin mRNA migrating as a band of 1 600 nucleotides is not the α-skeletal muscle form, but a cardiac-like actin mRNA. The degree of hybridization with the p91-200 probe which is specific for α-skeletal muscle actin mRNA would also suggest that in R984 myotubes

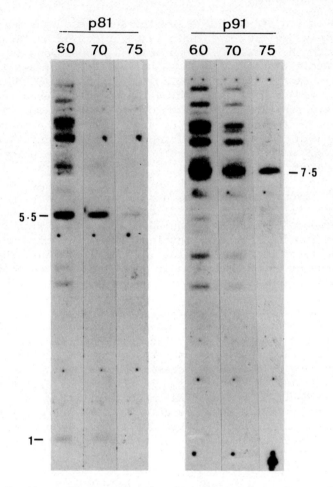

Figure 5. Southern blot analysis of actin genes. 5 μg of DNA prepared from mouse (129) embryos was digested with EcoRI and separated on 0.7 % agarose gels. The DNA was transferred to nitro cellulose (Southern, 1975) and hybridized with nick translated p81 or p91 DNA. The blots were washed successively in 0.1 x SSC under increasingly stringent conditions, 60, 70, 75°C, as indicated.

be less homologous to the other actin sequences, already gives one major band (5.5 Kb) and two minor bands (7.5 Kb, 1 Kb) after a 70° wash. The largest band represents slight cross-hybridization with the skeletal muscle gene. The major 5.5 Kb fragment corresponds to the 5' end of the cardiac gene. The small 1 Kb fragment is the 3' end of the same gene which is cut by EcoRI (Minty et al., in preparation). We interpret these results to indicate that p81 is homologous with a single gene coding for cardiac actin.

In the early stages of muscle fibre formation, mRNA for cardiac actin accumulates. Although it is not yet clear whether this mRNA is translated, the situation appears to be somewhat analogous to that observed for the embryonic light chain LC_{1emb} which is also found in heart muscle (Whalen et al., 1981). The physiological significance of cardiac isoforms in embryonic skeletal muscle remains to be demonstrated. From a regulatory point of view the co-expression of parts of two different phenotypic programmes at early stages of development poses some interesting problems. It suggests that the naive view that the programme for any phenotypic group of contractile proteins is regulated as a block is not necessarily correct. Indeed the diversity of myosin heavy chain isoforms expressed during skeletal muscle development, the presence, for example, of a new-born form not seen for the myosin light chains or actins (see Table 1), suggests that regulation of a developing muscle phenotype is complex.

CONCLUSIONS

In conclusion, the kind of experiment described here, based essentially on Northern blot analysis of accumulated mRNA, gives information both on the expression in different tissues of mRNAs of varying degrees of homology coding for different isoforms of a myosin or actin multigene family and on the co-ordination of accumulation of different contractile mRNAs during muscle cell differentiation.

It is clear from the cell culture experiments that mRNA accumulation and synthesis of the corresponding muscle protein are closely correlated. The Northern blot experiments confirm earlier attempts to look at this by comparison between in vivo and in vitro mRNA directed protein synthesis (Yablonka and Yaffe, 1977; Devlin and Emerson, 1979: Perriard et al., 1979; Daubas et al., 1981), and also by molecular hybridization experiments using cDNA synthesized from total RNA populations (Paterson and Bishop, 1977; Leibovitch et al., 1979; Affara et al., 1980; Zevin-Sonkin and Yaffe, 1980) or from fractions enriched for a single muscle mRNA (John et al., 1978; Dym et al., 1979; Benoff and Nadal-Ginard, 1979). Controversy as to whether untranslated forms of muscle mRNA accumulate prior to differentiation in the dividing myoblast population has now been resolved by the experiments with cloned probes. Whether muscle mRNAs accumulate in a ribonucleoprotein form immediately prior to their mobilisation into polysomes and whether unstable forms are present in myoblasts (see Buckingham, 1977), are questions which are not resolved by the Northern blot approach. However with cloned probes it should be possible to look at pulse labelled forms of nuclear and cytoplasmic muscle mRNAs, and at the conformational state of their chromatin by DNase I type analyses, during myogenesis.

It is important in assessing the results to consider the status of the dividing myoblast. In muscle cell lines the "myoblasts" dis-

Table 3. A comparison between part of the nucleotide sequence of the actin insertions in p81 and p91.

MUSCLE ACTINS

	5'								
91	GGG	ATC	CXX	XXX	XCC	ACC	TAC	AAC	AGC	ATC
81	GGC	ATT	CAT	GAA	ACT	ACT	TAC	AAC	AGC	ATC
	gly	ile	his	glu	thr	thr	|tyr|	asn	ser	ile
	272									
91	ATG	AAG	TGC	GAC	ATC	GAC	ATC	AGG	AAG	GAC
81	ATG	AAG	TAG	GAC	ATC	GAT	ATC	CGC	AAG	GAC
	met	lys	cys	asp	|ile|	asp	ile	arg	lys	asp
91	CTG	TAT	GCC	AAC	AAC	GTC	ATG	TCA	GGG	3'
81	CTG	TAT	GCC	AAC	AAT	GTC	CTX	TCT	GGX	
	leu	tyr	ala	asn	|asn|	val	met/leu	ser	gly
									300	

DNA sequencing was by the technique of Maxam and Gilbert (1980) as described in Minty et al. (1981). The residues characteristic of striated muscle are indicated.

other forms of actin accumulate. In new-born muscle small amounts of cardiac-like actin mRNA are presumably still present. DNA sequencing of the insertion in p81 demonstrates that it codes for cardiac actin. In Table 3 part of the p81 sequence is compared with that of p91. Amino acid positions 278, 286 and 296 are characteristic of muscle actin, while position 298 distinguishes cardiac from skeletal muscle actin. Comparison of the sequences of these two actin mRNAs indicates about 10 % divergence at the nucleotide level, compared with about 1 % (4/375) in the amino acid sequence for cardiac actin.

Although at the positions (89, 298, 375) plasmid p81 has residues characteristic of a cardiac sequence it may nevertheless be the product of a different "embryonic" gene. This has been looked at by Southern blot analysis (Fig. 5). Mouse genomic DNA, digested by EcoRI was separated on agarose gels, transferred to nitrocellulose paper (Southern, 1975) and hybridized with nick-translated plasmids p81 or p91. When the blots are washed at relatively low stringency (60°C, 0.1 x SSC) multiple actin bands are visible corresponding to different actin genes and probably pseudogenes. At higher stringencies the probe hybridizes only to its homologous genomic sequence. In the case of p91 the skeletal muscle actin gene is clearly distinguished as a band at 7.5 Kb after a 75° wash. Plasmid 81, which appears to contain part of the actin coding sequence and show tissue and species cross-reactivity. In contrast the non-coding sequence of p91, which has been sub-cloned to give p91 (200) is highly tissue- and species-specific under the Northern blot conditions used (see Figs. 3 and 4). It can be employed as a specific probe for mouse skeletal muscle α-actin.

(c) The Myosin Light Chains

Plasmid 161 cross-hybridizes with the messengers of the alkali myosin light chains LC_{1F} and LC_{3F} of adult fast skeletal muscle (Fig. 1). These peptides have a common C-terminal sequence of 141 amino acids. LC_3 has 8 residues at the N-terminus which differ from the 49 amino terminal sequence of LC_3 (Frank and Weeds, 1974). DNA sequence analysis of the p161 insertion shows that it corresponds to a non-coding fragment (Robert et al., 1981). This is a surprising result, since usually, as found even for highly conserved proteins like the actins, the non-coding sequence at the 3' end of the mRNA has diverged from the comparable sequences in other mRNAs of the multigene family. Conservation of the non-coding sequence may reflect utilisation of the same sequence of genomic DNA for the 3' region of the two mRNAs, possibly with a differential splicing at amino acid 141 in the coding sequence as suggested from the protein data by Matsuda et al. (1981). We are currently looking at the structure of the corresponding gene to investigate this possibility. Plasmid 161 shows some cross-hybridization with the LC_{1F}, LC_{3F} mRNAs (Fig. 1) from rat skeletal muscle. Given this cross-species hybridization we therefore looked at mRNA from the rat muscle cell line L_6, which expresses the embryonic light chain LC_{1emb} as the only major alkali light chain isoform (Whalen et al., 1978). A faint hybridization is seen with an mRNA of intermediate size (about 950 nucleotides). A similar mRNA is present in embryonic and to a lesser extent in adult mouse hearts. This probably corresponds to LC_{1emb} present in embryonic hearts and in adult heart atria (Whalen et al., 1981). These results show that the embryonic LC_1 mRNA is different, both in size and homology of the non-coding sequence, from the adult isoform.

THE EXPRESSION OF MYOSIN AND ACTIN mRNAs DURING MYOGENESIS IN A MOUSE MUSCLE CELL LINE

We have used the plasmid probes to look at muscle mRNAs in a mouse muscle cell line. This cell line, T-984 Cl 10, derived initially from the embryoid body of a mouse teratocarcinoma, forms differentiated muscle fibres in monolayer culture when the cells become confluent (Jakob et al., 1978). This is illustrated in Fig. 2 where the accumulation of creatine phosphokinase has been followed as a marker of differentiation. The synthesis of most muscle proteins, such as myosin heavy chain and α-actin, is detectable at 120 h and play a certain flexibility in their potential to differentiate. The T984 mouse line will give rise to fibroblast-like, adipose, or muscle cells according to the culture conditions (Darmon et al., 1981). One way of looking at this is to consider the mononucleated cells in these lines as further back in a differentiating pathway than most of the myoblasts of primary cultures. The advantage of the cell lines is that mononucleated populations are not contaminated by other cell types, nor by differentiated cells, as is the case in primary cultures. A combination of studies on muscle cell lines, primary cultures, and muscle tissue in the embryo is probably essential in order to describe the steps of muscle gene expression during myogenesis. Understanding the mechanisms by which a muscle phenotype is established will require knowledge of the muscle genes themselves, and the capacity to manipulate them in different biological contexts. In the muscle system it is now possible to envisage this perspective.

ACKNOWLEDGEMENTS

The laboratory is supported by grants from the Délégation Générale à la Recherche Scientifique et Technique, the Centre National de la Recherche Scientifique, the Institute National de la Santé et de la Recherche Medicale, the Commissariat à L'Energie Atomique, the Ligue Nationale Française contre le Cancer, the Fondation pour La Recherche Médicale Française and The Muscular Dystrophy Association of America.

M.C. was the recipient of a fellowship from the Swiss National Research Foundation and from I.N.S.E.R.M., A.M. from the Muscular Dystrophy Associations of America, P.D. from the Ligue Francaise contre le Cancer and S.A. from the D.G.R.S.T.

We are grateful to R. Williamson for providing the sample of human foetal muscle RNA.

REFERENCES

Affara, N. A., Robert, B., Jacquet, M., Buckingham, M., and Gros, F., 1980, Changes in gene expression during myogenic differentiation. I: Regulation of mRNA sequences expressed during myotube formation, J. Molec. Biol., 140:441.
Alwine, J. C., Kemp, D. J., Parker, B. A., Keiser, J., Renart, J., Stark, G. R., and Wahl, G. M., 1979, Detection of specific RNAs or specific fragments of DNA by fractionation in gels and transfer to diazobenzyloxymethyl paper, Methods in Enzymology, 68:220.
Benoff, S., and Nadal-Ginard, B., 1979, Most myosin heavy chain mRNA in L_6E_9 rat myotubes has a short poly(A) tail, Proc. Natl. Acad. Sci. U.S.A., 76:1853.

Buckingham, M. E., 1977, Muscle protein synthesis and its control during the differentiation of skeletal muscles in vitro, in: "International Review of Biochemistry, Biochemistry of Cell Differentiation II, Vol. 15," J. Paul, ed., University Park Press, Baltimore.

Caravatti, M., Perriard, J. C., and Eppenberger, H. M., 1979, Developmental regulation of creatine kinase isoenzymes in myogenic cell cultures from chicken. Biosynthesis of creatine kinase subunits M and B, J. Biol. Chem., 254:1388.

Caravatti, M., Minty, A., Robert, B., Montarras, D., Weydert, A., Cohen, A., Daubas, P., Gros, F., and Buckingham, M., 1981, Regulation of muscle gene expression: The accumulation of mRNAs coding for muscle specific proteins during myogenesis in a mouse cell line, Submitted for publication.

Darmon, M., Serrero, G., Rizzino, A., and Sato, G., 1981, Isolation of myoblastic, fibro-adipogenic, and fibroblastic clonal cell lines from a common precursor and study of their requirements for growth and differentiation, Expl. Cell Res., 132:313.

Daubas, P., Caput, D., Buckingham, M., and Gros, F., 1981, A comparison between the synthesis of contractile proteins and the accumulation of their translatable mRNAs during calf myoblast differentiation, Develop. Biol., 84:133.

Devlin, R. B., and Emerson, C. P., 1978, Co-ordinate regulation of contractile protein synthesis during myoblast differentiation, Cell, 13:599.

Devlin, R. B., and Emerson, C. P., 1979, Co-ordinate accumulation of contractile protein mRNA during myoblast differentiation, Develop. Biol., 69:202.

Dym, H. O., Kennedy, D. S., and Heywood, S. M., 1979, Subcellular distribution of the cytoplasmic myosin heavy chain mRNA during myogenesis, Differentiation, 12:145.

Fischman, D. A., 1970, The synthesis and assembly of myofibrils in embryonic muscle, Current Topics in Develop. Biol., 5:235.

Frank, G., and Weeds, A. G., 1974, The amino acid sequence of the alkali light chains of rabbit skeletal muscle myosin, Eur. J. Biochem., 44:317.

Garrels, J. I., 1979, Changes in protein synthesis during myogenesis in a clonal cell line, Develop. Biol., 73:134.

Jakob, H., Buckingham, M. E., Cohen, A., Dupont, L., Fiszman, M., and Jacob, F., 1978, A skeletal muscle cell line isolated from a mouse teratocarcinoma undergoes apparently normal terminal differentiation in vitro, Exp. Cell Res., 114:403.

John, H. A., Patrinou-Georgoulas, M., and Jones, K. W., 1977, Detection of myosin heavy chain mRNA during myogenesis in tissue culture by in vitro and in situ hybridization, Cell, 12:501.

Leibovitch, M. P., Leibovitch, S. A., Harel, J., and Kruh, J., 1979, Changes in the frequency and diversity of mRNA populations in the course of myogenic differentiation, Eur. J. Biochem., 97:321.

McMaster, G. K., and Carmichael, G. G., 1977, Analysis of single- and double-stranded nucleic acids on polyacrylamide and agarose gels by using glyoxal and acridine orange, Proc. Natl. Acad. Sci. U.S.A., 74:4835.

Matsuda, G., Maita, T., and Umegane, T., 1981, The primary structure of L-1 light chain of chicken fast skeletal muscle myosin and its genetic implication, FEBS Letters, 126:111.

Maxam, A. M., and Gilbert, W., 1980, Sequencing end-labeled DNA with base specific chemical cleavages, Methods in Enzymology, 65:499.

Merlie, J. P., Sobel, A., Changeux, J. P., and Gros, F., 1975, Synthesis of actylcholine receptor during differentiation of cultured embryonic muscle cells, Proc. Natl. Acad. Sci. U.S.A., 72:4028.

Minty, A. J., Caravatti, M., Robert, B., Cohen, A., Daubas, P., Weydert, A., Gros, F., and Buckingham, M. E., 1981, Mouse actin messenger RNA: construction and characterization of a recombinant plasmid molecule containing a complementary DNA transcript of mouse α-actin mRNA, J. Biol. Chem., 256:1008.

Nudel, U., Katcoff, D., Carmon, Y., Zevin-Sonkin, D., Levi, Z., Shani, Y., Shani, M., and Yaffe, D., 1980, Identification of recombinant phages containing sequences from different rat myosin heavy chain genes, Nucleic Acids Res., 8:2133.

Paterson, B. M., and Bishop, J. O., 1977, Changes in the mRNA population of chick myoblasts during myogenesis in vitro, Cell, 12:751.

Perriard, J. C., Perriard, E. R., and Eppenberger, H. M., 1979, Detection and relative quantitation of mRNA for creatine kinase isozymes in RNA from myogenic cell cultures and embryonic chicken tissues, J. Biol. Chem., 254:7036.

Robert, B., Weydert, A., Caravatti, M., Minty, A., Cohen, A., Daubas, P., Gros, F., and Buckingham, M., 1981, A cDNA recombinant plasmid complementary to mRNAs for both light chains 1 and 3 of mouse skeletal muscle myosin, Submitted for publication.

Shani, M., Nudel, U., Zevin-Sonkin, D., Zanut, R., Girol, D., Katcoff, D., Carmon, Y., Keiter, J., Fischauf, A. M., Yaffe, D., 1981, Skeletal muscle actin mRNA characterization of the 3' untranslated region, Nucl. Acids Res., 9:579.

Southern, E. M., 1975, Detection of specific sequences among DNA fragments separated by gel electrophoresis, J. Molec. Biol., 98:503.

Umeda, P. K., Sinha, A. M., Jakorcic, S., Merten, S., Hsu, H.-J., Subramanian, K. N., Zak, R., and Rabinowitz, M., 1981, Molecular cloning of two fast myosin heavy chain cDNAs from chicken embryo skeletal muscle, Proc. Natl. Acad. Sci. U.S.A., 78:2834.

Vandekerckhove, J., and Weber, K., 1978, Actin amino acid sequences: Comparison of actins from calf thymus, bovine brain, and SV transformed mouse 3T3 cells with rabbit skeletal muscle actin, Eur. J. Biochem., 90:451.

Whalen, R. G., Butler-Browne, G. S., and Gros, F., 1978, Identification of a novel form of myosin light chain present in embryonic muscle tissue and cultured muscle cells, J. molec. Biol., 126:415.

Whalen, R. G., Schwartz, K., Bouveret, P., Sell, S. M., and Gros, F., 1979, Contractile protein isozymes in muscle development: Identification of an embryonic form of myosin heavy chain, Proc. Natl. Acad. Sci. U.S.A., 76:5197.

Whalen, R. G., 1980, Contractile protein isozymes in muscle development: The embryonic phenotype, in:"Plasticity of Muscle," D. Pette, ed., Walter de Gruyter, Berlin, p.177.

Whalen, R. G., and Sell, S. M., 1980, Myosin from foetal hearts contains the skeletal muscle embryonic light chain, Nature, Lond., 286:731.

Whalen, R. G., Sell, S. M., Butler-Browne, G. S., Schwartz, K., Bouveret, P., and Pinset, I., 1981, the myosin heavy chain isozymes appear sequentially in rat muscle development, Nature, Lond., 292:805.

Whalen, R. G., Sell, S. M., Thornell, L. E., and Erickson, A., 1981, The myosin subunit in foetal and adult ventricles, atria, and Purkinje fibres, Submitted for publication.

Yablonka, Z., and Yaffe, D., 1977, Synthesis of myosin light chains and accumulation of translatable mRNA coding for light chain like polypeptides in differentiating muscle cultures, Differentiation, 8:133.

Yaffe, D., 1968, Cellular aspects of muscle differentiation in vitro, Current Topics in Devlop. Biol., 4:37.

Zevin-Sonkin, D., and Yaffe, D., 1980, Accumulation of muscle specific RNA sequences during myogenesis, Develop. Biol., 74:326.

GENES CODING FOR VIMENTIN AND ACTIN IN MAMMALS AND BIRDS

Wim J. Quax, Huub J. Dodemont, Johannes A. Lenstra,
Frans C. S. Ramaekers, Philippe Soriano*,
Marielle E. S. van Workum, Giorgio Bernardi* and
Hans Bloemendal[1]

Department of Biochemistry
University of Nijmegen
Geert Grooteplein N21
6525 EZ Nijmegen, The Netherlands

*Laboratoire de Génétique Moléculaire
Institut de Recherche en Biologie Moléculaire
F-75221 Paris, France

The past few years have seen an increasing interest in the cytoskeleton with regard to architecture and function. Among the structural components of the cytoskeleton, actin-filaments and microtubules are rather well-characterized. It has recently been recognised that the intermediate-sized filaments (IF) also play a role, presumably in mitosis and the maintenance of cell shape (Lazarides, 1980). Up till now five different classes of IF have been described in higher eukaryotic cells. One of them, the vimentin-type, as a rule exists in vivo only in cells of mesenchymal origin; our studies, however, on the lens cytoskeleton clearly showed that vimentin filaments also occur in the lenticular cells in spite of their epithelial origin (Ramaekers et al., 1980; Ramaekers and Bloemendal, 1981). Subcultured cells seem to synthesize vimentin irrespective of their embryological origin (Franke et al., 1979). Although intensive studies have been devoted to the elucidation of structural aspects of actin, tubulin, and their corresponding genes (Cleveland et al., 1980; Firtel, 1981; Kalfayan and Wensink, 1981; Ponstingl et al., 1981; Vandekerckhove and Weber, 1978), rather few data concerning the molecular biology of vimentin are available. The availability of a vimentin cDNA would help with the elucidation of the structure of proteins in IF (Steinert et al., 1980; Geisler and Weber, 1981) but would also make the study of developmental regulation and gene organization possible. This

---
[1] To whom correspondence should be addressed.

data should enable us to unravel the functions of IF which are still poorly understood.

In recent studies of transformation of cultured hamster lens cells (Bloemendal et al., 1980a) we described the isolation of cytoskeletal messenger RNAs including the species that code for vimentin and actin. In the present paper we wish to report for the first time on the molecular cloning of a mammalian vimentin cDNA. For comparison we also describe results obtained with an actin clone (β, γ) constructed from mRNA of the same source. From our studies two major aspects emerged;
1) The sequence encoding vimentin is, as in the case of actin, evolutionarily conserved in higher vertebrates.
2) Vimentin genes, in contrast to actin and tubulin genes, do not occur as a multigene family.

The construction and characterization of an SV40 transformed hamster lens cell line growing in suspension culture have been described elsewhere (Bloemendal et al., 1980a; Bloemendal et al., 1980b). Cell-free translation of the unfractionated poly(A)$^+$ RNA isolated from these cells suggests that the messages encoding cytoplasmic actin, vimentin and a hitherto unidentified 46 kd protein are present in appreciable amounts. The total mRNA was fractionated by sucrose density gradient centrifugation and the pooled mRNA from the 16-20 S region was converted into double-stranded cDNA. The products sizing from 1 200 to 2 600 base pairs were elctroeluted from agarose gels and inserted into the Pst I site of plasmid pBR322 by the polydC-polydG joining procedure (Bolivar et al., 1977). The hybrid DNA was used for transformation of E. coli strain HB101.

Poly(A)$^+$ RNA was isolated from the hamster lens cell culture (Bloemendal et al., 1980a) denatured with methylmercurihydroxide and fractionated on a sucrose gradient. The 16-20 S fractions were used as starting material for double-stranded cDNA synthesis. Double-stranded cDNAs, ranging from 1 200 to 2 600 base-pairs were oligo-dC tailed and cloned into Pst 1-cleaved(dG)-tailed pBR322. Hybrid molecules were used to transform E. coli strain HB101. From colonies with the expected phenotype (Tet^R, Amp^S) plasmid DNA was isolated (Birnboim and Doly, 1979) and 2 µg aliquots were denatured at 100°C for 5 minutes, bound to nitrocellulose and hybridized to 5 µg of the original 16-20 S hamster lens mRNA (Ricciardi et al., 1979). After extensive washings hybrids were melted out and the eluted mRNA was assayed for in vitro translation in the nuclease treated rabbit reticulocyte lysate (Pelham and Jackson, 1976). The translation products formed were electrophoresed in 10 % SDS-polyacrylamide and the gel was fluorographed.

One of the RNAs directed exclusively the synthesis of a product that co-migrated with vimentin. Another RNA directed exclusively the synthesis of actin when tested by in vitro translation (Fig. 1). Further characterization of the translation products was achieved by

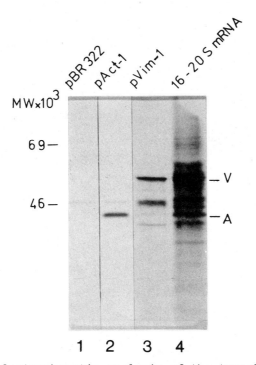

Figure 1. Gel electrophoretic analysis of the translation products whose synthesis has been directed by the hybrid-selected actin and vimentin mRNA.
Lane 1: Control incubation in which the 16-20 S mRNA was hybridized to immobilized pBR322 DNA. The endogenous band of the reticulocyte lysate is the only band visible.
Lane 2: Products synthesized under the direction of RNAs selected by the actin clone.
Lane 3: Products synthesized after translation of the RNA that was selected by the vimentin-specific plasmid. Besides the endogenous band of the reticulocyte lysate the 57 kDaltons vimentin subunit and its breakdown products can be observed (Gard and Lazarides, 1980).
Lane 4: Products synthesized formed by translation of the total 16-20 S hamster lens mRNA fraction.
V = vimentin; A = actin.

two-dimensional electrophoresis (Fig. 2). The product obtained in the case of vimentin had not only the correct isoelectric point, but also showed the steplike distribution of spots, due to specific breakdown products, which have been described as characteristic for vimentin (Fig. 2C) (Gard and Lazarides, 1980). With the actin colony the corresponding translation products co-migrated with β- as well as with γ-actin which had been added as unlabeled carriers (Fig. 2A, 2B).

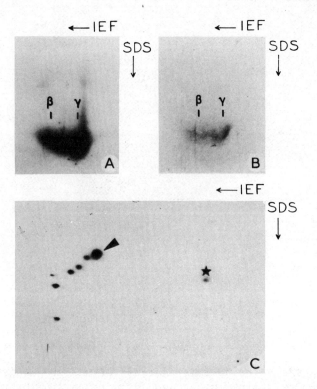

Figure 2. In order to establish the products formed in Fig. 1 (lanes 2 and 3), they were subjected to two-dimensional gel electrophoresis according to O'Farrell (1975). (A) corresponds to Coomassie Blue staining of added β- and γ-actin; (B) shows the fluorograph of the translation products obtained with the actin clone run on the same gel; (C) shows the fluorograph of vimentin (arrow) and a ladder of breakdown products specific of this protein. Unlabeled vimentin, extracted from the hamster lens cells, was run on the same gel and co-migrates exactly with the main spot of the fluorograph (not shown). The asterisk marks the endogenous product of the rabbit reticulocyte lysate.

The hybrid plasmid contains a vimentin-specific cDNA insert of about 1 700 base pairs. The ^{32}P-labeled, nick-translated probe was hybridized with a nitrocellulose filter, to which the initial hamster lens poly(A)$^+$ RNA had been transferred (Thomas, 1980) after electrophoresis on denaturing agarose gels. Only one hybridization band corresponding to a length of about 2 100 nucleotides could be detected. Identical results were obtained with vimentin-mRNAs derived from whole rat or calf lenses (data not shown). From the known molecular weight of vimentin (about 57 kd) a coding region of approximately 1 400 bases can be calculated for the corresponding message, leaving about 700

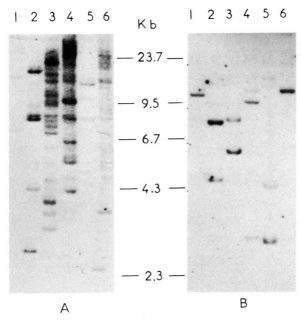

Figure 3. Detection of actin and vimentin sequences in genomic DNA: hybridization of EcoR1-digested DNA from different vertebrate species with the actin clone (A) and the vimentin clone (B). DNA sources were: 1) chicken erythrocytes 2) hamster liver 3) mouse liver 4) rat liver 5) calf thymus 6) human liver.

bases for the total non-coding regions. Since the inserted cDNA comprises 1 700 base pairs, it must contain a substantial portion, if not all of the coding sequence the latter which is to account for the hybridisation results.

The cloned vimentin probe was also hybridized to nitrocellulose filters on which Eco R1 restriction fragments of nuclear DNA from a variety of eukaryotic species had been blotted. High molecular weight genomic DNA was isolated from different sources and after digestion (5-10 µg) with Eco R1 (there is no Eco R1 site in the vimentin plasmid) electrophoresed on 0.6 % agarose gels and transferred to nitrocellulose paper (Southern, 1975). The vimentin plasmid was ^{32}P-labeled by nick translation.

Hybridizations were carried out at 42°C during 40 hours under the following conditions: 50 % formamide, 5 x SSC, 1 x Denhardt solution, 20 mM sodium phosphate (pH 6.8), 5 mM EDTA and 100 µg/ml single-stranded salmon sperm DNA. Washings were carried out at 42°C for 3 hours with the hybridization solution, at 55°C for 15 minutes in 2 x SSC, 0.5 % SDS and lastly at 55°C for 30 minutes in 0.1 x SSC,

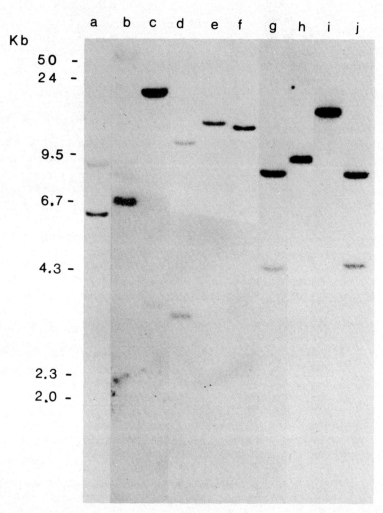

Figure 4. Estimation of the number of vimentin genes in mammalian genomes. Mouse, human and hamster DNAs were digested with different restriction enzymes and hybridized with the ^{32}P-labeled vimentin plasmid under the same conditions as in Fig. 3. Lanes a to d are mouse DNA digested with EcoR1 (a), Xhol (b), Bam H1 (c), and Hind III (d). Lanes e and f are human DNA digested with EcoR1 (e) and Bam H1 (f). Lanes g to j are hamster DNA digested with EcoR1 (g), Xhol (h), Bam H1 (i) and Hind III (j).

0.1 % SDS (Fig. 3B). All different species showed hybridizing restriction fragments, indicating that vimentin sequences are conserved throughout evolution in concert with their immunological cross-reactivity (Franke et al., 1978). Irrespective of the source of the DNA, a very simple fragment distribution is observed, quite different from the multiband pattern found in a parallel experiment with our actin probe which contains 600 nucleotides of the 3' coding part in addition to 600 nucleotides of the 3' non-coding segment (Fig. 3A) (Dodemont et al., 1981, submitted for publication). In order to estimate the number of vimentin genes, we digested nuclear DNA from hamster, mouse and man with different restriction enzymes (Fig. 4). After hybridization all lanes showed simple band patterns and in all species the vimentin hybridizing sequences can be reduced to one strongly hybridizing restriction fragment suggesting that the actual number of genes is one. Although the vimentin probe might not contain the total coding sequence, it is unlikely that genes from other parts of the genome have escaped detection for this reason. It is also unlikely that the strongly hybridizing restriction fragment contains a cluster of vimentin genes since in the case of calf DNA this fragment is about 3 kilobases (Fig. 3B, lane 5) whereas the corresponding messenger is 2 100 bases in size.

Although the elucidation of the details of the vimentin gene organization awaits further investigation it is obvious that, in marked contrast to actin, vimentin most probably is encoded by a single gene. Moreover in spite of 65 % homology between the 141 amino acid residues of the C-termini of vimentin and desmin (Geisler and Weber, 1981) one has to conclude from the present studies that this fact is not reflected by cross hybridization at the DNA-level.

ACKNOWLEDGEMENTS

We thank A. A. Groeneveld for maintaining the hamster lens epithelial cell line and Philippe Breton for the photographic work.

This work was supported in part by the Netherlands Foundation for Chemical Research (SON) and the Netherlands Organization for the advancement of Pure Research (ZWO). H.J.D. was recipient of a FEBS Fellowship. J.A.L. is research fellow of the Netherlands Society against Cancer (KWF).

REFERENCES

Birnboim, H. C., and Doly, J., 1979, A rapid alkaline extraction procedure for screening recombinant plasmid DNA, Nucl. Acids Res. 7:1513.
Bloemendal, H., Lenstra, J. A., Dodemont, H. J., Ramaekers, F. C. S., Groeneveld, A. A., Dunia, I., and Benedetti, E. L., 1980a, SV-40

transformed hamster lens epithelial cells: a novel system for
the isolation of cytoskeletal messenger RNAs and their translation products, Exp. Eye Res., 31:513.
Bloemendal, H., Ramaekers, F. C. S., Lenstra, J. A., Dodemont, H. J.,
Dunia, I., and Benedetti, E. L., 1980b, Neoplastic transformation
of lens cells in vitro, in:"Predictive value of short-term
screening tests in carcinogenicity evaluation," G. M. Williams
et al., ed., Elsevier/North Holland Biomedical Press, Amsterdam,
p.199.
Bolivar, F., Rodriguez, R., Greene, P., Betlach, M. C., Heyneker, H.
Heyneker, H. L., and Boyer, H. W., 1977, Construction and char-
acterization of new cloning vehicles. II. Multipurpose cloning
system, Gene, 2:95.
Cleveland, D. W., Lopata, M. A., MacDonald, R. J., Lowan, N. J.,
Rutten, W. J., and Kirschner, M. W., 1980, Number and evolution-
ary conservation of α- and β-tubulin and cytoplasmic β- and γ-
actin genes using specific cloned cDNA probes, Cell, 20:95.
Dodemont, H. J., Quax, W. J., Soriano, P., Ramaekers, F. C. S.,
Lenstra, J. A., Groenen, M. A. M., Bernardi, G., and
Bloemendal, H., 1981, manuscript in preparation.
Firtel, R. A., 1981, Multigene families encoding actin and tubulin,
Cell, 24:6.
Franke, W. W., Schmid, E., Osborn, M., and Weber, K., 1978, Different
intermediate-sized filaments distinguished by immunofluorescence
microscopy, Proc. Natl. Acad. Sci. U.S.A., 75:5034.
Franke, W. W., Schmid, E., Winter, S., Osborn, M., and Weber, K.,
1979, Widespread occurrence of intermediate-sized filaments of
the vimetin-type in cultured cells from diverse vertebrates,
Exp. Cell Res., 123:25.
Gard, D .L., and Lazarides, E., 1980, The synthesis and distribution
of desmin and vimentin during myogenesis in vitro, Cell, 19:263.
Geisler, N., and Weber, K., 1981, Comparison of the proteins of two
immunologically distinct intermediate-sized filaments by amino
acid sequence analysis: Desmin and vimentin, Proc. Natl. Acad.
Sci. U.S.A., 78:4120.
Kalfayan, L., and Wensink, P., 1981, α-tubulin genes of Drosophila,
Cell, 24:97.
Lazarides, E., 1980, Intermediate filaments as mechanical integrators
of cellular space, Nature, Lond., 283:249.
O'Farrell, P. H., 1975, High resolution two-dimensional electrophor-
esis of proteins, J. biol. Chem., 250:4007.
Pelham, H. R. B., and Jackson, R. J., 1976, An efficient mRNA-
dependent translation system from reticulocyte lysates, Eur. J.
Biochem., 67:247.
Ponstingl, H., Krauhs, E., Little, M., and Kempf, T., 1981, Complete
amino acid sequence of α-tubulin from porcine brain, Proc. Natl.
Acad. Sci. U.S.A., 78:2757.
Ramaekers, F. C. S., Osborn, M., Schmid, E., Weber, K.,
Bloemendal, H., and Franke, W. W., 1980, Identification of the
cytoskeletal proteins in lens-forming cells, a special epithel-

ioid cell type, Exp. Cell Res., 127:309.

Ramaekers, F. C. S., and Bloemendal, H., 1981, Cytoskeletal and contractile structures in lens cell differentiation, in:"Molecular and Cellular Biology of the Eye Lens," H. Bloemendal, ed., Wiley, New York, p.85.

Ricciardi, R. P., Miller, J. S., and Roberts, B. E., 1979, Purification and mapping of specific mRNAs by hybridisation-selection and cell-free translation, Proc. Natl. Acad. Sci. U.S.A., 76:4927.

Southern, E. M., 1975, Detection of specific sequences among DNA fragments separated by gel electrophoresis, J. Mol. Biol., 98:503.

Steinert, P. M., Idler, W. W., and Goldman, R. D., 1980, Intermediate filaments of baby hamster kidney (BHK-21) cells and bovine epidermal keratinocytes have similar ultrastructure and subunit domain structures, Proc. Natl. Acad. Sci. U.S.A., 77:4534.

Thomas, P. S., 1980, Hybridisation of denatured RNA and small DNA fragments transferred to nitrocellulose, Proc. Natl. Acad. Sci. U.S.A., 77:5201.

Vanderkerckhove, J., and Weber, K., 1978, At least six different actins are expressed in a higher mammal: an anlysis based on the amino acid sequence of the amino-terminal tryptic peptide, J. Mol. Biol., 126:783.

THE UNUSUALLY LONG RAT CRYSTALLIN αA$_2$ mRNA IS MONOCISTRONIC

R. Moormann, H. Dodemont*, J. G. G. Schoenmakers
and H Bloemendal*

Department of Molecular Biology
University of Nijmegen
The Netherlands

*Department of Biochemistry
University of Nijmegen
The Netherlands

INTRODUCTION

The structural proteins of the eye lens, called crystallins, are particularly suited to the study of differentiation and development (Papaconstantinou, 1967; Clayton, 1974; Bloemendal, 1977). Although we have acquired substantial knowledge concerning the structure and biosynthesis of lens proteins during the last decade (Bloemendal, 1981) nothing is known so far concerning the regulation of crystallin gene expresion at the genome level.

CRYSTALLIN CLONES

Recently the first reports on δ-crystallin gene organization have been published by Piatigorsky and his colleagues (Bhat et al., 1980) and by Clayton and co-workers (Bower et al., 1981). Our group described the construction of plasmid clones bearing duplex DNA sequences complementary to rat crystallin mRNAs (Dodemont et al., 1981). These recombinant plasmids containing cDNA sequences encoding the majority of α-, β- and δ-crystallins could be detected by positive hybrid-selection and translation. In Fig. 1 a typical analytical polyacrylamide gel electrophoretic pattern is depicted showing the identification of the translation products obtained with the different clones.

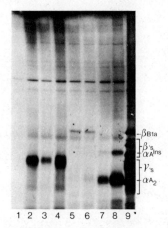

Figure 1. Identification of rat lens crystallin clones by hybridization translation. The products were analyzed by SDS polyacrylamide gel electrophoresis and fluorography. The numbers refer to the type of probe used for hybridization. Lane 1 is a control; Lanes 2, 3, 4: α-clones; Lanes 5, 6: β-clones; Lanes 7, 8: γ-clones; Lane 9: translation products of total lens mRNA.

THE αA_2 MESSAGE IS MONOCISTRONIC

At present the solution of several problems with regard to α-crystallin awaited the availability of crystallin cDNA and genomic clones.

Firstly there is the problem of the difference that exists between the genetic messengers directing the synthesis of the two kinds of α-crystallin polypeptide chains αA_2 and αB_2. Both chains have the same molecular weight (about 20 000 daltons; they reveal approximately 60% homology in amino acid sequence, but are encoded by messengers of remarkably different size. The αB_2 messenger sediments at 10 S whereas αA_2 mRNA has a sedimentation value of about 14 S. In general terms this 14 S mRNA contains enough nucleotides to encode two polypeptides of 2 000 daltons. In fact we envisaged this possibility when we first isolated this unusually long mRNA species (Berns et al., 1972). Later on Chen and Spector (1977) reported on the presumed bicistronic nature of αA_2. However, biosynthesis experiments with cap analogues carried out in our laboratory did not support this assumption, in that our results strongly suggested the presence of only one initiation site on the 14 S mRNA (Asselbergs et al., 1978). Molecular cloning of the αA_2 cDNA eventually led to the definitive solution. The major part of the nucleotide sequence of αA_2 mRNA present in two of our plasmid clones unequivocally excluded a bicistronic nature of the 14 S α-crystallin messenger (Moormann et al., 1981).

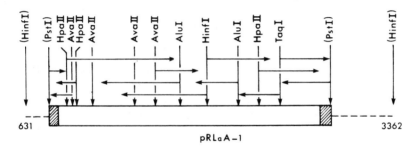
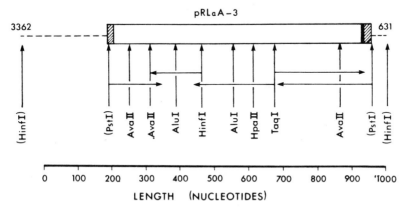

Figure 2. Restriction map of 2 αA_2-crystallin clones. The restriction sites used for nucleotide sequencing are shown. The cDNAs were inserted in the vector in opposite direction. Open boxes = mRNA sequences; Hatched boxes = poly (dG-dC) tracks; Black box = polyA residues.

Fig. 2 shows the restriction maps of these two clones. The position of the various restriction enzyme cleavage sites found in each map indicate that both DNA inserts represent overlapping sequences. It can be seen that one clone has a more extended 5' region whereas the other one is larger at the 3' end. The deduced nucleotide sequence is shown in Fig. 3. The overlap comprises 580 nucleotides whose sequences are identical in both clones.

Comparison with the known amino acid sequence of αA_2 crystallin (Van der Ouderaa et al., 1973; de Jong et al., 1975) reveals that the nucleotide sequence corresponds to a polypeptide fragment ranging from position 53 to the C terminal amino acid at position 173.

THE UNUSUAL LENGTH OF THE 14 S αA_2 RNA IS DUE TO NONCODING SEQUENCES

The coding frame shown is the only reading frame which can be

```
                        20                         40
         TC TTC CGC ACA GTG TTG GAC TCC GGC ATC TCT GAG GTC CGA TCT GAC CGG GAC AAG
            Phe Arg Thr Val Leu Asp Ser Gly Ile Ser Glu Val Arg Ser Asp Arg Asp Lys
                60                     80                    100
         TTT GTC ATC TTC TTG GAT GTG AAG CAC TTC TCT CCT GAG GAC CTC ACC GTG AAG GTA
         Phe Val Ile Phe Leu Asp Val Lys His Phe Ser Pro Glu Asp Leu Thr Val Lys Val
                120                    140          → pRLαA-3
         CTG GAA GAT TTC GTG GAG ATC CAT GGC AAA CAC AAC GAG AGG CAG GAT GAC CAT GGC
         Leu Glu Asp Phe Val Glu Ile His Gly Lys His Asn Glu Arg Gln Asp Asp His Gly
                180                    200                    220
         TAC ATT TCC CGT GAA TTT CAC CGT CGC TAC CGT CTG CCT TCC AAT GTG GAC CAG TCC
         Tyr Ile Ser Arg Glu Phe His Arg Arg Tyr Arg Leu Pro Ser Asn Val Asp Gln Ser
                240                    260                    280
         GCC CTC TCC TGC TCC TTG TCT GCG GAT GGC ATG CTG ACC TTC TCT GGC CCC AAG GTC
         Ala Leu Ser Cys Ser Leu Ser Ala Asp Gly Met Leu Thr Phe Ser Gly Pro Lys Val
                300                    320                    340
         CAG TCT GGC TTG GAT GCT GGC CAC AGC GAG AGG GCC ATT CCC GTG TCA CGG GAG GAG
         Gln Ser Gly Leu Asp Ala Gly His Ser Glu Arg Ala Ile Pro Val Ser Arg Glu Glu
                        360                    380                    400
         AAG CCC AGC TCG GCA CCC TCG TCC TGA GC AGGCCTCGCC TTGGTTGTCC CCTGATGCCC CTG
         Lys Pro Ser Ser Ala Pro Ser Ser
                        420                    440                    460
         ATCCATC TGCCCAGGGG CCACAGCAAA GAGTCTGCCT TCCTGACTTC TTTTCTTTCT CTTTGTTTCC T
                        480                    500                    520                    540
         TTCCACTTT CTCAGAGGGC TGAGGATTTG AGAGAGTGGC TTAAAGAGCT TGGGGGGTCT TGGCCTGAGA
                        560                    580                    600
         TGGCTGCGGG TTCAGGGTGA CCCAGGCTCA ACACCAGCCG GTCAGAGGGA ATGATGGCAT TGAACTCTT
                        620                    640                    660
         A AGATTTCCTG TCCTCCTGGA AAGTGGCATC GAGCTCTGCC AAAGGCAGAG TGAATGGTGG CTAACCA
                680                    700                    720                    740
         ACC CCAAGAGCCC TCTGCCAAGC CCCTGGATGG CAGCCTCCCA CCCCCTTTGC CCACACTTAC CGCAG
         pRLαA-1 ←  760                    780                    800
         GCGTA TATGCTGGGC TCCAACAGTC CGCTTCTCTC ATGCCCTCTT CCTGTGACTT TCTCTACTAT GTA
                820                    840                    860                    880
         GTATCGC TCCTGGGGAC CCTGATCACC CATGAGAATG GGCCCCTGGC AGAC AATAAA GAGCAGGTGA

         CAAGCAAAAA AA
```

Figure 3. Nucleotide seqeunce of αA$_2$-crystallin cDNA, deduced from two "overlapping clones" (1-759 and 146-892, respectively). Note the extremely long non-coding region at the 3' end.

deduced to direct the synthesis of a polypeptide of the required length. The other two possible reading frames comprise several termination codons. The coding sequences in the latter two frames would only give rise to very short polypeptides. The excess region at the 3' terminal end comprises about 550 nucleotides which might theoretically encode a protein of about 180 amino acid residues. However,

```
  1                              10
Ac-Met-Asp-Val-Thr-Ile-Gln-His-Pro-Trp-Phe-Lys-Arg-Ala-Leu-Gly-Pro-Phe-Tyr-Pro-
  20                             30
-Ser-Arg-Leu-Phe-Asp-Gln-Phe-Phe-Gly-Glu-Gly-Leu-Phe-Glu-Tyr-Asp-Leu-Leu-Pro-Phe-
  40                             50
-Leu-Ser-Ser-Thr-Ile-Ser-Pro-Tyr-Tyr-Arg-Gln-Ser-Leu-Phe-Arg-Thr-Val-Leu-Asp-Ser-
  60       63
-Gly-Ile-Ser-Glu         Leu-Met-Thr-His-Met-Phe-Val-His-Met-Asn-Gln-Pro-Pro-Gly-
                                                         64              70
-Ala-His-Lys-Asn-Asn-Pro-Gly-Lys       Val-Arg-Ser-Asp-Arg-Asp-Lys-Phe-Val-Ile-
                    80                             90
-Phe-Leu-Asp-Val-Lys-His-Phe-Ser-Pro-Glu-Asp-Leu-Thr-Val-Lys-Val-Leu-Glu-Asp-Phe-
                    100                            110
-Val-Glu-Ile-His-Gly-Lys-His-Asn-Glu-Arg-Gln-Asp-Asp-His-Gly-Tyr-Ile-Ser-Arg-Glu-
                    120                            130
-Phe-His-Arg-Arg-Tyr-Arg-Leu-Pro-Ser-Asn-Val-Asp-Gln-Ser-Ala-Leu-Ser-Cys-Ser-Leu-
                    140                            150
-Ser-Ala-Asp-Gly-Met-Leu-Thr-Phe-Ser-Gly-Pro-Lys-Val-Glu-Ser-Gly-Leu-Asp-Ala-Gly-
                    160                            170
-His-Ser-Glu-Arg-Ala-Ile-Pro-Val-Ser-Arg-Glu-Glu-Lys-Pro-Ser-Ser-Ala-Pro-Ser-Ser-OH
```

Figure 4. The insertion found in an additional rat αA_2-like polypeptide named αA^{Ins}. (Amino acid sequence between residues 63-64 of the normal αA_2 polypeptide chain.)

in each of the three possible reading frames the ATG triplets present are followed by several stop codons at distances of at most 40 triplets. This means that the 3' region of the αA_2 messenger in spite of its having more than 50% of the total nucleotide content has no coding capacity.

THE SELECTED CLONES ARE NOT OF THE αA^{Ins} TYPE

Cohen et al. (1978a) observed the existence of an additional gene product closely related but not identical to αA-crystallin named αA^{Ins} in the lenses of rodents. This extra chain, which occurs as a rather low percentage of α-crystallin, is characterized by the insertion of 22 amino acid residues (Fig. 4) between residue 63 and 64 of αA_2. The occurrence of three methionine residues in this polypeptide fragment is an unusual feature.

Analysis of our clones revealed that they all were of the "normal" αA_2 type (see Fig. 5) but not of the αA^{Ins} type. Cohen et al. (1978b) were able to show that the synthesis of both αA^{Ins} and αA_2 was directed by 14 S messengers. Prolonged electrophoresis of total rat mRNA followed by blotting and detection with the aid of our nick-translated αA_2 clone indicated that in fact the 14 S population consists of two discrete messenger species.

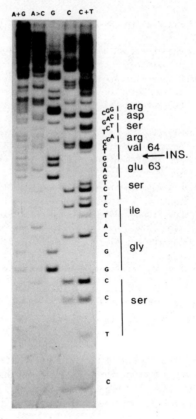

Figure 5. Autoradiograph of a sequencing gel showing that the cDNA clones described lack the theoretically necessary nucleotide track of 66 residues required for coding of αA^{Ins}.

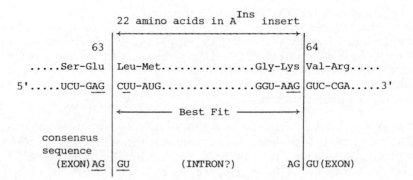

Figure 6. Is αA$_2$ polymorphism due to splicing distortion? Tentative interpretation of the possible origin of αA^{Ins} by splicing distortion.

SPECULATIVE INTERPRETATION OF THE ORIGIN OF THE αA_2, AND αA^{Ins} mRNAs

It might well be that the rodent αA crystallin products originate from two different genes. However, taking into account the known amino acid sequence of the insertion and data from the literature which revealed the nature of splice points in the pre-messenger (Lerner et al., 1980), one may think in terms of a distorted splicing phenomenon. This tentative explanation is illustrated in Fig. 6.

Forthcoming analysis of the α-crystallin gene organization will eventually provide us with the definite answer.

REFERENCES

Asselbergs, F. A. M., Peters, W. H. M., van Venrooij, W. J., and Bloemendal, H., 1978, Inhibition of translation of lens mRNAs in a messenger-dependent reticulocyte lysate by cap analogues, Biochim. biophys. Acta, 520:577.

Berns, A. J., M., Strous, G. J. A. M., and Bloemendal, H., 1972, Heterologous in vitro synthesis of lens α-crystallin polypeptide, Nature New Biol., 236:7.

Bhat, S. P., Jones, R. E., Sullivan, M. A., and Piatigorsky, J., 1980, Chicken lens crystallin DNA sequences show at least two α-crystallin genes, Nature, Lond., 284:234.

Bloemendal, H., 1977, The vertebrate eye lens, Science, 197:127.

Bloemendal, H., 1981, The lens proteins, in:"Molecular and Cellular Biology of the Eye Lens," H. BLoemendal, ed., John, Wiley & Sons, Inc., New York, p.1.

Bower, D. J., Errington, L. H., Wainright, N. R., Sime, C., Morris, S., and Clayton, R. M., 1981, Cytoplasmic RNA sequences complementary to cloned chick δ-crystallin cDNA show size heterogeneity, Biochem. J., 201:339.

Chen, J. H., and Spector, A., 1977, The bicistronic nature of lens α-crystallin 14 S mRNA, Proc. Natl. Acad. Sci. U.S.A., 74:5448.

Clayton, R. M., 1974, Comparative aspects of lens proteins, in:"The Eye," H. Davson and L. T. Graham, eds., Academic Press, New York, p.399.

Cohen, L. H., Westerhuis, L. W., de Jong, W. W., and Bloemendal, H., 1978a, Rat α-crystallin A chain with an insertion of 22 residues, Eur. J. Biochem., 89:259.

Cohen, L. H., Westerhuis, L. W., Smits, D. P., and Bloemendal, H., 1978b, Two structurally closely related polypeptides encoded by 14 S mRNA isolated from rat lens, Eur. J. Biochem., 89:251.

Dodemont, H. J., Andreoli, P. M., Moormann, R. J. M., Raemakers, F. C. S., Schoenmakers, J. G. G., and Bloemendal, H., 1981, Molecular cloning of mRNA sequences encoding rat lens crystallins, Proc. Natl. Acad. Sci. U.S.A., 78:5320.

De Jong, W. W., van der Ouderaa, F. J., Versteeg, M., Groenewoud, G., van Amelsvoort, J. M., and Bloemendal, H., 1975, Primary

structures of the α-crystallin A chains of seven mammalian species, Eur. J. Biochem., 53:237.
Lerner, M. R., Boyle, J. A., Mount, S. M., Wollin, S. L., and Steitz, J. A., 1980, Are snRNPs involved in splicing?, Nature, Lond., 283:220.
Moormann, R. J. M., van der Velden, H. M. W., Dodemont, H. J., Andreoli, P. M., Bloemendal, H., and Schoenmakers, J. G. G., 1981, An unusually long non-coding region in rat lens α-crystallin messenger RNA, Nucleic Acids Res., 9:4813.
Van der Ouderaa, F. J., de Jong, E. W., and Bloemendal, H., 1973, The amino acid sequence of the αA_2 chain of bovine α-crystallin, Eur. J. Biochem., 39:207.
Papaconstantinou, J., 1967, Molecular aspects of lens cell differentiation, Science, 156:335.

COMMITMENT: A MULTI-STEP PROCESS DURING

INDUCED MELC DIFFERENTIATION

    Richard A. Rifkind, Zi-xing Chen, Judy Banks and
Paul A. Marks

    Memorial Sloan-Kettering Cancer Center
New York
NY 10021

Murine erythroleukemia cells (MELC) are virus transformed erythroid precursors which maintain their capacity, under in vitro conditions, for self-renewal (Marks and Rifkind, 1978; Friend et al., 1971; Liao and Axelrod, 1975; Hawkins et al., 1978; Wendling et al., 1981). Till and co-workers (1964) proposed that in differentiating systems the "decision" for loss of self-renewal (commitment to terminal cell divisions) and expression of differentiated characteristics, involves a random-event component. The factors which determine the probability for commitment to terminal erythroid differentiation (or self-renewal) are unknown. In MELC it is clear that inducing agents can markedly increase the probability of the event(s) leading to terminal cell division and differentiation (Marks and Rifkind, 1978; Fibach et al., 1977; Gusella et al., 1976).

In MELC unsynchronized with respect to the cell cycle and cultured with various inducers, commitment to terminal cell division(s) and expression of erythroid characteristics are detectable by as early as 12 to 16 hrs (Nudel et al., 1977; Marks et al., 1979). Commitment is defined as the capacity of the cells to differentiate and enter terminal cell division in the absence of inducer, after a period of exposure to inducer. Commitment is assayed at the single cell level, by culture of MELC in suspension with an inducer, then transfer of the cells at intervals to inducer-free, semi-solid cloning medium; the colonies, derived from single cells, are scored 5 days later for proliferation (colony size) and differentiation (benzidine reactivity indicating hemoglobin content). In this assay, commitment is dependent on the nature of the inducer, as well as on the duration and concentration of inducer. By this assay, commitment of MELC cultured with 5 mM hexamethylene bisacetamide (HMBA) can be detected as early as 16 hrs and is essentially complete by 48 hrs. Fully differentiated colonies are small, containing fewer than 64 cells, suggesting that a committed cell is limited in proliferative capacity to fewer than 5 to 6 cell divisions. MELC cultured with butyrate (BA) accumulate committed cells at the same rate as observed for HMBA, but fewer cells are committed (over 95% of cells are committed with HMBA, while only about 70% with BA). MELC cultured with dimethylsulfoxide (Me_2SO) show initial commitment by 16 hrs and achieve over 80% commitment by about 48 hrs after onset of culture. With HMBA, BA or Me_2SO, the first accumulation of α-globin is detected at about the same time as commitment to terminal cell division; the accumulation of β-globin lags behind that of α-globin by 12 to 18 hrs.

Hemin has a quite different effect on expression of characteristics of differentiation from the three inducers described above. Hemin causes an increase in newly synthesized α- and β-globin mRNA, as early as 6 hrs after onset of culture, but commitment to terminal cell division either does not occur or occurs very late (after 72 hrs or more) and even then only in a relatively small proportion of the population (Nudel et al., 1977; Marks et al., 1979; Gusella et al., 1980). These findings form part of the evidence suggesting that events implicated in commitment to terminal cell division may be distinct from events leading to expression of erythroid-specific genes such as the globin structural genes. During induction of differentiation by inducers such as HMBA, there appears to be coordinate expression of the program of erythroid differentiation, including commitment to terminal cell division, expression of globin genes, increased activity of heme synthetic enzymes, increased accumulation of spectrin, globin and hemoglobins (Marks and Rifkind, 1978); but coordinate expression is clearly not essential for induction of some components of the program of erythroid differentiation.

Tumor promoters, such as 12-O-tetradecanoyl-phorbol-13-acetate (TPA), are potent inhibitors of inducer-mediated differentiation of MELC and other cells (Fibach et al., 1979; Diamond et al., 1977; Cohe et al., 1977). TPA suppresses the onset of terminal cell division, the accumulation of newly synthesized globin mRNA, and hemoglobin formation. By the transfer of MELC from medium containing HMBA and RPA to medium without these agents, it has been shown that MELC retain, for a period, a "memory" of prior exposure to HMBA (Fibach et al., 1979). This "memory" is expressed upon removal of the phorbol ester.

The steroid, dexamethasone, also suppresses or inhibits the effects of the inducer, HMBA. Previous studies from several laboratories have demonstrated that dexamethasone is a potent inhibitor of inducer-mediated terminal erythroid differentiation in MELC, includin suppression of terminal cell division and of accumulation of globin mRNA, globin and hemoglobins (Tsiftsoglou et al., 1979; Lo et al., 1978; Scher et al., 1978; Santoro et al., 1978; Chen et al., 1982).

Table 1. Effect of dexamethasone on accumulation of newly synthesized globin mRNA in nuclear and cytoplasmic fractions prepared from MELC cultured with HMBA or HMBA plus dexamethasone.

Addition	^{3}H-globin mRNA*			
	Cytoplasm		Nucleus	
	$\dfrac{cpm}{10^6 \text{ cells}}$	% total ^{3}H-RNA	$\dfrac{cpm}{10^6 \text{ cells}}$	% total ^{3}H-RNA
HMBA, 4 mM	301	0.4	94	0.02
HMBA, 4 mM plus dexamethasone, 10^{-6} M	54	0.06	14	0.002

*Determined after 65½ hrs in culture with HMBA or HMBA plus dexamethasone. For the last 30 mins of culture, ^{3}H-uridine was added. Data are from Chen et al. (1982). In each experiment, cells cultured with HMBA accumulated over 90% benzidine reactive cells, while those cultured with HMBA plus dexamethasone accumulated less than 5% benzidine reactive cells, as assayed 4 d after initiation of cultures.

Dexamethasone suppresses the accumulation of benzidine reactive cells when MELC are cultured with Me$_2$SO, HMBA or actinomycin D (another powerful differentiation-inducer in these cells; Terada et al., 1978). The suppression of accumulation of benzidine reactive cells is associated with a marked decrease in accumulation of newly synthesized cytoplasmic globin mRNA as well as nuclear globin mRNA sequences (Table 1). These studies were interpreted as suggesting that dexamethasone acts to suppress transcription or processing of globin mRNA, in agreement with the studies of Mierendorf and Mueller (1981).

Further to define the parameters of "memory" for inducer-mediated commitment, MELC were cultured with HMBA and, at intervals up to 72 hrs, aliquots were transferred to inducer-free, semi-solid medium without or with 4×10^{-6} M dexamethasone. By 30 hrs of exposure to HMBA over 50% of MELC are committed; however, if dexamethsone is included in the cloning medium, little or none of that commitment is apparent. By 40 hrs in culture with HMBA 80% of the cells are committed, but if transferred into dexamethasone, only about 30% of the colonies show evidence of that commitment. By 50 hrs in HMBA approximately 95% of cells are committed, and these display their commitment whether cloned with or without the inhibitor. These studies suggest that changes induced by MBA which lead to commitment are initiated within 20 hrs, but expression of commitment requires up to

Table 2. Effect of hemin on cultures in which dexamethasone has suppressed induction.[+]

Addition	Cell density* ($\times 10^6$/ml)	Commitment** (%)	Hb* (μg/10^6 cells)
None	1.82	1	0.71
HMBA (5 mM)	2.12	94	6.70
HMBA + dexa (4×10^{-6} M)	2.04	21	1.14
HMBA + dexa + hemin (0.1 mM)	1.78	12	5.40
Hemin	2.00	3	4.71
HMBA + hemin	1.99	89	10.54
Hemin + dexa	1.97	6	4.02

*Assayed at 120 hrs after initiation of culture. Hb, hemoglobin content.
**Assayed at 72 hrs after initiation of culture.
[+]From Chen et al. (1982).

40-50 hrs before it is resistant to suppression by dexamethasone. These findings further suggest that commitment involves a multi-step process, and expression of some early step(s) can be suppressed by dexamethasone.

Additional evidence has been obtained consistent with this interpretation. MELC were cultured in suspension with both HMBA and dexamethasone; at intervals up to 72 hrs, aliquots were transferred to inducer-free semi-solid cloning medium without or with dexamethasone. In the first case, cells transferred after 30, 40, 50 and 72 hrs to medium without dexamethasone showed progressively increasing proportions of committed colonies. In the second case, cells transferred into medium with dexamethasone produced few or no committed colonies. These results suggest, as for TPA, that a memory for past exposure to HMBA can be established even in the presence of agents which suppress expression of the committed state. It seems reasonable that commitment involves several steps, and that some later or final steps are not suppressible by dexamethasone. It is also suggested that an early inducer-mediated step in commitment involves changes with an operationally finite half-life. The molecular mechanism of such a finite memory might involve a molecule of limited stability or a marginal concentration, diluted with each cell division.

The relationship between inducer-mediated commitment and expression of globin structural genes was examined in studies on the effect of hemin during dexamethasone-mediated suppression of HMBA-induced differentiation. Hemin, as noted, rapidly initiates α- and β-globin mRNA accumulation but not commitment to terminal cell division. MELC were cultured with HMBA alone, HMBA plus dexamethasone, and HMBA plus dexamethasone plus hemin (Table 2). Under these conditions, hemin permits hemoglobin formation, but does not reverse the suppression of commitment to terminal cell division. Pulse-chase experiments have shown that the rate of accumulation of newly synthesized α-globin mRNA is the same in cultures with HMBA alone and in cultures containing HMBA, dexamethasone and hemin; under both these conditions, the mRNA has similar stability.

In summary, inducer-mediated commitment to terminal erythroid differentiation appears to involve several steps. Some are responsible for the acquired "memory" for exposure to inducer; this "memory" decays with a distinct half-life and appears to be insensitive to dexamethasone and probably to RPA as well. A more distal step, implicated in expression of the committed state, appears to be the principal target of these inhibitory agents. The observed requirement for DNA synthesis for inducer-mediated commitment of MELC (Rifkind et al., this volume) may reflect the need to restructure chromatin during this process.

ACKNOWLEDGEMENTS

These studies were supported, in part, by Grants CA-13696, CA-31768, and Core Grant CA-08748-16 from the National Cancer Institute, and Grant CH-68A from the American Cancer Society.

REFERENCES

Chen, Z. X., Banks, J., Rifkind, R. A., and Marks, P.A., 1982, Inducer-mediated commitment of murine erythroleukemia cells to differentiation: A multi-step process, Proc. Natl. Acad. Sci. U.S.A., in press.

Cohen, R., Pacifici, M., Rubenstein, N., Biehl, J., and Holtzer, H., 1977, Effect of a tumor promoter on myogenesis, Nature, Lond., 266:538.

Diamond, L., O'Brien, T. G., and Rovera, G., 1977, Inhibition of adipose conversion of 3T3 fibroblasts by tumor promoters, Nature, Lond., 269:247.

Fibach, E., Reuben, R. C., Rifkind, R. A., and Marks, P. A., 1977, Effect of hexamethylene bisacetamide on the commitment to differentiation of murine erythroleukemia cells, Cancer Res.,

37:440.
Friend, C., Scher, W., Holland, J.G., and Sato, T., Hemoglobin synthesis in murine virus-induced leukemic cells in vitro: Stimulation of erythroid differentiation by dimethylsulfoxide, Proc. Natl. Acad. Sci. U.S.A., 68:378.
Gusella, J., Geller, R., Clarke, B., Weeks, V., and Housman, D., 1976, Commitment to erythroid differentiation by Friend erythroleukemia cells: a stochastic analysis, Cell, 9:221.
Gusella, J. F., Weil, S. C., Tsiftsoglou, A. S., Volloch, V., Neumann, J. R., Keys, C., and Housman, D., 1980, Hemin does not cause commitment of murine erythroleukemia (MEL) cells to terminal differentiation, Blood, 56:481.
Hawkins, W. D., Kost, T. A., Koury, M. J., and Krantz, S. B., 1978, Erythroid bursts produced by Friend leukemia virus in vitro, Nature, Lond., 276:506.
Liao, S. K., and Axelrad, A. A., 1975, Erythropoietin-independent erythroid colony formation in vitro by hemopoietic cells of mice infected with Friend virus, Int. J. Cancer, 15:467.
Lo, S. C., Aft, R., Ross, J., and Mueller, G. C., 1978, Control of globin gene expression by steroid hormones in differentiating Friend leukemia cells, Cell, 15:447.
Marks, P. A., Rifkind, R. A., Bank, A., Terada, M., Gambari, R., Fibach, E., Maniatis, G., and Reuben, R., 1979, Expression of globin genes during induced erythroleukemic cell differentiation, in: "Cellular and Molecular Regulation of Hemoglobin Switching," G. Stamatoyannopoulos and A. W. Nienhuis, eds., Grune and Stratton, New York, p.437.
Migrendorf, R. C., and Mueller, G. C., 1981, Role of dexamethasone in globin gene expression in differentiating Friend cells, J. Biol. Chem., 256:6736.
Nudel, U., Salmon, J., Fibach, E., Terada, M., Rifkind, R. A., Marks, P. A., and Bank, A., 1977, Accumulation of alpha and beta globin messenger RNAs in mouse erythroleukemic cells, Cell, 12:463.
Santoro, M., Benedetto, A., and Jaffe, B. M., 1978, Hydrocortisone inhibits DMSO-induced differentiation of Friend erythroleukemia cells, Biochem. Biophys. Res. Commun., 85:1510.
Scher, W., Tsuei, D., Sassa, S., Price, P., Gabelman, N., and Friend, C., 1978, Inhibition of dimethyl sulfoxide-stimulated Friend cell erythrodifferentiation by hydrocortisone and other steroids, Proc. Natl. Acad. Sci. U.S.A., 75:3851.
Terada, M., Epner, E., Nudel, U., Salmon, J., Fibach, E., Rifkind, R. A., and Marks, P. A., 1978, Induction of murine erythroleukemia differentiation by Actinomycin D, Proc. Natl. Acad. Sci. U.S.A., 75:2795.
Till, J. E., McCulloch, E. A., and Siminovitch, L., 1964, A stochastic model of stem cell proliferation, based on the growth of spleen colony-forming cells, Proc. Natl. Acad. Sci. U.S.A., 51:29.
Tsiftsoglou, A. S., Gusella, J. F, Volloch, V., and Housman, D., 1979, Inhibition by dexamethasone of commitment to erythroid

differentiation in murine erythroleukemia cells, Cancer Res., 39:3849.

Wendling, F., Moreau-Gachelin, F., and Tambourin, P., 1981, Emergence of tumorigenic cells during the course of Friend virus leukemia, Proc. Natl. Acad. Sci. U.S.A., 78:3614.

STUDIES OF THE RELATIONSHIP BETWEEN DNA METHYLATION

AND TRANSCRIPTION OF THE RIBOSOMAL RNA GENES

Adrian P. Bird, Donald Macleod and Mary H. Taggart

MRC Mammalian Genome Unit
West Mains Road
Edinburgh   EH9 3JT

INTRODUCTION

Like many cellular macromolecules, DNA is often subject to post-synthetic modification. In animals this modification involves methylation of cytosine to give 5-methyl-cytosine (5mC), most commonly in the dinucleotide CpG (see Razin and Riggs, 1980 for references). From the point of view of developmental biology, interest in DNA methylation springs from two properties. First, it is known that in bacteria the presence or absence of 5mC can determine whether a protein (i.e. a restriction endonuclease) interacts productively or unproductively with its recognition sequence on DNA, and this raises the possibility that eukaryotic methylation is also involved in switching DNA-protein interactions (Riggs, 1975; Holliday and Pugh, 1975). Secondly, there is clear evidence that the pattern of methylation in the DNA of a cell can be inherited by its daughter cells (reviewed by Wigler, 1981). The idea of a heritable switch that is covalently attached to DNA, but which is potentially reversible has obvious attractions.

Is there a discernable link between DNA methylation and the pattern of transcription in a cell? Although definitive evidence is lacking, there is indeed a widespread inverse correlation between the level of methylation in or near a gene, and its expression (reviewed by Razin and Riggs, 1980). Our recent research has concentrated on the ribosomal RNA genes. These genes are of particular interest because they have been well studied (e.g. Moss et al., 1980; Sollner-Webb and Reeder, 1979) and because in different animals they can be found heavily methylated, unmethylated or both (Bird and Taggart, 1980; Bird et al., 1981a). The following account shows that, in spite of this wide variation, there are correlations between local-

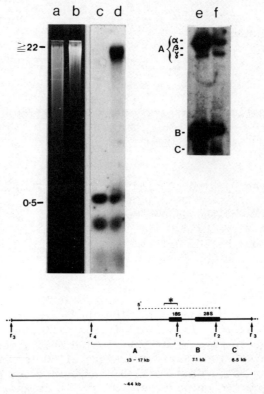

Figure 1. Methylated and unmethylated rDNA in mice. DNA from Balb/c liver was digested with MspI (lanes a and c), or HpaII (lanes b and d) and fractionated on a 2% agarose gel. Lanes a and b show the gel stained with ethidium bromide to localise total DNA. The comparative resistance to HpaII is due to CpG methylation. Lanes c and d are a and b after blotting and hybridisation to P^{32}-labelled rDNA (plasmid pX1r101; a gift from R. H. Reeder, Hutchinson Cancer Center, Seattle, Washington). Note HpaII-resistant hybridisation at the top of lane d. Lanes e and f demonstrate that this methylated fraction is rDNA. Balb/c DNA was digested with EcoRI to give bands A, B and C as indicated in the map (bottom; Cory and Adams, 1977). Band A includes some non-transcribed spacer and is heterogeneous in length, giving 3 bands in Balb/c (α, β and γ). Lane f shows that double digestion with SmaI and EcoRI leaves a proportion of bands A, B and C at the full sized position. Unmethylated rDNA is digested to low molecular weight fragments as in lane d (not shown). Note that the EcoRI-A band is not degraded by SMaI (i.e. is mostly methylated), while α and β are mostly unmethylated.

ised or generalised hypomethylation and rDNA expression.

VARIABLE METHYLATION OF rDNA AMONGST ANIMALS

The standard method used for detecting CpG methylation is to compare digestion of genomic sequences with the isoschizomers HpaII and MspI. MspI cleaves at CCGG regardless of CpG methylation (Waalwijk and Flavell, 1978), whereas HpaII will only cleave at CCGG if CpG is unmethylated (Bird and Southern, 1978; Mann and Smith, 1978). When genomic DNA from a variety of animals is digested with each enzyme and hybridised to an rDNA probe (Southern, 1975), the results are of two kinds (Bird and Taggart, 1980). Either the bulk of rDNA is heavily methylated at these and other sites (e.g. fish and amphibia), or it is not detectably methylated (invertebrates, reptiles, birds and mammals). The finding of predominantly unmethylated rDNA in reptiles, birds and mammals is particularly striking, because genomic DNA as a whole is highly methylated in these organisms.

METHYLATED AND UNMETHYLATED rDNA IN THE MOUSE

Although the majority of mammalian rDNA showed no signs of methylation, a few species showed additional hybridisation to a high molecular weight fraction that was resistant to HpaII, but not MspI (Bird and Taggart, 1980). This result was followed-up in the mouse, by showing that these HpaII-resistant sequences are cleaved with EcoRI to give fragments characteristic of mouse rDNA (Fig. 1). This and other experiments led to the conclusion that mouse cells contain two fractions of rDNA: a major fraction that is unmethylated, and a smaller fraction (between 5% and 30% in different inbred lines) that is heavily methylated (Bird et al., 1981a). There was no evidence for sparsely methylated rDNA intermediate between these extremes, and this agrees with a previous suggestion that sequences become methylated de novo in blocks, rather than at individual CpG sites (Bird et al., 1979). In addition, the results showed that methylated and unmethylated rDNA fractions were present in unchanging proportions in a range of somatic tissues, and in sperm. Thus there was no evidence for a developmental change in rDNA methylation.

We approached the question of rDNA transcription in mouse cells by comparing the DNase I sensitivity of methylated and unmethylated genes (Bird et al., 1981a). The rationale for this kind of experiment comes from the earlier demonstration by Weintraub and Groudine (1975) that the DNA of transcriptionally active chromatin is hypersensitive to DNase I. We tested liver nuclei from Balb/c mice, because this strain has rDNA spacers of three distinguishable lengths, and only one of these is heavily methylated (Fig. 1). Digestion with DNase I depleted the predominantly unmethylated spacer fragments, but the heavily methylated fragment was relatively resistant to digestion.

A DEVELOPMENTAL CHANGE IN rDNA METHYLATION IN XENOPUS

The apparently straightforward correlation between a stable lack of methylation and transcription that is observed for mouse rDNA cannot easily extend to the rDNA of amphibia and fish, because rDNA in these species is heavily methylated (Dawid et al., 1970). Detailed analysis of somatic rDNA methylation in Xenopus laevis (Bird and Southern, 1978; Bird et al., 1981b) has detected 250 out of an estimated 1000 CpGs in each repeat. Most CpGs are methylated to a level of 99% (i.e. there is a 1% probability that a CpG is in the unmethylated form), but this high level of methylation is interrupted at two regions: (1) a GCGC sequence in the 28 S gene that is 50% methylated at the CpG; (2) a pair of regions (UM-1 and UM-2) in the non-transcribed spacer that contain at least one unmethylated CpG in 99% of repeat units (Fig. 2). Examination of rDNA from different tissues showed that UM-1 and UM-2 are present in all the somatic tissues tested, but are absent in sperm. The transition from sperm pattern to somatic pattern takes place during the first day of embryonic development, and it is during this period that rRNA synthesis is initiated (Brown and Littna, 1964). Thus there is a correlation between the localised demethylation of otherwise heavily methylated rDNA, and the onset of rRNA synthesis. In the light of the correlations between hypomethylation and transcription mentioned above, it is tempting to suggest that the removal of spacer methylation is a necessary prerequisite for transcription. The same kind of contrast between heavy methylation in non-expressing tissues has been noted for several protein-coding genes (reviewed by Razin and Riggs, 1980).

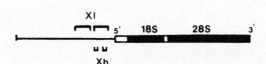

Figure 2. Hypomethylated sites in the rDNA of Xenopus laevis and Xenopus borealis. One repeat unit is shown, comprising a transcribed region, which codes for rRNA precursor (thick line), and a non-transcribed spacer of variable length (thin line). The nucleotide sequence of transcribed and non-transcribed spacers differs greatly between X. borealis and X. laevis. Brackets above the line (X1) show the hypomethylated regions seen in X. laevis somatic rDNA. Brackets below the line (Xb) show the hypomethylated regions seen in X. borealis. The transcribed portion of the repeat unit is about 8 kb in length.

UNCOUPLING OF TRANSCRIPTION AND HYPOMETHYLATION IN XENOPUS HYBRIDS

Interspecies hybrids between Xenopus laevis and Xenopus borealis have allowed us to study in more detail the putative relationship between DNA methylation and transcription (Macleod and Bird, manuscript in preparation). The hybrid system was chosen because the cells of hybrid tadpoles show only one nucleolus (Cassidy and Blackler, 1974), and biochemical studies have detected synthesis of only laevis-type rRNA precursor (Honjo and Reeder, 1973). Thus transcription of borealis rDNA is suppressed. We have verified this result by S1-protection experiments (Berk and Sharp, 1977) which show that 97-98% of hybrid tadpole rRNA is laevis-type. In addition, it was found that laevis rDNA is uniformly hypersensitive to DNase I compared with borealis rDNA. DNA methylation, however, showed no obvious correlation with suppression of the borealis genes. Both parent species have hypomethylated sites in corresponding regions of the spacer (Fig. 2), and these are also present in the hybrid. Thus, in this case localised hypomethylation and DNase I hypersensitivity are uncoupled. This argues against the idea that hypomethylation triggers the DNase I-sensitive chromatin structure. It does not, however, discount a less intimate relationship between DNA methylation and transcription. For example, it is possible that hypomethylation, or lack of methylation, is an obligatory precondition for gene activity, but it is not sufficient to ensure that transcription takes place. A "permissive" role of this kind is consistent with the data published so far, and has more in common with the developmental process of "determination", than with the final resolution of the determined state as seen at "differentiation".

ACKNOWLEDGEMENTS

This work was supported by the Medical Research Council (London).

REFERENCES

Berk, A. J., and Sharp, P. A., 1977, Sizing and mapping of early adenovirus mRNA by gel electrophoresis of S1-digested hybrids, Cell, 12:721.

Bird, A. P., and Southern, E. M., 1978, Use of restriction enzymes to study eukaryotic DNA methylation: I. The methylation pattern in ribosomal DNA from Xenopus laevis, J. molec. Biol., 118:27.

Bird, A. P., and Taggart, M. H., 1980, Variable patterns of total DNA and rRNA methylation in animals, Nucleic Acids Res., 8:145.

Bird, A. P., Taggart, M. H., and Smith, B. A., 1979, Methylated and unmethylated DNA compartments in the sea urchin genome, Cell, 17:889.

Bird, A. P., Taggart, M. H., and Macleod, D., 1981, Loss of rDNA methylation accompanies the onset of ribosomal gene activity in early development of X. laevis, Cell, 26 (in press).

Bird, A. P., Taggart, M. H., and Gehring, C., 1981b, Methylated and unmethylated ribosomal RNA genes in the mouse, J. molec. Biol. (in press).

Brown, D. D., and Littna, E., 1964, RNA synthesised during the development of Xenopus laevis, The South African clawed toad, J. molec. Biol., 8:669.

Cassidy, D. M., and Blackler, A. W., 1974, Repression of nucleolar organiser activity in an interspecific hybrid of the genus Xenopus, Develop. Biol., 41:84.

Cory, S., and Adams, J. M., 1977, A very large repeating unit of mouse DNA containing the 18 S, 28 S, 5.8 S rRNA genes, Cell., 11:795.

Dawid, I. B., Brown, D. D., and Reeder, R. H., 1970, Composition and structure of chromosomal and amplified ribosomal DNAs of Xenopus laevis, J. molec. Biol., 51:341.

Holliday, R., and Pugh, J. E., 1975, DNA modification mechanisms and gene activity during development, Science, 187:226.

Honjo, T., and Reeder, R. H., 1973, Preferential transcription of Xenopus laevis ribosomal RNA in interspecies hybrids between Xenopus laevis and Xenopus mulleri, J. molec. Biol., 80:217.

Mann, M. B., and Smith, H. O., 1978, Specificity of HpaII and HaeII DNA methylases, Nucleic Acids Res., 4:4211.

Moss, T., Boseley, P. G., and Birnstiel, M. L., 1980, More spacer sequences of X. laevis rDNA, Nucleic Acids Res., 8:467.

Razin, A., and Riggs, A. D., 1980, DNA methylation and gene function, Science, 210:604.

Riggs, A. D., 1975, X inactivation, differentiation and DNA methylation, Cytogenet. Cell Genet., 14:9.

Sollner-Webb, B., and Reeder, R. H., 1979, The nucleotide sequence of the initiation and termination sites for ribosomal RNA transcription in X. laevis, Cell, 18:485.

Southern, E. M., 1975, Detection of specific sequences among DNA fragments separated by gel electrophoresis, J. molec. Biol., 98:503.

Wallwijk, C., and Flavell, R. A., 1978, MspI, an isoschizomer of HpaII which cleaves both unmethylated and methylated HpaII sites, Nucleic Acids Res., 5:3231.

Weintraub, H., and Groudine, M., 1976, Chromosomal subunits in active genes have an altered conformation, Science, 193:848.

Wigler, M. H., The inheritance of methylation patterns in vertebrates, Cell, 24:285.

TISSUE SPECIFIC EXPRESSION OF MOUSE α-AMYLASE GENES

Ueli Schibler, Otto Hagenbüchle, Richard A. Young[1], Mario Tosi and Peter K. Wellauer

Swiss Institute for Experimental Cancer Research
Chemin des Boveresses
CH - 1066 Epalinges
Switzerland

SUMMARY

Two distinct α-amylase genes, Amy-1a and Amy-2a, are expressed in the mouse strain A/J. Amy-1a and Amy-2a are interrupted by 10 and 9 introns, respectively. With the exception of the first Amy-1a intron, which has no counterpart in Amy-2a, introns are located at analogous positions within the two genes. Comparable exons of Amy-1a and Amy-2a are more highly conserved in sequence than analogous introns.

Amy-2a specifies pancreatic α-amylase mRNA. Two apparently identical copies of this gene exist in the haploid mouse genome.

The single copy of Amy-1a is expressed in a tissue specific fashion in the salivary gland and the liver. It specifies α-amylase mRNA with identical translated and 3' nontranslated sequences but different 5' nontranslated sequences in the two tissues. These different mRNAs are generated by tissue specific splicing events. S_1 nuclease mapping of nuclear transcripts from salivary gland and liver suggests the presence of at least two promotors of different strength in Amy-1a. A strong promoter appears to be active in the salivary gland exclusively, while a weak promoter is apparently used in both the salivary gland and the liver. The data suggest that regulation of Amy-1a expression occurs primarily at the transcriptional level.

---

1. Present address: Dept. of Biochemistry, Stanford University School of Medicine, Stanford, California, 94305.

INTRODUCTION

A more profound understanding of cell differentiation requires the elucidation of the molecular basis for the regulation of tissue-specific gene expression. Although a considerable amount of information has been accumulated on the structure of eucaryotic genes during the past few years, the mechanisms which determine the activity of cell type specific genes are still poorly understood. The mouse α-amylase genes, Amy-1^a and Amy-2^a (Sick and Nielsen, 1964), offer a system of choice to investigate such problems. Amy-2^a specifies pancreatic α-amylase mRNA, while Amy-1^a is expressed in the salivary gland and in the liver. The α-amylase mRNAs accumulate to highly different amounts in these three tissues (Schibler et al., 1980). Using recombinant DNA technology we have studied the structure of the two α-amylase genes and of their transcripts (Schibler et al., 1980, 1981; Hagenbüchle et al., 1980, 1981; Young et al., 1981; Tosi et al., 1981).

RESULTS AND DISCUSSION

Structural comparison of Amy-1^a and Amy-2^a

Genomic recombinant DNA libraries were constructed by insertion of partial EcoRI restriction fragments of salivary gland and pancreas nuclear DNA into a λ vector. DNA fragments containing Amy-1^a and Amy-2^a sequences were obtained by screening pancreas and salivary gland DNA libraries with the cloned amylase cDNAs of recombinant plasmids pMPa21 and pMSa104, respectively (Schibler et al., 1980). Genomic α-amylase sequences were identified by sequence comparison of exons with mRNA sequences (Hagenbüchle et al., 1980,1981). Restriction mapping and elctron microscopic analysis of RNA/DNA hybrids allowed us to orient overlapping DNA fragments with respect to each other. In Figure 1 the maps of cleavage sites for several restriction enzymes are superimposed on the intron-exon maps of Amy-1^a and Amy-2^a. The precise location of cap- and polyadenylation sites was determined by sequence analysis. Amy-1^a and Amy-2^a contain about 22 and 11 kb of DNA, respectively. The large size difference between the two genes is mostly due to different numbers and lengths of introns. The first Amy-1^a intron, which separates the tissue-specific leader segments (S5', L5') from common salivary gland and liver exons (a-j), has no counterpart in Amy-2^a. The first Amy-2^a exon (a') is homologous to the second Amy-1^a exon (a). The 9 remaining introns are located at analogous positions within the two genes (Schibler et al., 1981). While exon sequences of the two genes exhibit 89% sequence homology, comparable introns are considerably less well conserved. This is demonstrated by different lengths and restriction maps of Amy-1^a and Amy-2^a introns (Fig. 1) and by comparative sequence analysis (Schibler et al., 1981).

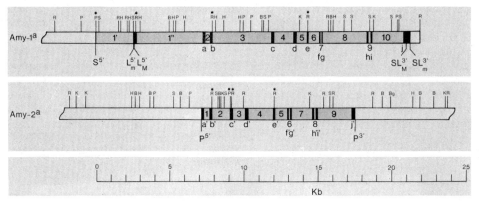

Figure 1. Architecture of Amy-1ᵃ and Amy-2ᵃ. Data from a variety of sources (see text and Schibler et al., 1980) are summarized in this map of two segments of the mouse genome. Filled portions designate exons for the salivary gland (S), liver (L) and pancreatic (P) alpha-amylase mRNAs. The CAP and polyadenylated sites for these mRNAs are indicated by the superscripts (5') and (3'), respectively. (M) and (m) subscripts represent sequences for the termini of the major and minor mRNA species, respectively. Restriction endonuclease cleavage sites are shown above each of the two genes; these include EcoRI (R), PstI (P), SacI (S), HindIII (H), BamHI (B) and KpnI (K). Lower case letters designate comparable exons in the two genes. Introns are numbered.

Quantitative Southern blot hybridization experiments reveal that Amy-1ᵃ is a single copy gene while Amy-2ᵃ exists as two identical copies in the haploid mouse genome. It is conceivable that two active Amy-2ᵃ genes are required for the high production of pancreatic α-amylase (Bloor et al., 1981). Molecular cloning experiments and genomic Southern blot experiments demonstrate the presence of several additional α-amylase related DNA sequences in the mouse genome. At the present time it is unclear whether these sequences represent functional genes or pseudogenes.

Tissue-specific transcription of Amy-1 in salivary gland and liver

The α-amylase mRNAs accumulate to highly different amounts in the salivary gland (10^4 molecules/cell) and the liver (10^2 molecules/cell) (Schibler et al., 1980). Sequence analysis of amylase mRNAs from these tissues and of cloned Amy-1ᵃ DNA segments indicates that different leaders which contain exclusively 5' nontranslated sequences are spliced in a tissue-specific fashion onto those sequences which are common to both mRNAs (Hagenbüchle et al., 1981; Young et al., 1981). While mRNAs with liver type leaders are present in similar concentrations in the liver and the salivary gland, no salivary gland type transcripts could be detected in the liver. None of these amylase mRNAs have been found in other tissues such as lung, heart-muscle, spleen, brain and kidney (Tosi et al., unpublished observations). The sequences specifying the tissue-specific leaders are preceded by TATA motifs, which are thought to be part of transcription initiation signals (Goldberg, 1979). The cap sites for salivary gland and liver α-amylase mRNAs are therefore likely to coincide with sites of transcription initiation. This notion is supported by S_1 nuclease mapping of nuclear amylase transcripts from salivary gland and liver. These experiments also reveal that the relative cellular concentration of nuclear transcripts is about 100-fold higher in the salivary gland than in the liver.

Our results are best explained by a model in which two promoters of different strength are used in a tissue-specific manner in Amy-1^a. According to this model, a weak promoter would be active in both the salivary gland and the liver. A strong promoter would be active in addition to the weak one in the salivary gland exclusively. It should be emphasized that our model has to be tested by kinetic analysis of transcription rates. The observation of Derman et al. (1981) that the relative abundance of several liver mRNAs corresponds well to the relative rate of synthesis of these RNAs, supports our model in which different promoter strengths may play an important role for the differential regulation of gene activity.

ACKNOWLEDGEMENTS

We thank A. C. Pittet and R. Bovey for excellent technical assistance and S. Cherpillod for typing of the manuscript. This work was supported by a grant from the Swiss National Science Foundation.

REFERENCES

Bloor, J. J., Meisler, M. H., and Nielsen, J. T., 1981, Genetic determination of amylase synthesis in the mouse, J. Biol. Chem., 256:373.
Derman, E., Kranter, K., Walling, L., Weinberger, C., Ray, M., and Darnell, J. E., 1981, Transcriptional control in the production of liver specific mRNAs, Cell, 23:731.
Goldberg, M., 1979, Ph.D. thesis, Stanford University, Stanford.
Hagenbüchle, O., Bovey, R., and Young, R. A., 1980, Tissue-specific expression of mouse α-amylase genes: nucleotide sequence of isoenzyme mRNAs from pancreas and salivary gland, Cell, 21:179.
Hagenbüchle, O., Tosi, M., Schibler, U., Bovey, R., Wellauer, P. K., and Young, R. A., 1981, Mouse liver and salivary gland α-amylase

mRNAs differ only in 5' nontranslated sequences, Nature, Lond., 289:643.
Schibler, U., Tosi, M., Pittet, A.-C., Fabiani, L., and Wellauer, P. K., 1980, Tissue-specific expression of mouse α-amylase genes, J. molec. Biol., 142:93.
Schibler, U., Pittet, A.-C., Young, R. A., Hagenbüchle, O., Tosi, M., Gellman, S., and Wellauer, P. K., 1981, The mouse α-amylase multigene family: Sequence organization of members expressed in the pancreas, salivary gland and liver, manuscript submitted for publication.
Sick, K., and Nielsen, J. T., 1964, Genetics of amylase isozymes in the mouse, Hereditas, 51:291.
Tosi, M., Young, R. A., Hagenbüchle, O., and Schibler, U., 1981, Multiple polyadenylation sites in a mouse α-amylase gene, Nucleic Acids Res., 9:2313.
Young, R. A., Hagenbüchle, O., and Schibler, U., 1981, A single mouse α-amylase gene specifies two different tissue-specific mRNAs, Cell, 23:451.

REGULATION OF THE RELATIVE ABUNDANCES OF mRNAs

IN HEPATOMA AND LIVER

G. D. Birnie, H. Jacobs, Rosemary Shott
and P. R. Wilkes

Beatson Institute for Cancer Research
Switchback Road
Glasgow G61 1BD

SUMMARY

Comparison of the polysomal poly(A)+ RNAs of normal rat liver and hepatoma (HTC) cells have shown that, while few sequences are specific to either hepatocytes or hepatoma cells, some of the abundant liver mRNAs (perhaps 20% by weight) are much rarer in the polysomal RNA from HTC cells. In contrast, these mRNA sequences are at quite similar abundances in nuclear poly(A)+ RNA from hepatocytes and hepatoma cells. The use of cloned cDNAs to measure the relative abundances of individual liver mRNA sequences in hepatocytes and HTC cells has shown that these changes occur in both directions, and indicated that post-transcriptional modulations, which are cell-type-specific and sequence-specific, play a significant role in establishing steady-state levels of polysomal poly(A)+ mRNAs.

To investigate post-transcriptional regulation of mRNA abundance, a cell-free system consisting of isolated HTC-cell nuclei has been developed. This system supports the in vitro processing and/or transport of rRNA and a complex mixture of poly(A)+ RNA sequences which resembles polysomal poly(A)+ RNA from HTC cells. The pattern of relative abundances of the sequences in released poly(A)+ RNA is intermediate between that of polysomal and nuclear poly(A)+ RNAs, and indicates that some degree of sequence-specific selection of processing and/or transport is maintained in isolated nuclei. Sequence-specific selection of individual poly(A)+ RNA sequences in vitro has been detected with cloned cDNAs.

INTRODUCTION

The phenotype of a mammalian cell is dependent not only on which mRNA sequences there are on its polysomes but also on the quantities of these sequences relative to each other (reviewed in Minty and Birnie, 1981). Three basic facts regarding polysomal poly(A)+ mRNAs have led to this conclusion. First, all cells contain a wide spectrum of mRNAs ranging in abundance from several thousand to less than 10 copies per cell although most genes coding for mRNAs are present once (or, at most, a few times) in the haploid genome (for example, Bishop et al., 1974; Birnie et al., 1974; Levy-W. and McCarthy, 1975; Williams and Penman, 1975; Young et al., 1976; Parker and Mainwaring 1977; Getz et al., 1977; Wilkes et al., 1979; Jacobs and Birnie, 1980). Second, cells at different stages in development, or in different lineages, share many mRNA sequences in common, but the abundances of these common sequences relative to each other frequently differ (by factors of 100 or more) between cells of different phenotype (Young et al., 1976; Hastie and Bishop, 1976). Dramatic changes in phenotype, such as those associated with differentiation (Parker and Mainwaring, 1977; Minty et al., 1978), growth induction (Williams and Penman, 1975; Getz et al., 1976; Wilkes et al., 1979), and transformation (Getz et al., 1977; Rolton et al., 1977; Williams et al., 1977) are more often marked by quantitative alterations in the relative abundances of mRNAs than by qualitative changes in the complement of mRNA sequences. Third, a direct correlation between the relative abundances of cellular proteins, or their rates of synthesis, and mRNA abundance has been demonstrated in a number of instances (Hastie and Held, 1978; Jacobs and Birnie, 1980).

The importance of the relative abundances of mRNAs in the establishment and maintenance of cellular phenotype poses two questions: what mechanisms generate the disparities in mRNA abundances in a cell, and, in which way do these mechanisms vary to cause changes in the relative abundances of the same mRNA sequences in different cells and thus be responsible (in large part at least) for changes in cellular phenotype? It has been argued that large disparities in mRNA abundances in a cell could arise from a combination of quite small differences in rates of transcription, rates of nuclear processing and nucleocytoplasmic transfer, and stabilities of mature mRNAs in the cytoplasm (Tobin, 1979). However, while there is evidence implicating each of these mechanisms in the generation of disparate mRNA abundances, their relative significance is still in dispute.

REGULATION BY TRANSCRIPTIONAL AND POST-TRANSCRIPTIONAL MECHANISMS

Evidence directly implicating transcriptional mechanisms in the control of mRNA abundance is now beginning to emerge. For example, the genes for moderately abundant mRNA sequences specific to mouse liver are transcribed at a much lower rate (if at all) in nuclei from mouse brain and hepatoma (Derman et al., 1981). Also, it has been shown that the 100-fold difference between the α-amylase mRNA concentrations in salivary gland and liver corresponds to a similar difference in the rate of transcription of the α-amylase gene in these two tissues (Schibler et al., 1981). Finally, the relative abundance of six out of nine randomly selected mRNA sequences in CHO cells are explicable in terms of the relative rates at which the genes are transcribed; however, that of the other three cannot be accounted for in this way, thus implicating post-transcriptional controls in the establishment of mRNA abundance (Harpold et al., 1979).

There is also good evidence from experiments with cDNAs transcribed from populations of mRNAs that, in a variety of cell types, post-transcriptional events (processing, transport, mRNA stability) are largely responsible for at least modulating if not fully determining, the relative abundances of a large proportion of mRNA sequences. This evidence is exemplified by comparisons of the polyadenylated RNA sequences in normal and 16-hour regenerating rat liver and rat hepatoma (HTC) cells (Wilkes et al., 1979; Jacobs and Birnie, 1980). Saturation hybridization of labelled unique DNA with polysomal poly(A)+ RNA showed that these RNAs are qualitatively very similar in all three tissues: if any sequences are unique to one cell type they must constitute less than 4% of the total number (about 25 000) represented on the polysomes of these cells. However, quantitative differences among the polysomal poly(A)+ RNA sequences common to all three tissues were clearly demonstrated by comparisons of the kinetics of homologous and heterologous poly(A)+ RNA-cDNA hybridization reactions. The difference between normal and regenerating liver is relatively small: some mRNAs which are very abundant in normal liver become less abundant in regenerating liver while some that are rare in normal liver become more abundant (Wilkes et al., 1979). These changes in relative abundances of mRNAs parallel the change in the balance between free and membrane-bound ribosomes which follows partial hepatectomy, and presumably reflect the shift in the balance of protein synthesis towards intracellular proteins which accompanies the earliest stages of regeneration (Braun et al., 1962).

The quantitative differences between the polysomal poly(A)+ RNAs of normal liver and HTC cells are more dramatic, particularly among those mRNA sequences which are most abundant in liver (Jacobs and Birnie, 1980). On average, the abundant HTC-cell mRNAs are about 5-fold less abundant on liver polysomes while most abundant liver mRNAs are about 100-fold rarer in HTC-cell polysomal RNA; some are even rarer. However, these differences are not reflected in the nuclear poly(A)+ RNA sequences from these cells: even the abundant liver polysomal mRNA sequences are at very similar concentrations in the nuclear poly(A)+ RNAs of the two tissues - at most, there is a 5-fold difference between them. These data, obtained from experiments using cDNA probes for populations of poly(A)+ RNA sequences, have been confirmed for some individual sequences from measurements of their relative concentrations using cloned cDNAs from a rat-liver polysomal cDNA plasmid library (Birnie et al., 1981). Thus a major cause of the depletion of some abundant liver polysomal mRNAs in HTC cells must be sequence-specific alterations in one or more post-transcriptional regulatory mechanism.

REGULATION BY NUCLEAR PROCESSING MECHANISMS

Whether post-transcriptional modulations in mRNA abundance are regulated by nuclear processing mechanisms, or changes in cytoplasmic mRNA stability, or both, cannot be assessed from data such as those summarized above. However, it would appear that differential stability of mRNAs alone is not sufficient to explain all post-transcriptional modulations of mRNA abundance. Although some correlation between the stabilities of cytoplasmic mRNAs and their abundances has been demonstrated (Lenk et al., 1978; Meyuhas and Perry, 1979), it has also been shown that the difference in concentration between the most and the least abundant classes of mRNAs in sea-urchin embryos (Galau et al., 1977) and HeLa cells (Lenk et al., 1978) is at least an order of magnitude greater than the differences in mRNA stability. Moreover, there is evidence in favour of a role for nuclear processing mechanisms in the determination of mRNA abundances: this comes from studies of globin mRNA synthesis in developing chick embryos (Chan, 1976) and of viral mRNA synthesis in adenovirus-infected cells (Shaw and Ziff, 1980), and from measurements of the relative abundances of mRNA sequences in polyadenylated and non-polyadenylated nuclear RNA in Friend cells (Balmain et al., 1980).

To determine the extent to which nuclear processing events are sequence-selective and, thus, contribute to the steady-state concentrations of polysomal mRNAs, we have developed an in vitro system which supports the processing and transport of RNA from isolated HTC-cell nuclei (Jacobs and Birnie, 1981). While the methods of isolating and incubating the nuclei were based on some described by others, particular attention was paid to measuring the extent of, and minimising, artefactual release of RNA in vitro. By far the largest proportion of the RNA released during a 75 min incubation of the nuclei consisted of mature ribosomal RNA, but about 0.5% was polyadenylated RNA which, by a number of commonly accepted criteria (size, labelling characteristics and organization into mRNP-like particles) was mRNA-like. No more than 10% of this poly(A)+ RNA could be accounted for by non-specific leakage of nuclear RNA, while a further 10-15% was attributable to release of cytoplasmic RNA which adhered to the nuclei when the cells were lysed and which resisted removal by washing in a neutral detergent. Thus 75-80% of the poly(A)+ RNA released from the nuclei appears to result from nuclear processing and transport mechanisms still operative in vitro (Jacobs and Birnie, 1981).

The kinetics of the hybridization reaction between the released

Table 1. Ratios of mole fractions of individual RNA sequences in
poly(A) RNAs from HTC-cell polysomal and nuclear RNAs,
RNA released in vitro from HTC-cell nuclei, and normal
liver polysomal RNA.

Ratios of mole fractions of poly(A)+ RNAs in:	pRR 117	pRR 133	pRR 83	pRR 5B
HTC polysomal/HTC nuclear	6.4	19.0	0.5	0.5
HTC released/HTC nuclear	5.0	2.0	0.7	0.7
HTC released/HTC polysomal	0.8	0.1	1.3	1.4
HTC polysomal/liver polysomal	4.0	1.4	0.01	0.01

Recombinants were isolated from a cDNA library prepared by blunt-end ligation of double-stranded cDNA, transcribed from rat-liver polysomal poly(A)+ RNA, into the Bam H1 site of plasmid pAT 153. Filter-bound plasmid DNAs were hybridized with ^{32}P-cDNAs transcribed from the poly(A)+ RNAs, and the proportions of input cDNA which hybridized were measured (Kafatos et al., 1979).

poly(A)+ RNA and the cDNA transcribed from it showed that the released RNA consited of a complex mixture of sequences at heterogeneous abundances resembling poly(A)+ polysomal RNA from HTC cells. Moreover, comparisons of the rates and extents of the reactions between polysomal poly(A)+ RNA and the cDNAs to poly(A)+ RNA from nuclear RNA, in-vitro-released RNA, and polysomal RNA indicated, first, that the bulk of the poly(A)+ RNA released in vitro was qualitatively very similar to HTC-cell polysomal RNA and, second, that the relative abundances of the sequences in released poly(A)+ RNA were intermediate between those of the same sequences in polysomal and nuclear poly(A)+ RNA. Thus, some nuclear poly(A)+ RNA sequences have been processed and/or transported more efficiently than others during incubation of these nuclei (Jacobs and Birnie, 1981).

This conclusion was confirmed by measurements of the relative abundances of four individual mRNA sequences in the poly(A)+ RNA from released RNA, polysomal RNA and nuclear RNA of HTC cells, which also showed that some degree of sequence-specific selection among RNA sequences destined for export occurs in HTC-cell nuclei in vitro (Table 1). The concentration of one sequence (corresponding to clone pRR 117) was also increased in the released RNA as compared to nuclear RNA, though to a smaller extent than in polysomal RNA. In contrast, the concentrations of two sequences (pRR 83 and pRR 5B) in polysomal and released RNA appeared to be decreased relative to their concentrations in nuclear RNA. Interestingly, these two sequences are at least 100-fold less abundant in HTC-cell polysomal RNA than in liver polysomal RNA.

CONCLUSIONS

While it seems that the relative steady-state abundances of some polysomal mRNA sequences are determined by the relative rates at which their gene sequences are transcribed, it also appears that those of some mRNAs are dependent, to a greater or lesser extent, on other mechanisms. The data from the experiments with isolated nuclei suggest that intranuclear post-transcriptional mechanisms are capable of modulating the relative abundances of mRNAs - a conclusion already reached from the studies of the steady-state populations of nuclear and polysomal poly(A)$^+$ RNAs (Jacobs and Birnie, 1980). These data are also consistent with the view that post-transcriptional modulation of mRNA abundance is not wholly the result of intranuclear processing and/or transport, but, to some extent, such modulation must also occur in the cytoplasm as a consequence of sequence-specific differences in the rates of turnover of mRNAs. However, it is not yet clear to what extent the final concentrations of mRNA sequences depend on regulation at a single stage in their metabolism, and to what extent multi-stage regulation (as discussed by Tobin, 1979) occurs. Once this is known it is possible that we may be able to classify genes according to whether their expression is regulated transcriptionally, or by intranuclear post-transcriptional mechanisms, or by mRNA stability, or by a combination of two (or more) of these processes. In this case it will be interesting to search for correlations between such classes of genes and the functions of the proteins they specify.

ACKNOWLEDGEMENTS

The Beatson Institute is supported by grants from C.R.C. and M.R.C. We are grateful to many colleagues for invaluable, critical discussion of the work from this laboratory.

REFERENCES

Balmain, A., Minty, A. J., and Birnie, G. D., 1980, Frequency distribution of pre-messenger RNA sequences in polyadenylated and non-polyadenylated nuclear RNA from Friend cells, Nucleic Acids Res., 8:1643.
Birnie, G. D., MacPhail, E., Young, B. D., Getz, M. J., and Paul, J., 1974, The diversity of the messenger RNA population in growing Friend cells, Cell Differentiation, 3:221.
Birnie, G. D., Balmain, A., Jacobs, H., Shott, R., Wilkes, P. R., and Paul, J., 1981, Post-transcriptional control of messenger

abundance, Molec. Biol. Rep., 7:159.

Bishop, J. O., Morton, J. G., Rosbash, M., and Richardson, M., 1974, Three abundance classes in HeLa cell messenger RNA, Nature, Lond., 250:199.

Braun, G. A., Marsh, J. B., and Drabkin, D. L., 1962, Synthesis of plasma albumin and tissue proteins in regenerating liver, Metabolism, 11:957.

Chan, L-N. L., 1976, Transport of globin mRNA from nucleus to cytoplasm in differentiating embryonic red blood cells, Nature, Lond., 261:157.

Derman, E., Krauter, K., Walling, L., Weinberger, C., Ray, M., and Darnell, J. E., 1981, Transcriptional control in the production of liver-specific mRNAs, Cell, 23:731.

Galau, G. A., Lipson, E. D., Britten, R. J., and Davidson, E. H., 1977, Synthesis and turnover of polysomal mRNAs in sea-urchin embryos, Cell, 10:415.

Getz, M. J., Elder, P. K., Benz, E. W., Stephens, R. E., and Moses, H. L., 1976, Effect of cell proliferation on levels and diversity of poly(A)-containing mRNA, Cell, 7:255.

Getz, M. J., Reiman, H. M., Siegal, G. P., Quinlan, T. J., Proper, J., Elder, P. K., and Moses, H. L., 1977, Gene expression in chemically transformed mouse embryo cells: selective enhancement of the expression of C type RNA tumor virus genes, Cell, 11:909.

Harpold, M. M., Evans, R. M., Salditt-Georgieff, M., and Darnell, J. E., 1979, Production of mRNA in Chinese hamster cells: relationship of the rate of synthesis to the cytoplasmic concentration of nine specific mRNA sequences, Cell, 17:1025.

Hastie, N. D., and Bishop, J. O., 1976, The expression of three abundance classes of messenger RNA in mouse tissues, Cell, 17:1025.

Hastie, N. D., and Held, W. A., 1978, Analysis of mRNA populations by cDNA-mRNA hybrid-mediated inhibition of cell-free protein synthesis, Proc. Natl. Acad. Sci. U.S.A., 75:1217.

Jacobs, H. J., and Birnie, G. D., 1980, Post-transcriptional regulation of messenger abundance in rat liver and hepatoma, Nucleic Acids Res., 8:3087.

Jacobs, H., and Birnie, G. D., 1981, Isolation and purification of rat-hepatoma nuclei active in the transport of messenger RNA in vitro, Eur. J. Biochem., (in press).

Kafatos, F. C., Jones, C. W., and Efstratiadis, A., 1979, Determination of nucleic acid sequence homologies and relative concentrations by a dot hybridization procedure, Nucleic Acids Res., 7:1541.

Lenk, R., Herman, R., and Penman, S., 1978, Messenger RNA abundance and lifetime: a correlation in Drosophila cells but not in HeLa, Nucleic Acids Res., 5:3057.

Levy-W., B., and McCarthy, B. J., 1975, Messenger RNA complexity in Drosophila melanogaster, Biochemistry, 14:2440.

Meyuhas, O., and Perry, R. P., 1979, Relationship between size, stability and abundance of the messenger RNA of mouse L-cells, Cell, 16:139.

Minty, A. J., Birnie, G. D., and Paul, J., 1978, Gene expression in Friend erythroleukaemia cells following the induction of haemoglobin synthesis, Expl. Cell Res., 115:1.

Minty, A. J., and Birnie, G. D., 1981, Messenger RNA populations in eukaryotic cells - evidence from recent nucleic acid hybridization experiments bearing on the extent and control of differential gene expression, in: "Biochemistry of Cellular Regulation, Vol III: Development and Differentiation," M. E. Buckingham, ed., CRC Press, Boca Raton, Fla.

Parker, M. G., and Mainwaring, W. I. P., 1977, Effect of androgens on the complexity of poly(A) RNA from rat prostate, Cell, 12:401.

Rolton, H. A., Birnie, G. D., and Paul, J., 1977, The diversity and specificity of nuclear and polysomal poly(A)+ RNA populations in normal and MSV-transformed cells, Cell Differentiation, 6:25.

Schibler, U., Hagenbüchle, O., Young, R. A., Tosi, M., and Wellauer, P. K., 1982, Tissue-specific expression of mouse α-amylase genes, in this book.

Shaw, A. R., and Ziff, E. B., 1980, Transcripts from the adenovirus-2 major late promoter yield a single early family of 3' coterminal mRNAs and five late families, Cell, 22:905.

Tobin, A. J., 1979, Evaluating the contribution of post-transcriptional processing to differential gene expression, Develop.Biol., 68:47.

Wilkes, P. R., Birnie, G. D., and Paul, J., 1979, Changes in nuclear and polysomal polyadenylated RNA sequences during rat-liver regeneration, Nucl. Acids Res., 6:2193.

Williams, J. G., and Penman, S., 1975, The messenger RNA sequences in growing and resting mouse fibroblasts, Cell, 6:197.

Williams, J. G., Hoffman, R., and Penmam, S., 1977, The extensive homology between mRNA sequences of normal and SV40-transformed human fibroblasts, Cell, 11:901.

Young, B. D., Birnie, G. D., and Paul, J., 1976, Complexity and specificity of polysomal poly(A)+ RNA in mouse tissues, Biochemistry, 15:2823.

SUMMARY OF DISCUSSION ON QUANTITATIVE

REGULATION OF GENE EXPRESSION

Bloemendal said that there is a considerable difference in the length of the mRNAs for two related proteins of very similar size, αA_2 and αB_2 crystallins, and this difference is conserved in evolution.

Schibler mentioned that a similar conservatism holds for the length of the non-coding sequences of muscle and non-muscle actin. Harrison suggested that the detection of crystallin gene non-coding sequences in extra-lenticular tissue should indicate whether lens crystallins or isoforms were present. Buckingham suggested that isoforms might be sought by Southern blots using crystallin probes and Bloemendal replied that this was now being undertaken. McDevitt asked whether the inserted sequence in the rodent A^{Ins} chain was found in species related to rodents such as lagomorphs. Bloemendal replied that it was found in the rat, mouse, gerbil, desert rat, and hamster, but not in the guinea pig or rabbit. The insertion was unusual in having three methionines in a sequence of twenty two amino acids. A splice signal at the 3' end but not at the 5' end suggested that its origin might be a defect in the splicing mechanism. Truman asked whether expressed introns might have a high methionine content and Schibler said that introns do not generally contain long coding sequences, but did in yeast petite mutants.

The relationship between the cell cycle, the state of the chromatin, as judged for example by DNAaseI sensitivity, and gene expression was discussed. Buckingham asked Rifkind whether he thought that the time of the cell cycle at which a gene was replicated depends on whether it is activated. Rifkind said there was some generalised evidence which suggests that early S phase is important for the cell: BUdR being particularly toxic to differentiating systems if applied in early S. α and β globin are transcribed early but not synchronously. Yaffe described experiments on muscle cells which were grown in medium permitting proliferation and changed to medium permitting differentiation. It was found that it was cells which were in late S phase when the medium was changed which could complete one division and then differentiate. Exposing cells to medium change for 6 hours only in lag phase also showed that cells which fused later had been in late S during this period. Rifkind commented that in some systems there was only 1 hour between early and late S phase. Schibler asked whether DNAase I sensitivity had been examined for different genes during the cell cycle. Rifkind replied that such work was now being undertaken in several laboratories. Harrison reported experiments of Dr Conkie on Friend cells blocked in G_1 at the non-permissive temperature. Cells blocked in the presence of inducer do not differentiate later when inducer is removed and the block is released, but cells taken through a similar protocol at the permissive temperature do differentiate. Owen asked Rifkind how he could relate data presented earlier in the symposium on tumour cells to data on the cell cycle. Rifkind replied that the consensus is that the transformed cells which are able to form large colonies are frozen between the BFUE and CFUE stages. Owen said that in both B and T cells there is a halt in G_1 following the proliferative phase, and the FC receptors, surface immunoglobulins and 1A antigens appear at this stage. However there is a further proliferative phase following stimulation by antigen. Rifkind thought that if, after induction, cells entered a small fixed number of divisions and then stop, these final divisions constituted the post-induction phenotype and these divisions should not be regarded as equivalent to division in cells which divide without such limitations.

There was some discussion of the role of isoforms of proteins. Buckingham suggested that their significance might sometimes be not so much physiological as related to the need for regulation between tissues. In the case of globin there was clear physiological significance but the actins had very few amino acid differences in sequence between cardiac and skeletal muscle actin. Their physiological differences were likely to be slight, and molecular recombinations can be made between cardiac and skeletal actins and cardiac and skeletal myosins. Both forms were detectable in embryonic skeletal cells. Yaffe pointed out that even subtle differences might be relevant in the longer term in vivo. Cardiac muscle had to function continuously for perhaps 80 years or so. Rifkind enquired whether ontogenic changes occurred in a single cell, or whether there was a replacement of cells. Buckingham replied that the situation was complex in vivo. The fibre type changes, and new fibres are formed but many investigators consider that the same fibre could express embryonic and adult forms at different times. In tissue culture fused myotubes first synthesise the LC1 embryonic form of myosin and then this is phased out and replaced by the LC1 adult form. Yaffe pointed out that evidence for the physiological significance of isoforms is implied by the observation that electrical stimulation of fast muscle can cause it to switch to slow, and Buckingham agreed that in two types of change, embryonic to adult and fast to slow, the same fibre was involved. Clayton asked whether differences between embryonic and adult could be related to function, and whether the signal was known which could lead to the transition. Was it external, or did the breakdown of the earlier form provide such a signal? Yaffe replied that denervation led to muscle degeneration, and there was a relationship between nerve stimulation, the type of muscle activity and the state of the protein, and that the slow movements of the embryo differed from the movements of the new born and adult. John pointed out that slow activity was a stimulus for further development and was necessary for creatine synthesis. Buckingham said that there was a question as to whether the genes were regulated in a block or arranged in a disperse manner. There was no evidence in mammals, but in Drosophila, hybridisation showed that actin genes are dispersed on the chromosomes. The question must be whether they are grouped according to the proteins expressed together in the phenotype. Moscona asked whether vertebrate muscle forming cells had a clonal origin as in tunicates and Truman said that limb bud cells were a population that could differentiate into cartilage, muscle and bone. Owen asked about muscle cells in tetraparental mice and Yaffe replied that Mintz found that in most tissues there were two isozymes but in muscle, two isozymes and the hybrid form, showing that the syncytium originates from both genotypes.

There was extensive discussion of the significance of methylation of DNA in regulation of differentiation, in particular whether methylation might be involved early on in determination. The possibility of inheritance of the pattern of methylation makes this a tempting model to explain the stability of differentiation. Yaffe cited several cases in the literature, pointing out that one X chromosome was stably inactivated, but 5 azacytidine used in cell culture could activate genes on the inactive X, while Weintraub found that an oncogenic virus in a cell line which did not express it was expressed after 5 azacytidine treatment, implying a permanent block of expression by methylation. Bird pointed out that there are many instances of genes being switched on and off regardless of the state of methylation of the DNA. He also warned against interpretations of experiments involving 5-azacytidine. Though this functions as an analogue of cytidine which cannot be methylated, methylation is drastically reduced even at very low levels of incorporation into DNA, and there are also a large number of toxic side effects on cells. Schibler pointed out that hormonal induction of mRNA synthesis was rapid and did not require cell division and would not be likely to involve methylation, and Buckingham cited the work of Mandel in Chambon's laboratory in which it was possible to separate temporally methylation, DNAaseI sensitivity and transcription of the ovalbumin gene following hormonal induction. In the oviduct the ovalbumin gene is already undermethylated while still DNAaseI insensitive. The first exposure to the hormone induces DNAaseI sensitivity and transcription. After withdrawal of the hormone ovalbumin transcripts are no longer detectable, but the gene remains DNAaseI sensitive and undermethylated.

Paul said that while it had been suggested that gene methylation might provide the signal for the changes leading to DNAaseI sensitivity, there was also the possibility that changes which lead to DNAaseI sensitivity might enable the demethylation enzyme to have access to the DNA. Bird said that it was not possible rigorously to exclude this possibility, but there was evidence against it. For example borealis genes were not methylated but were also not DNAaseI sensitive.

There were questions concerning the similarity of the amy-1 and amy-2 loci, and Schibler replied that no differences had been found by restriction mapping which had extended over a 30 kb region. Asked by Yaffe whether any loci were methylated or DNAaseI sensitive he replied that there were only 4 methylation sites in 15 kb and the possibility of heterogeneity in the cell population made measurements of DNAaseI sensitivity difficult. By S1 mapping of transcripts through the 'CAT' region to the promoter it could be seen that the promoter site known from sequence data and transcripts from other tissues was not used in liver. If it were assumed that the cell population was not heterogeneous, the Weintraub model of wrapping the promoter in chromatin was not appropriate since all the region was transcribed so that chromatin changes do not account for changes in DNAaseI sensitivity. He also pointed out that there were only 3-4 kb between promoters on the amy chromosome.

Iscove asked whether the introns described by Buckingham and Yaffe were evolutionarily stable in sequence. Yaffe replied that the stable regions were flanking sequences involved in splicing: the intron seuqences were not conserved. Drosophila has an intron in position minus 13 in the non-translated part of the message. This position required explanation - did this intron have a function? It might be that introns could remain if they did not interfere with function but if they arose within a domain, they would interfere and would be lost. Schibler asked Bloemendal about the αA^{Ins} chain - which was primary, the insertion or the deletion? Bloemendal thought that usually of 2 proteins the normal was the longer one and was found at a higher concentration, but αA^{Ins} was at a lower concentration. Referring to the difference in size of the αA and αB mRNAs he said that the difference was highly conserved. Schibler said that salivary gland α-amylase was nine nucleotides and three amino acids longer than in liver. The coding region for these 9 nucleotides corresponded to a non-coding region in the other gene at about residue 450. This was not a question of exons and introns. In dihydrofolate reductase there were multiple polymerase sites and mRNAs with identical coding groups had non-coding regions of about 200-1300 nucleotides inside the transcribed region.

THE DIFFERENTIATED STATE AND ITS REGULABILITY:

FINAL DISCUSSION

J. Paul

In the absence of having specific ideas from people I sat down and thought about some of the questions that still trouble me. I am going to split this into discussing cellular problems and molecular problems. It seems to me that the phenomenon we have been discussing fairly extensively during the past few days has really been regeneration in relation to differentiation. I thought the thing that might be quite interesting to set out first was how these systems are related and what lessons one system can give the other. Some of us were having a discussion last night about the universal set of slides, the set of 12 slides that covers every subject in science. I haven't got mine with me but I have drawn one of them to illustrate some of my points.

We discussed four systems, namely, the formation of blood tissue, regeneration of the lens, myogenesis and development of nervous tissue. If we look at these we realise that the haemopoietic system is continuously regenerating throughout the entire life of the animal. That is its characteristic. The lens is a system which sometimes regenerates in some creatures. Muscle is rather similar to nervous tissue; it is a tissue which practically never regenerates except in

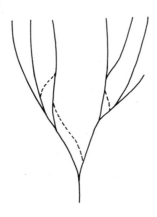

a very few lower animals. I was wondering if there are any elements in common in these or whether they are totally different. When we think of the haemopoietic system we think in terms of a stem cell pattern. I have some questions about that I will bring up in a minute. It seems to me that one thing that we are repeatedly asking ourselves is whether the traditional notion of ON/OFF switching which I introduced at the beginning has any validity at all or whether we are dealing with a much more flexible system. We should not forget that regeneration is a very common feature in most living creatures, in platyhelminths, for example, and plants. Regeneration in those is due to the fact that the organism, as I recollect, carries some of its stem cells along with it, the meristematic cells in plants, and neoblasts in platyhelminths, from which the entire organism can regenerate. In both plants and platyhelminths of course these cells form quite a high proportion of the organism but is it possible that we also possess stem cells for most activities as a minor population. Is it possible, moreover, that these are disguised as other kinds of cells.

When we think of differentiation we think of a situation in which decisions are taken at branching points and we tend to be rather hypnotised by some of the old models. There is that very fine picture in Waddington's book which you all remember of the ball rolling down the little hill; falling into troughs on the way. Of course when it has gone down the hill it can never come back up. I am just wondering how true that is. I know we have been over this ground in the past but is it not possible in fact that a differentiation pathway can jump across from one branch to another as indicated by dotted lines in Fig. 1? We should maybe be thinking in terms of networks as an alternative to branches.

When a cell is at a branch point it has to make some kind of decision and there are several kinds it can make. There are three that we have been considering during the past few days. One is a purely directive decision, as I think Norman Iscove called it. When the cell comes to a branch point it is directed along one branch. The second possibility is a purely stochastic decision - as is suggested by that old model of Waddington's - when the ball is rolling down the hill and gets to a branch point it has a fifty-fifty chance of going one way or the other. And then the third possibility of course is a stochastic model associated with selection such that when the ball rolls down one channel it drops into a hole and doesn't go any further. Any of these mechanisms may be involved at decision points but what is the possible molecular significance?

Are stem cells really universal and are they sometimes disguised as other cells which have the capacity to jump from one branch to another, i.e. transdifferentiate? This is a question which is seen in different ways by different people. The haematologist for example, tends to think in terms of a stem cell which follows one branch and then another. It has no alternative decisions. However, it is quite possible that there isn't one stem cell, but several. In early embryonic development mammals go through what is called regulative development which in the early stages, certainly up to blastocyst development, is very flexible indeed. One wonders to what extent this flexibility persists into the development of tissues in the adult. To what extent, for example, do we have not single stem cells but several cells which are already partially committed within the blood system and which switch over from one pathway to another? There is a certain amount of evidence that this can happen. So let me then put some of these questions to the group here.

Are stem cells universal? What is a stem cell?

Is it possible that stem cells can also be cells which have developed a set of characteristics but can switch over to another set in the way we describe as transdifferentiation?

These are fairly important things to discuss because of the implications they have for the molecular mechanisms. If we have a rather inflexible system of devlopment it has one set of implications; if we have a very flexible system it has other implications. What's your feeling about the stem cell, Brian?

B. Lord

It depends on where you put the limit on a stem cell, of course. I had just jotted your last comment down to take you up on because when we are talking about a spleen colony-forming cell, there is no question at all that the spleen colony is not a clone. Unique chromosome markers have shown that the colony is quite definitely derived from one cell and what is more, further spleen colony-forming cells are reproduced within that colony. When the colony is transplanted into further mice, the colony-forming cells in it will produce further colonies containing the normal spectral range of blood cells and yet more CFU-S. Thus, one cell is quite definitely giving rise to a variety of cell types. It is not the committed daughters of the stem cells that are all giving rise to colonies. The stem cell (or CFU-S) is giving rise to cells which can go off into different lines of commitment. You will recall, however, that I discussed in my talk the possibility of an age structure in the CFU-S compartment and that the direction of commitment may depend on the location of the pluripotent CFU-S in that structure. In that sense, if this idea is correct, commitment may arise from specifically defined (though still nominally) pluripotent CFU-S.

J. Paul

Yes, but is the first stage of commitment irrevocable, that is really the question, or is the first stage of commitment one that still allows it to go in different directions?

B. Lord

It is impossible to answer that as yet. I don't think there is any good evidence on that point at all but one would tend to feel that once it is committed it is going to stay committed. You certainly would never get a granulocyte/macrophage committed cell going into the erythroid line: not that we would necessarily see it in our experimental conditions. It doesn't mean to say it is not there because we haven't seen it. Perhaps the techniques are not right for seeing it but the methods of culturing these cells are now really quite refined with conditions defined for growing specific cell types or mixed colony cultures. But to my knowledge, no-one has ever seen a granulocyte/macrophage precursor cell going into erythroid cell production.

J. Paul

That may already be too far along the way of course. You see one of the things that intrigued me about your talk was the fact that your cells are limited to 200 divisions. Now that is fine if you are a mouse, but it is not very good if you are a man.

B. Lord

No! That 200 divisions is the average number in a mouse's lifespan but we know that the stem cell population is, proliferatively, just as good in an old mouse as it is in a young mouse and with serial transplantation this can easily be extended to 600 or 1000 divisions possibly many more. Now the average turnover time for blood precursor cells in the human is very long compared with the mouse and one can similarly expect that a mouse's stem cell turnover time of about 5 days is probably equivalent to weeks or months in the human. Those 200 divisions can go a long way then.

J. Paul

Yes, you don't know there do you?

THE DIFFERENTIATED STATE AND ITS REGULABILITY

R. Clayton

John, I'm bothered by this, because, we are already discussing the minutiae of stem cell properties in one system and I feel we ought perhaps to ask ourselves first to what extent a finding from any particular area is going to be something we can extrapolate widely. There may be a whole lot of different mechanisms and we must not be tempted to decide that just because one has been worked out to a certain degree, it has to explain everything in every system. The second thing I feel is that there is a tendency to accept as given something which is in part, I suspect, a product of our technical ability in detection and nothing to do with the real biological situation. What I have in mind is this: let us say for the sake of argument that the entry into a differentiated state and the capacity to be displaced out of it have a stochiometric element and that some cells are fairly readily pushed into various alternative pathways and that others are less readily so, so that an effective procedure for redirecting the more stable types of cell has yet to be found. If that is so we have to ask ourselves: what is it which characterises a cell which is de facto relatively stable? Now one of the things we already know is that cell-cell association and cell association with basement membranes tend to stabilize cells, at least in some systems such as transdifferentiation of ocular tissues. This must mean that they are necessarily irrevocably fixed provided you detach them from the information that they get from their neighbours. Another possibility is that when a cell such as a still-nucleated muscle or lens fibre, is crammed full of some superabundant spatially organised product, so that it cannot divide, and there is not much room to accomodate anything else without turning over what is already there, then the odds are that it is to all intents and purposes an irreversible terminal cell. This seems to be the case for cells like a lens fibre, before it's actually lost its nucleus (although there is a recent report, Iwig et al. (1978), which I think needs very careful corroboration, that at least early lens fibre cells may be caused to reversibly dedifferentiate and then divide). A lot of tissues were once thought to be totally stable but now there are techniques for destabilizing them and so they have become disqualified as examples of stability, and you say, ah! there must be a stem cell there after all. If we are to look to the future there may be a number of other cell types which people will discover how to reverse or to redifferentiate. For example, we don't know what is the nature of the black pigment in the lens epithelial cells, after we have treated them once with MNNG (and these black pigmented cells which Charles Patek found didn't appear until the second cell culture generation, so that was a long time after the treatment and many cell divisions). However, until we obtained these it was thought that lens cells at least were an end point and couldn't really redifferentiate into something else. I suspect that to a large extent we are prisoners of the manipulations accessible to us, and finally we are the prisoners of the limitations on our ability to discover if anything has happened - because so long as we rely on obvious markers like striated myofibrils or crystallin antigens, we may indeed not know how much noise and heterogeneity there is in any given group of cells. In other words, I feel that we are in danger of deciding that certain phenomena exist when (a) they may not be universal, or (b) we may be mistaken because we have not yet either searched for or have the means to discover how much variability there is, and, finally we perhaps have not yet got for every possible cell type the means of pushing it out of its pathway. If it turns out that a very high proportion of cells can, by one means or another, be deflected (whether they flop back or not is another matter), then it may be that this desperate search for stem cells everywhere - even asking them to be disguised by being totally jet black and so on, would simply be an interesting aspect of scientific history. Perhaps where you get systems whose whole function is to be constantly renewed, as in the blood or skin you do need stem cells that are constantly pushing forward, but in other systems perhaps you don't need to have stem cells at all, but that doesn't mean that you can't manipulate the cells in some way.

R. Rifkind

But Ruth, you know it is the minutiae that define the range of the possible; we don't have any other tool but to look at minutiae. For example, take an example that does not really address the stem cell. Let us take the Friend cell. It doesn't have multi-potentiality so it doesn't qualify, but it is a self-replicating cell which can make a decision to differentiate or not under special circumstances. You made the statement that there is an obvious truth, that, if a cell is filled with its final product it can't divide. The fact of the matter is that the Friend cell can fill up with haemoglobin, which has got to be one of the grossest examples of an over-stuffed cell, and yet it is still a self-replicating cell, if you induce it with haemin. Now surely that is an artificial differentiation, yet it defines the range of the possible - stuffing a cell with haemoglobin is not tantamount to determining its lifespan. It is still a precursor cell in the Friend cell's sense. It is not what I or Brian would call a stem cell but it's what a Friend cell would call a stem cell.

R. Clayton

I would agree with you actually that it isn't maybe a question of cell division, but a question of how easy is it to offer options. Maybe it's not the space for products so much as the space for messages.

THE DIFFERENTIATED STATE AND ITS REGULABILITY

R. Rifkind

If we knew as much about the Friend cell as some people know already about K562, I would know that there is an agent that in fact can take the cell that is already accumulating small amounts of haemoglobin and probably making spectrin, certainly knows it's a red cell and turn it into a macrophage by giving it other inducers. Now these inducers are not physiological but they define the range of the possible for that cell.

J. Paul

The thing I was trying to achieve here, of course, was to point out that we are dealing with phenomena which are not so very different, but we certainly look at them in very different ways, because the person who thinks in terms of stem cells in the haemopoietic system thinks of an irrevocable future for a cell which has gone in a given direction. Now you think of this rather differently, Ruth, you think that if you just have the right buttons to push you make certain cells into quite different cells.

R. Clayton

No, I think there is a whole range of degrees of facility, ranging from something which is very easily pushed to something else to cells which are only redirected with immense difficulty (or perhaps, in the extreme, not at all), but if we have got a continous spectrum, which we very likely have, then the true question surely becomes: what is it that defines where the cell is on the spectrum?

P. Harrison

I think that in molecular terms it is hard to think of what distinctions you can make. You are asking how you can get from one gene set being expressed to another: in a way the difference between the various viewpoints concerns whether transdifferentiation or transdetermination occurs per se, whether the stem cell itself shows a certain differentiated phenotype or whether the differentiated cell has somehow to begin to replicate and revert to a stem cell before gene sets can change. The difference reduces, in molecular terms, as to whether there is a cell cycle involvement in changes in gene expression. We don't want to go over that old ground again - but if you try to think of what the difference in these hypotheses might mean in molecular terms then this is one fairly clear distinction between your view and that of someone like Holtzer. Do you mean to imply in what you say that you would get gene sets being switched from one cell type to another within a single cell (for example, from a lens containing large amounts of crystallins to a red blood cell containing haemoglobin) by quantitative variations in the leaky regulation of the expression of the relevant genes?

R. Clayton

No.

P. Harrison

Without anything, in a single cell, is that not an extreme case?

R. Clayton

If we look at what is reported in the literature, one gets the impression that using, say, 5-azacytidine or using MNNG or whatever agent or conditions are effective, such as cell dissociation, what you may get out of your initial cell population is not unlimited potential. It is not a complete range of everything. In the cases where transdifferentiations have been obtained, there is apparently a limited range of possible pathways, so that the status of the cell, at the time that it is treated, must, in some sense limit what can happen thereafter. Fibroblasts can go into fat cells or they can go into muscle cells, but nobody has yet got them going into liver or lens cells, as far as I know.

P. Harrison

But again that could be interpreted in terms of multipotency or that there is a limited stem cell population which is triggered to proliferate and differentiate in various ways according to the conditions.

R. Clayton

I don't think you have to say stem cell. I think you have to say that any kind of cell may have some limited set of options open to it which are defined by its current status.

P. Harrison

You could have an alternative. You could have a partly committed stem cell which has only a limited range of options and that what you are really doing by treating with mutagens, etc., is equivalent to what happens to the haemopoietic stem cell when it is irradiated.

R. Clayton

If you want it to be a stem cell, then Aron Moscona's experiments suggest that virtually every single glial cell in the retina is also a stem cell. That is a bit much isn't it?

P. Harrison

Well in that case it comes down to the question as to whether a single cell can shift from expression of one gene set to another as a single cell. That is the critical test of the two hypotheses.

R. Clayton

Well then you are asking for sets to be switched on and switched off absolutely and not quantitatively regulated - and that is one of the problems we are running into.

B. Lord

How can we say that every single cell is a stem cell?

J. Paul

I think we want to exclude that idea. That is impossible. We should perhaps clarify this because what we are thinking of is a population of cells which have the capability of developing into other cells and these other cells are terminally differentiated. I don't think any of us envisage the possibility of a neurone becoming an erythrocyte or an erythrocyte becoming a neurone. That is quite out of the question, but the question we are asking is about a population of flexible cells and how far up does it go in the tree. You see we have two very rigid views which are quite opposite to each other. One is the view of the haematologist that the flexibility of the cells is restricted at a very early stage, that they don't have the ability to switch over from one line of development to another and once they are committed their future is determined, whereas we have this view of the transdifferentiationists that cells which are quite advanced in their development are nevertheless fairly flexible. Now I am just wondering whether we can see some sort of convergence in our thinking. The reason why we have rather fixed ideas about the erythroid system may be because we don't know very much about the primitive stem cells themselves except by inference - we can't see them but the question

I am asking is: Are these retina lens cells which can go on and differentiate into lens cells really in the same sort of category: Are they cells which are going in one pathway but still have the flexibility to go back and enter into another pathway and is this a fairly universal phenomenon?

A. Moscona

John, is it inconceivable that a cartilage cell may transform into a muscle cell, assuming that the genome still contains the necessary, latent, information?

J. Paul

Well I think one does not want to enter into a detailed argument on this because in some cases one has to say almost certainly the whole genome is there and in other cases we don't really know. No-one has taken a brain nucleus and put it into an egg and made it grow into a complete organism. There is a lot of ignorance in this area.

A. Moscona

The outcome of John Gurdon's experiments was that you can take a nucleus from a differentiated cell and put it into an oocyte and get it to develop into a tadpole.

P. Harrison

As I understand it, now it is perhaps not even sure that Gurdon's experiments can be interpreted in the strict sense in which they were originally because it is quite conceivable that the nuclei giving rise to tadpoles at very low frequency were not derived from differentiated cells as such but rather from (say) stem cells in the population.

A. Wylie

I think there are two problems here. One is the orginal set of experiments where multiple transfer was required to get the tadpoles and I think that everyone is agreed there is a potential hole in the argument there. But I think the second set of experiments where products were looked for in oocytes into which nuclei had been transplanted shows that there is tremendous plasticity in the ability of the Xenopus oocyte to allow read off of new products, but although there is some question as to whether you can build a complete tadpole,

THE DIFFERENTIATED STATE AND ITS REGULABILITY

the general question of whether you can read a completely new set of products off an inserted nucleus remains very definitely established.

J. Paul

Well, I don't think that has been answered yet because these products of course are demonstrated essentially by 2D-electrophoresis. As I said at the beginning you see only about 10% of the total gene products. It may well be that the nucleus that you put in already has the ability to make a lot of those products but at a low level and that what you are simply doing is gearing them up to produce these products at a higher level.

D. de Pomerai

That is precisely what we have been saying about the transdifferentiation of retina into lens.

J. Paul

Exactly - could be the same and I think that is perhaps Ruth's view in fact.

R. Clayton

Well, yes, we think that there is at least a parallel, a very strong parallel, and we haven't so far found exceptions to it, between the possession of crystallin message in the intermediate abundance class and the ability to push that into the high abundance class, given the right provocation, so that the cell ends up as being, de facto a lens cell. On the other hand, tissues where it might conceivably be present in one copy per cell probably include all the tissues that don't transdifferentiate. So there might be quantitative factors which limit a cell's potential.

M. Buckingham

I was just going to make a more general comment that if one takes certain cells and puts them in culture, maybe it is important to stress that, then they do show varying degrees of flexibility. For example, in the muscle system or more generally in the teratocarcinoma system, it is clear that one can cultivate cells which have varying degrees of flexibility and the amount of flexibility they show and the range of cell types they will go to seems to be rather restricted, as you were implying with the neurone/erythrocyte transformation. The mouse muscle cell line we use will, if you push it, go to fibroblasts or go to fat cells, but normally it goes to muscle, and the problem is that in molecular terms I don't think one has any idea what that means. By Northern blotting we do not detect transcripts of the muscle genes in the myoblasts and David Yaffe's results suggest that the genes aren't DNase I accessible. Our earlier experiments with enriched cDNA fractions also suggested that the muscle genes aren't DNase I accessible and that no muscle transcripts are present in myoblasts of the mouse muscle cell line. The methylation experiments haven't been done but I would be surprised if all the genes that can eventually be expressed in the different cell types are under-methylated. Neither the idea that the genes are being transcribed at low levels, nor the idea that they are DNase I accessible, nor the idea, probably, that they are under-methylated can explain the fact that there does exist for these cells a limited range of differentiated possibilities. I think it is very difficult to envisage in molecular terms what this kind of restricted flexibility for different cell types means, or what, more generally, committment means during development.

D. McDevitt

I am not so sure that, although it sounds a bit heretical, it is correct to try to make some of the systems we have heard spoken about in this conference, namely the haematopoietic system, exactly in line with the lens regeneration or the transdifferentiation system. I think they are different and that we have to realise that. I think there is information to be gained, certainly, from the lens regeneration or the transdifferentiation system but I myself don't see, with all the talk about stem cells, what cells one would call a stem cell in transdifferentiation or regeneration. I guess I always think back to the source of the cell and whether it is withdrawn from the cell cycle as a criterion. I don't see that we are dealing with each system starting out with a stem cell.

M. Buckingham

The problem is that those cells have to undergo division and perhaps become something else before they demonstrate the transdifferentiation.

G. Chader

Much depends really on the length of time the systems we are discussing (e.g. erythropoietic, transdifferentiation, etc.) have been studied. We know a great deal about some systems but are relatively ignorant of even the basic phenomena in others. This is not the ignorance of "stupidity" but of simple lack of knowledge. We must not

forget for example that it is only a few years ago that Drs. Okada
and Eguchi defined "transdifferentiation" or at least described its
overt characteristics. We can then talk rather definitively of stem
cells in the erythropoietin system or specific signals and gene products in normal lens differentiation but yet haven't the foggiest
idea about these factors in transdifferentiation. In this latter
system, we are yet at the phenomenological point, basically describing and documenting the characteristics of the system in culture.
Moreover, different things are known about different systems. We do
know quite a bit, for example, about the signal(s) that turn on haemoglobin synthesis in the erythroid cell and that cyclic AMP enhances
pigmentation in the pigment epithelial (PE) cell. On the other hand,
we know little about the signals that make a muscle cell differentiate
or that cause a PE cell to "transform" into a cell with lentoid characteristics, although Dr. de Pomerai and I now have found that various
components of the culture medium definitely do affect lentoid body
formation. Similarly, Dr. Bloemendal's lovely presentation demonstrates the advanced state of the m-RNA/cDNA art in the lens system -
an art which is yet in the most rudimentary stages in the transdifferentiation process in PE cells, neural retina and lens. We thus have
to weigh what we know about a system versus what we would _like_ to
know about it before making a judgement as to its validity. At this
stage, therefore, I would suggest that the term "transdifferentiation"
as originally proposed by Drs. Okada and Eguchi be used if only in
an operational sense. A logical sequence of events can be envisioned
however in proposing a model for this phenomenon. One could first
envision the presence of a Primary Stem Cell which contains the entire
DNA complement, is actively dividing and in which many mRNAs are made
if only at a low level. This cell is exposed to a series of signals
(cell-cell interactions, substrata, humoral/hormonal, environmental,
etc.) which seek out and amplify mRNAs for a limited number of proteins possibly by action at a genomic switch point and poduces a
Secondary Stem Cell. It is interesting in this regard that
Dr. Clayton has found similar mRNAs in at least intermediate abundance
levels in retina, PE and lens tissues. Tertiary "fine tuning" of the
system then could further restrict the choice of pathways for the
Secondary Cell, allowing for "final" differentiation into a retinal,
PE or lenticular cell-type but not muscle, liver, etc. The "final"
differentiated state _in vivo_ may be "final" however, only in that
proper signals are not forthcoming to allow for further changes. One
might even consider this to be a natural form of "arrested development". Placing the retina in culture, for example, may supply these
signals or remove blocks in the differentiation pathways, freeing a
restricted genetic domain and attendent mRNA complement for reversible
redirection ("transdifferentiation") of cell morphology and function,
possibly through the Secondary Stem Cell Intermediate.

D. McDevitt

We still haven't defined a stem cell.

J. Paul

I am not sure that we need to define a stem cell, that was just to keep the conversation going. What we are really trying to do here is to try to identify mechanisms and I think that we all start with the premise that common mechanisms act in all cells. This is where we have to try to reconcile what happens in erythropoiesis and what happens in lens differentiation. You can say they are two different processes. Do you suggest that therefore in these two processes entirely different molecular mechanisms operate?

D. McDevitt

No. What I would say - by saying two different processes - they are both types of differentiation but to try to ally the two so as to almost artificially obtain agreement I don't think is informative. I think the one system is extremely valuable, namely the transdifferentiation system, but admittedly it is an exception. I don't think there are entirely different molecular mechanisms involved but I think the steps differ leading up to the point where the transdifferentiation begins to take place, then from then on I would say probably they use the same mechanism. I am not sure what causes that switch, whether it has something to do with cell replication or not. I can tell you one point that hasn't come out in the conference so far is - again I go back to the in situ situation with lens regeneration from the dorsal iris - it takes six cell cycles in the dorsal iris to reach the stage when crystallin synthesis is first detectable by whatever means. The ventral iris, which if you remember is not capable in situ of regenerating a lens but is capable of doing so in culture, only goes through four cell cycles, so that there is a shorter cell cycle time in cells which to all intents and purposes are the same in the dorsal iris and the ventral iris. That brings about the interesting possibility or hypothesis that it is the shortened cell cycle time in a given unit of time that forces these cells in the dorsal iris into the pathway of making a lens rather than as with the ventral iris cells, which repigment themselves and go back into being the same old melanocytes. I am not sure but in that region of the initial events, this suggests there might be some difference; after that I am sure it's the same.

J. Paul

The erythroid system may not be so different except that it has more maturation divisions. There are many maturation divisions in erythrocyte maturation and a number of experiments suggest that different globin genes are switched on during different maturation divisions so that if you only go part of the way you have more immature genes expressed than if you go through with the full set of divisions.

D. McDevitt

Well that is certainly a common feature....

D. de Pomerai

But if you attempt to interpret Wolffian lens regeneration in terms of stem cells you still have the problem that all the initial iris cells have withdrawn from the cell cycle and are all pigmented, therefore your stem cell has to be some kind of wolf in sheep's clothing that is superficially differentiated, in other words it is making melanin and to all intents and purposes it looks like an ordinary pigment cell. Nevertheless, you are saying that somehow it is different: it is a stem cell really.

J. Paul

That's right. You see in the erythroid system it's not so different because the actual stem cell is quiescent most of the time.

B. Lord

So quiescent in fact that Dr. Rifkind asked the question earlier this week whether under normal conditions that stem cell is ever really contributing anything.

J. Paul

And we don't know the answer. For all we know the stem cell may be a pigmented cell lying down there in the bone marrow somewhere.

R. Clayton

I think you are going in circles on these pigmented cells. Can we back-track to what Dave McDevitt said. He said something that I would not quite go along with. First of all he said that Wolffian regeneration is a limited phenomenon, it is not found in mammals. That is absolutely true in the _in vivo_ situation but _in vitro_ it looks as if in every vertebrate that has been tested, the iris or retina cells do have transdifferentiation capacity, so a particular in vivo inability must be due to some local restriction imposed upon a fairly basic kind of potential. As to whether this is a unique phenomenon confined to this set of tissues, I don't know: I still think it is open to consideration that some of the sorts of phenomena that John Ratcliffe told us about are parallel phenomena, that not all of them can be explained away by cells migrating someplace and there expressing a potential that they have always had. In some of these tissues there isn't any evidence that neural crest cells migrate. There is even some evidence from le Douarin and others against it. Nevertheless, certain cells can suddenly under some circumstances, when they get the provocation, begin to produce at superabundant level a product which normally characterises a totally different cell system in a quite different location. If we look at some of the recent evidence over the last two or three years, at least some of the sorts of tissues which can, for instance, start producing chorionic gonadotrophin in large and unacceptable amounts, include I believe some of those in which it is normally present at very low levels. That has only been found recently because the sensitive techniques detecting them are relatively modern and have only recently been applied here. I don't think that transdifferentiation in the eye as we have studied it, is unique but I think that it's been accessible, and it's been worked on, but it may be a paradigm for events and changes elsewhere. The last thing Dave mentioned was the cell cycle. The numbers of cell cycles that have to take place before transdifferentiation is something that is open to manipulation because Goro Eguchi has shown that you can change it by some sorts of manoeuvre, and Aron Moscona also has shown as well you can get transdifferentiation to take place after a very short period of time, yet previously it took a long time. The number of cell cycles that is required must be characteristic of the experimental situation and may be the number needed to get a sufficient number of cells in contact. It is not necessarily immutable.

D. McDevitt

Is it done in situ or in culture?

R. Clayton

In culture.

D. McDevitt

You see I am talking about the in vivo situation.

THE DIFFERENTIATED STATE AND ITS REGULABILITY

N. Iscove

I would like to add the view of someone interested in stem cell problems. When one looks at transdifferentiation, one asks whether a cell which expresses a particular set of genes associated with "end-stage" function can later express a different set. Because of the stem cell argument, experiments involving populations of growing cells clearly can not be decisive on this issue. Experiments on single cells, on the other hand, would be decisive. If a single cell possesses particular "end-stage" markers (state 1) and then gives rise to a population of cells possessing another set of markers (state 2), the point is made. Of course, one would then wish to know whether the proportion of state 1 cells which can do this is high or low. So I think Dr. Eguchi's experiments are the right ones to do. I would only caution not to generalize too quickly. What holds in avians may not hold in mammals, for example, and there is no substitute for doing the right experiments in each system.

R. Clayton

Eguchi and Okada did it.

N. Iscove

It was a clone, but did the experiment begin with one cell?

G. Eguchi

Yes.

J. Paul

That is why there isn't a question of whether people consider it a stem cell. Let us get the definition of differentiation sorted out. Differentiation means becoming different, that's all.

G. Birnie

I cannot help wondering whether we underestimate the importance of something mentioned by Dr. Buckingham earlier, that is, putting a cell from a tissue into culture, either alone or in company with other cells, simply makes the cell more flexible, since it is then no longer constrained by intercellular influences which may be important in vivo.

M. Buckingham

And also because it can divide, so in fact you are looking at a conversion which doesn't take place in the original cell because there you would have to demonstrate that without division, the pigment cell for example goes to another type, but in fact it divides and it loses its pigmented properties and then it becomes another cell.

B. Pessac

Why should not a stem cell be pigmented? Can it be that there are stem cells that express differentiation markers that are pigmented and are still stem cells? Margaret Buckingham is very very right, I tried to point out this point earlier and I said that transdifferentiation would be a differentiated cell that goes from state A to state B without division.

T. Okada

I do not like to pretend that I can follow all the discussion so far, since it runs very often so fast and is much too complicated. However, I have got an impression that more people seem to be more comfortable to think of differentiation starting from a stem cell and are less willing to accept the idea of a sudden change from state A to state B, in order to interpret the results of a series of experiments on the phenomenon which we call "transdifferentiation". This is a situation which I had predicted to some extent. Perhaps, the existence of changes which we call transdifferentiation does not readily fit a general logical framework for understanding the process of differentiation at the molecular level. However, there is the fact, which is now proved to be quite sound, that the progenies of retinal pigment cells (which are actually black) can differentiate into lens and the progenies of cells partially differentiated in a neuronal direction in neural retina cells can also differentiate into lens. Do you still prefer to call such pigment cells as "black stem cells" with bipotentiality to differentiate into black cells and lens cells? As we all know, differentiated somatic cells of plants can differentiate into a complete set of cell types to make up a complete plant. Then, do you state that a plant body consists of stem cells? Of course, this is just a matter of terminology. However, it is certainly outside of our common usage of the word "stem cell".

There are several reasons why the differentiation of stem cells should not be categorically mixed up with the transdifferentiation of once specialized cells with well defined differentiative phenotypes. a) Before transdifferentiation of lens, for instance, black cells loose their pigment granules. But these so-called dedifferentiated cells are not equivalent to multipotential progenitors present

in early embryos. They are programmed only to differentiate into
either pigment cells to lens cells, and into nothing else. b) It is
very important to emphasize that retina and lens are well separated
at the very early stages of cell lineage in embryonic development.
"Stem-like" progenitors to have the potential to differentiate into
both retinal types (pigmented and neural) are not expected to be
present in normal development. Therefore, I have to emphasize again
that dedifferentiated retinal pigment cells are not equivalent to such
"stem-like" cells, if any, at all. Transdifferentiation is not mediated by the reversible process of differentiation to "stem-like"
cells, but is a forward process. c) Finally, the transdifferentiation
may not be always preceeded by mitotic cycles, as indicated in this
symposium by Moscona and perhaps by myself. If you would like to see
another example in relation to this specific topic, I only mention
here the beautiful demonstration of the transdifferentiation of
ciliated non-striated muscle cells from striated muscle cells of
medusa. Dr. Schmid (1975, 1980) showed the presence of "mixed" cells,
a striated muscle cell with cilia, in the transitory stage of this
type of transdifferentiation.

Taken altogether, it is a fact that there are examples of a
shift of differentiated state A to B. Three points which I have
raised may make it very unnatural to interpret such process in terms
of the differentiation from "stem-like" cells. If you do not like
still to face these facts frankly and you still consider the differentiation of "stem-like" cells as a sole model of differentiation,
I dare to say, it is simply dogmatic.

I feel it is sad that we are asked by not a few people whether
we had the demonstration of transdifferentiation starting from a
single cell in clonal cell culture. Our approach to the problem in
question was initiated by the success of such clonal cell culture
work. These were published already (Eguchi and Okada, 1973; Okada,
1977; Yasuda et al., 1980). Without having had such experiment at
the initial stage of the research, we would not continue to study
this problem.

H. Bloemendal

I would like to make the following comment. If I understood
Dr. Courtois correctly, he made the observation that one can force
a differentiated cell to regain its original programme and in this
respect I am thinking more in particular of epithelial lens cells.
When you bring these cells in culture they cease to synthesize crystallins which are the predominant proteins in situ: apparently they
lose something which, according to Dr. Courtois, has to come in from
the retina. This observation means that indeed one can restore a
cell's programme presumably lost by artificial growth conditions. It
also means that "neighbouring" cells need certain factors to carry

out and maintain their protein biosynthetic programme and my question
is: could one think in terms of the following picture. The farther
we go in differentiation the more factors we are removing, factors
which are required to maintain a certain programme. In view of the
consideration that the retina can force the lens fibre to carry out
its crystallin programme again, a second question arises: can we
travel in the opposite direction, can we force the fibre to make, say,
rhodopsin or another retinal protein? At any rate Dr. Courtois'
experiment seems to suggest that cells which are very close in terms
of distance may affect each other in one or two directions. Hence
it may be that the more remote a tissue is localized from the original
stem cells the more easily it loses just simply factors that would
promote the expression of a particular gene product.

J. Paul

I think that helps us to move forward to thinking about mechanisms. We have been over the whole question of stem cells thoroughly
now and perhaps we haven't come to any conclusions except that possibly the transdifferentiationists should think a little bit more
about stem cells, and the stem cell people should think a little more
about flexibility in early stem cells. But I want to turn now to a
discussion of some general phenomena related to mechanisms before we
talk about molecules. And perhaps we can consider some of the things
that Dick Rifkind and Norman Iscove talked about yesterday and particularly look at the question of whether, when a decision is made
it is a directive decision. In other words does the cell institute
a new programme because of some environmental factor or is it a stochastic event whereby the cell follows an inbuilt programme, unaffected
by the environment and the effect of the environment subsequently is
selection of the cells which have already been determined. Now this
is a very important distinction because in one case we are saying
that the cell has a number of possible fates arrived at randomly and
the eventual outcome depends on which has been selected. On the other
hand we are saying that the cell is the subject of its environment
and can be directed. Have people got any very strong views about
that?

R. Clayton

I don't think they are in opposition. I think you could have
both operating.

J. Paul

I think they imply different mechanisms. One says that if a
cell which has the possibility of going towards A or B receives an

external signal it then goes to A or goes to B depending on the signal.

R. Clayton

And it can still after that be selected.

J. Paul

Ah, yes, but that is by the way. The other says that no matter what you do with it environmentally you cannot alter this decision. The cell goes on its way and is then selected. That is a different thing.

R. Clayton

Can we not have a compromise between the two - perhaps it is neither the one nor the other. The cell may have an intrinsic sort of stoichiometric system built into it with several potentialities but still be constrained by its neighbours, to go in one direction only: in other words the identical cell might move inflexibly in one way in one cellular environment and you would then say this has got an inbuilt pathway to go along, but nevertheless in another environment you might discover it may have other options. I think your distinction is artificial, John.

J. Paul

I am sorry we are differing in this, but what I am saying is that phenomenologically the two things are identical but from a mechanistic point of view they are very different.

N. Sueoka

I have been wondering about whether or not there are some cases in mammalian development where one stem cell divides into two different cell types (A and B), namely dichotomous branching at an individual stem cell level. In our rat neurotumor RT4 cell culture system I think we have eliminated this pattern as the regular rule. When cell type conversion occurs, we do not see two different derivative cell types show up simultaneously. Instead only one type of cell shows up as a patch. Multi-potential stem cells, however, can generate more than one derivative cell type at the population level, thus forming a branch point of, for example, neuronal-type and glial-type cells. This is a probabilistic view.

J. Paul

We are all familiar with Holtzer's notion of the quantal cell-cycle in which an asymmetric division gives rise to one stem cell and one differentiated cell. The question Dr. Sueoka is asking is whether you ever get a stem cell giving rise to two different differentiated cells? I think actually it depends what you mean by differentiation.

A. Moscona

A possible example is the special kind of cell in the cuticle of Drosophila which in 1 to 2 divisions gives rise to a bristle, a socket, and a neurone. The question is, is it a stem cell?

R. Clayton

Yes, it is a stem cell. Mutants affecting the bristle-socket patterns such as forked surely show this.

N. Sueoka

Another point I would like to make is about the case where the majority of seemingly undifferentiated cells become differentiated by some external agent like DMSO. This should be classified as maturation of already determined cells and should not be treated equally with differentiation of the multi-potential stem cell. I am not against the use of the term, "stem cell", for this situation but want to point out that there are at least two conceptually different classes of cells which are called stem cells. This confuses the general discussion in the field.

Y. Courtois

Recently I heard about work performed by M. Darmon (1981) in G. Sato's laboratory showing that by using growth factor and different media they could direct teratocarcinoma cells in various pathways.

B. Pessac

I think he started with the embryonic carcinoma line. In serum-free medium the cells shift to neuronal differentiation: in serum-containing medium they remain undifferentiated.

Y. Courtois

I think thus that environmental signals can participate in modulations of the differentiation of the cells strains; and that the discovery of these signals (i.e. growth factors, extracellular matrix) and their use in tissue culture will allow us to investigate directly how they can participate in stabilizing the cells in different stages of differentiation.

B. Pessac

If Dr. Sueoka's hypothesis is true it means that during all life there is a population of stem cells which will be carried over and give rise to newer stem cells, for if a stem cell always divides giving rise to a stem cell and a neurone....

J. Paul

Provided you have those cells alive.

B. Pessac

I mean that it seems difficult to imagine that for neurone cells, though some people claim actually that neurone cells can arise from other types.

A. Radbruch

I just want to make the point that the selection of totally differentiated cell populations is very evident in the immune system. I mean, that is the way it works, and to my mind it is very hard to imagine that this is an entirely new achievement and it has to have evolved somehow. If we look at it a little bit closer then more and more factors in the immune system that propagate proliferation of different specialised cell types are described − factors for certain types of T-cells or for B-cells − and the literature is growing and growing. These factors selectively propagate growth of very narrow, small populations of specialised cells.

D. de Pomerai

Returning to your point about the branching: I think all the evidence from determination of Drosophila imaginal discs suggests that groups of cells take a decision to form a particular adult tissue type as a group. It is not taken by single cells independently. That would suggest that neighbouring cells do have an influence on the direction taken at a particular branch point.

U. Schibler

Let's go back to the immune system. If one looks at the end result of B-cell differentiation superficially, one might easily conclude that a given antigen triggers differentiation of a very limited number of lymphocyte populations which produce antibody molecules with high affinity to the antigen. What is partiuclarly nice in this system is that differentiation can be monitored by looking at rearrangements of immunoglobulin gene constant and variable regions. Since in many cases lambda light chain producers have been found to contain rearranged kappa light chain genes, it seems likely that, at least in this case, differentiation of specific lymphocytes occurs according to a stochastic trial-error system. That is to say that lymphocyte precursor cells try out different possibilities until a successful one is found. So, the antigen does not at all stimulate a unidirectional differentiation pathway of antibody producer cells, but rather selects B-cells manufacturing antibodies with high affinity to itself among a large repertòir of B-cells, most of which are producing nonsense antibodies.

R. Clayton

One of the things that has come out of the work on the Drosophila discs on the basis of electrophoretic separations is how amazingly similar they are. Most of the differences between them are quantitative; in the relative balance of components and not in anything unique (Rodgers and Shearn, 1977; Arking, 1978; Seybold and Sullivan, 1978), so perhaps it is not very difficult for them to slip into particular directions in transdetermination experiments.

If the basis of change in a system is on balance stochastic then committment might involve the choice of cells with the best fit (or a sufficient fit) from a varied population. However, in systems like imaginal discs or ocular tissues, which demonstrate a range of overlapping future options and also express a molecular overlap, there is no reason to suppose that only those cells survive which have the best fit for the local requirements. You would expect to see cell death associated with differentiation - if an immunoglobulin type of model is appropriate. Although cell death is seen in some systems, like the formation of digits or semi-circular canals, it is not general, it is also spatially organised. However, even in cells which seem to overlap in potential and in molecular characteristics, there is probably a point along a pathway of increasing quantitative divergence when there is a real committment, a binary choice. Whether we recognise it correctly depends on available technology.

The state of differentiation might also affect possible tumour formation and responses to carcinogens. Tumours can not be really dedifferentiated, since pathologists can designate their tissue of origin. The effect of a carcinogen apparently depends on the state of differentiation of the cells attacked: for example neuroblastomas are produced by exposing embryo rats to ethyl nitrosourea at the right stage of development. In our data, the effect of MNNG on transdifferentiation depends on when it is applied. Early treatment has a more drastic effect than late treatment. The cells differ at these stages; for example, there are fewer mRNAs each in larger amounts in later cultures.

The initial effect of a carcinogen might be to superimpose a range of changes on a cell population with certain molecular restrictions. If any such cells then differentiate terminally or divide at the normal or a lesser rate they will not be noticed, but if some of them divide faster, they may be seen as a tumour. They will be selected as tumours by their rate of mitosis just as antibody cells are selected by antigens for the Ig chains on their surface. Both models are stochastic. Substances synthesised or accumulated in a long cycle may not appear in a rapid cycle and the cells will appear to dedifferentiate, or products favoured by a short cycle may have to predominate, suggesting "retrodifferentiation". If growth is slowed down the cells should potentially "redifferentiate". Yves (Courtois) and colleagues found an inverse relationship between visible differentiation and mitotic interval in the effect of EDGF on cornea cell cultures. Changes such as chromosomal losses accumulated by the descendants of cell lines started by carcinogen exposure are a separate problem.

R. Balazs

I only wanted to comment on problems related to differentiation of cells in neural tissue. As our repertoire of specific markers for the different classes of differentiated neural cells increases it is becoming evident that, depending on the stage of development, overlap between gene expression is relatively common. The earliest examples have been obtained by the examination of cell lines derived from neural tumors. These express simultaneously certain properties which in the mature nervous system are selectively associated with either oligodendrocytes or astrocytes. However, overlaps of this kind are not restricted to transferred cells. It seems that during development cells which can be identified as nerve cells express transiently molecular characteristics that in the mature nervous system are exclusively detected in glial cells. Furthermore, expression of completely 'alien' genes has also been detected. For example, astrocytes in the immature brain contain vimentin, an intermediate filament protein originally isolated from fibroblasts.

N. Sueoka

I do not imply that all the markers of, for example, neuronal vs glial cells separate clearly at one developmental branch point. However, in the RT4 system, it is already clear that two glial markers (S100 and GFA proteins) are separated together from a neuronal marker (veratridine stimulated Na^+-influx). We are examining more neural markers on this point. Another point I have to bring up is the following. In the case of mouse hypothalamus development it has been shown that both a neuronal marker (14-3-2 protein) and a glial marker (S100 protein) are co-expressed in the precursor cells (ependymal cells) (DeVitry et al., 1980), a situation similar to the RT4 system. However, when one of our graduate students, Kurt Droms, examined the rat superior cervical ganglia at the newborn stage, catecholamines were clearly detectable whereas S100 protein was not. Only a few days later S100 protein was seen but already glia and neurones were segregated. This suggests that the determination at the gene or chromatin level may not necessarily be accompanied by phenotypic expression of the gene products which constitute the marker properties of the fully expressed cells. What is this determination? That seems to be the key question for the study of branch-determination at the molecular level.

A. Moscona

We may have reached a certain consensus, in spite of the diversity of views which have been expressed. Are we agreed that stem cells are in fact, differentiated cells? In other words, that a stem cell which gives rise to haemopoietic cells is different from a stem cell which gives rise to keratinising cells, and from one that gives rise to primordial germ cells? These stem cells are, in fact, committed cells. Can we agree on that?

Chorus

Yes, certainly.

B. Lord

No! What evidence do you have that it is not merely the environment a stem cell finds itself in which defines the direction of its committment?

A. Moscona

Then, we agree also that a problem exists with the conventional

THE DIFFERENTIATED STATE AND ITS REGULABILITY

definition of a stem cell. A stem cell is really a "determined" cell committed to a certain line of development. The issue to a developmental biologist is what commits that stem cell to being a progenitor of A, B or C cell line. As to the point raised by Dr. Sueoka: In the early embryonic neural retina, cells diverge into glial cells and neurones, and these undergo progressive specializations. I find it very difficult to consider a mature, differentiated glial cell a stem cell, no matter how much potential developmental plasticity it still retains.

J. Paul

Sir Michael Woodruff has accused me of simply substituting a potentiometer for a switch but I think that is an important distinction. If we talk about mechanisms, what are people thinking about in molecular terms. There are two extreme kinds of models which I might just summarise for you before we move the discussion on. Firstly, there is the strictly digital model, which is that we go from a minus state to a plus or a minus state. This is summarised in its extreme form by the postulates which are made about what happens when a DNA molecule replicates. It is at this point of course wrapped up in a

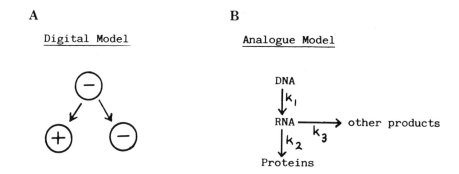

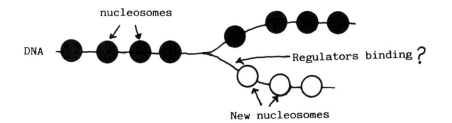

nucleosome that has regulatory proteins associated with it. When the
DNA strands replicate there is enough protein to reconstitute one
nucleosome but the other DNA molecule is temporarily naked. The proposal is that when the chromosome is dividing the structure of the
parent chromosome is perpetuated in one of the daughter chromosomes
but at the point when DNA is temporarily naked non-histone proteins
and other regulators may bind and a different kind of chromosome
is constructed. And this, of course, Dick Rifkind would tell us, is
what happens during that crucial cell division. This is one model
which implies that two different chromosomes result, one of which is
in the 'plus' condition and one of which is in the 'minus' condition.
That is one kind of model. That is the switch model.

The other kind of model is what we might describe as the analogue
model. The gene gives rise to RNA, according to a reaction which is
governed by a rate constant (k_1), the RNA has two fates, it can either
go on to give rise to protein, again there is a rate constant involved (k_2), or else it can become degraded, perhaps by ineffectual processing. The main mechanism is probably deciding whether to process
or not and so you get another rate constant here (k_3) which gives rise
to degradation products. Now in this model - and of course a series
of genes is expressed - you can postulate that these constants are
affected by the products of other genes so what you then have is a
considerable network. This results in a metastable state within the
cell which can be perturbed by changing any of the elements eventually
involved in control.

These are two extreme models. The digital is a rather rigid
model: It is the sort of model which you might expect in cells which
took stochastic decisions and then were selected out to be one kind
of cell or another. The analogue one is the kind of flexible model
which you might see in cells which diverged gradually and which might
be switched from one differentiated form to another. It seems to me
that the haematologists by and large are thinking in terms of the
digital, stochastic model and the transdifferentiationists tend to
think in terms of the analogue model. Perhaps this would be a useful
time to consider some of the molecular evidence for these different
models. Ruth, do you think that transdifferentiation can be explained
in terms of this model here?

R. Clayton

I think that I wouldn't like in fact to be very rigid about this
because what we know is derived from a cell population. One can show
that the amount of heterologous message increases at first very slowly
and then very steeply and then it flattens off again, it plateaus,
but really to distinguish digital and analogue models....on the face
of it it would look more like the stoichiometric (analogue) model but
actually one needs to know what goes on on a per cell basis. There

is evidence that the initial population may well be heterogeneous (Okada et al., 1979), and we also suspect that the content of crystallin mRNA per cell and the number of cells with sufficient of it can depend on the age of the embryo (Thomson et al., 1979, 1981), but we really have to wait until we have done a whole lot of in situ measurements and assess the heterogeneity of the cell population.

U. Schibler

To my mind there is not one single gene which does not require both models for its regulation. The digital model may explain selection, but the analogue model is necessary for modulation of gene activity. Selection of a gene is not enough. You have to tell the gene how much to make.

J. Paul

My question is do you select a gene, because if you look at the saturation levels that we were hearing about the other day about 40% of the DNA in liver was being transcribed. Is that right?

N. Sueoka

RNA complexity measurement of the total nuclear RNA show that in rats, for example, brain, liver, and kidney nuclear RNAs have unique genome complexities of 30%, 20% and 10% respectively, in a nested fashion, in other words, 10% of kidney nuclear RNA complexity is included in the 20% of liver nuclear RNA complexity and 20% of liver RNA complexity is included in that of the brain (Chikaraishi et al., 1978). Cytoplasmic RNA complexities are 20% in brain and 5% in liver (Beckman et al., 1981). However, we are dealing with a large number of different RNAs (typically 1% of RNA complexity means roughly 10 000 gene transcripts). So unless you have enormous change in the spectrum of transcription during differentiation, it is impossible to detect small differences in RNA complexity. However, if you have specific gene clones, you should obviously be able to measure the average numbers of their transcripts per cell during differentiation.

U. Schibler

One might get more conclusive answers if one looks at specific genes. The question is: e.g. do neurones and myotubes produce significant amounts of globin mRNA?

J. Paul

Well, this has been claimed you know.

U. Schibler

Most of these claims were solely based on relatively simple hybridization experiments. Hybridization signals are certainly not sufficient for identification of a bonafide gene product.

J. Paul

That's right.

P. Harrison

And also that the initiation of the transcript is specific rather than it being merely a consequence of run-through from another gene, for example.

U. Schibler

That's what I meant.

J. Paul

I think that all that you say is correct, and therefore it is incumbent on the people who believe in the ON/OFF situation to prove that these antibodies explain the findings, because the fact of the matter is that we do see low levels of specific products in other tissues.

R. Clayton

Intuitively I would feel happier with a stoichiometric (analogue) model, as matching rather better what we see, with the proviso that in fact there is no critical data as yet until people relate what happens in the cell population to what happens in the single cell.

M. Buckingham

If you have low levels of the message then you don't really know whether there are one or two cells in the population which happen to be "leaky" or whether every cell in the population is expressing at a low level. I would have thought, because one is limited by the few tools one has for looking at this, that the DNase I tool, and maybe the methylation tool too, would tend to suggest that in the majority of tissues the majority of genes which would express differentiated products not characteristic of that tissue, are in an inactive conformational state whatever that means. I think you have to talk about the DNase I experiments, and maybe the methylation experiments as well, because these do suggest there is a conformational difference.

J. Paul

I challenge the DNase I experiments. I think the DNase I experiments are fine in so far as they apply to globin gene, ovalbumin genes and things like that, these are what I describe as in flagrante transcripto but what about other genes, which are transcribed perhaps twice in an hour, this is certainly true of most genes, there is no question about that.

M. Buckingham

If you stimulate a system with a hormone, for example, then the genes which have been stimulated by that hormone will remain in a DNase I-accessible state even if you can't detect any transcripts coming off them at all.

D. de Pomerai

But ovalbumin mRNA can be detectably induced by hormone in tissues like liver where it is expressed only at very low levels.

R. Clayton

That's right.

M. Buckingham

What I am saying is that if John Paul's model is correct, which indeed it would seem to be, that a gene may only be transcribed once in an hour or something like that, then probably the experimental evidence would suggest that that gene is always in the active DNase I conformation.

J. Paul

Well, I only know of one experiment that has been done to test this: it is an experiment that we did some years ago in which we took total HnRNA from cells and looked at DNA from DNase I-digested and undigested nuclei. In that case certainly the evidence suggested that virtually all active genes were rather more sensitive to DNase but it didn't say they all were. The evidence suggested that there is a great spectrum of sensitivity.

D. de Pomerai

Could it be, not so much that these gene sites are differentially susceptible to DNase I but rather that they are sensitive only some of the time, perhaps while they are actually being transcribed, after which they close up again?

M. Buckingham

I think all the evidence is against that and I think the ovalbumin and vitellogenin experiments, for example, show that that is not true.

A. Bird

What is the evidence for a spectrum of DNase I sensitivity?

P. Harrison

In an erythroid cell, in general the genes outside the globin gene region are not sensitive whereas the globin gene cluster is moderately sensitive and the globin genes which are actually being transcribed are even more sensitive. That is quite apart from hypersensitive sites (usually to the 5' side of transcribed genes) which might be trigger points for chromosome folding.

G. Birnie

With DNase I sensitivity experiments we have to remember that the answer obtained is not a simple yes or no. The genes which are being transcribed are much more susceptible to digestion than the rest which are not, but they are also susceptible to digestion, given enough DNase and enough time.

THE DIFFERENTIATED STATE AND ITS REGULABILITY

J. Paul

Yes, in fact if you look at the kinetics of digestion of several genes in a tissue, some of which are active and some of which have been active, you find a spectrum of digestibilities. There are two elements to the DNase I digestion not everyone of you will be familiar with. One is the total digestibility of the gene and the other is the very high sensitivity of specific sequences, which is another question. It's not one I am addressing myself to.

G. Harrison

There is one other point as well which I am sure Dr. Birnie would agree with. He has tried to look at it in some ways and so are we. In the past, for purely technological reasons the genes coding for super-abundant messengers in any system have been studied most intensively, like globin or ovalbumin or whatever. Now what we need to know is whether the same principles apply to the tissue-characteristic but less abundant gene products. Davidson's model of gene regulation makes such a distinction, for example.

M. Buckingham

The muscle messages for example are not superabundant messages. They are between 0.5 and 2%.

U. Schibler

We are looking at a gene, (α-amylase gene Amy-1^a) which produces vastly different amounts of mRNA's in two different tissues. In the parotids and liver α-amylase mRNA accounts for 2% (10^4 molecules/cell) and 0.02% (10^2 molecules/cell) respectively. Quantitative S_1 nuclease mapping experiments suggest that nuclear precursor mRNA colinear with the genomic Amy-1^a sequence are also 100 fold more abundant in salivary gland than in liver. Assuming that the sites of transcription initiation coincide with the cap sites, these results indicate that the salivary specific promoter of Amy-1^a is either silent or extremely inactive in liver. Therefore, it appears likely that, at least in the case of tissue-specific α-amylase production, transcriptional control plays an important role.

J. Paul

Well, I think possibly that's right. There are several ways that can work, of course. One way you could inactivate your active promoter would be by putting it into some sort of chromosomal conformation that made it inaccessible. That's one. The second possibility is that you can have a diffusible repressor which would bind nearby, that's another one. A third possibility is that you could have a site between two promoters that was particularly sensitive to a processing enzyme which is present in one case and not in another.

U. Schibler

But we know that the whole area upstream is transcribed, even at low frequency. In those cases the region upstream from that is transcribed at low frequency. Aside from that I did not make any speculations right now on how this is accomplished. The only thing I said is it's digital.

J. Paul

Yes, I think we are getting into a little bit of philosophical point about when this analogue becomes digital. I want to avoid that one, actually. Let us consider that according to the analogue model you eventually in the extreme get to the digital situation.

R. Rifkind

The digital doesn't imply that there is no product under one switch and product under the other. It doesn't have to be 'none' and 'all'.

J. Paul

Well, I think that's the implication.

R. Rifkind

It doesn't have to be. The mechanism you have created can actually modulate between levels. It can modulate between a low, inconsequential level of one per minute or one per hour transcription and one in which there are polymerases along the whole thing. That could be the modulation. It is a digital mechanism.

J. Paul

That I would put in the analogue category.

R. Rifkind

But that is arbitrary. It is still made by exactly the same mechanism you proposed.

J. Paul

What I am talking about here essentially is the mechanisms which people have proposed and envisaged. This is a kind of old-fashioned one, but it is still around. The notion here is this sort of thing that you actually have different cofigurations.

R. Rifkind

But the difference does not have to be between 'on' and 'off'.

J. Paul

I think that raises another question because I think the proposals which have been around do say that, in fact. Does it happen? That is really the question I am raising. So I thought I would go along this way and say that it doesn't have to be 'on' or 'off'. And I then suggested it might involve an equilibrium situation. If whereas in the 'digital model' you don't have that, you simply have two different states of chromatin and the question is how strong is evidence for this one.

Y. Courtois

In this context, I want to ask you if differentiation could be associated with gross modifications in chromatin structure. You know the experiments by Mary Weiss using cell hybrids where she shows that you can correlate the differential expression of albumin with different size in the repeat unit of the DNA.

So my question is: is this a general mechanism that could be observed in Friend cells or in any other described differentiating system? If this is the case and if there is between stage A and stage B a progressive change in the whole structure of the chromatin, could this be the mechanism which allows a set of genes to be expressed at a time?

J. Paul

It is certainly a good question, but I think the hybrid situation is one that we haven't raised. It is quite interesting in that in many hybrid cells you can detect the expression of genes which are not expressed in the cell from which the genome is derived.

P. Harrison

But not just <u>any</u> gene. There again, which genes are expressed is related to the parental cell types. One can activate genes expressed in one or both of the parental cells but not anything. Sometimes both parental gene sets coding for the same protein become active. The genes which are "silent" in a hybrid cell initially still seem to have some sort of "memory" as to whether they were active in the parental cells prior to fusion, and often they are re-expressed during culture of the hybrid cell due to a change in the chromosome balance or whatever.

Y. Courtois

What about mitosis? Is that not related to the structure of the chromatin?

J. Paul

I think the question then is whether it is a matter of introducing molecules which seize the opportunity to get into the chromatin or is it a matter of introducing molecules which perturb the equilibrium. That is the sort of question I am asking and the reason I am asking it is because in the context of this meeting one could envisage that this might be a rather more labile situation which might fit the transdifferentiation case. Perhaps the two are not mutually exclusive.

P. Harrison

With allosteric-type effects one cannot distinguish the two because although you have two states in equilibrium it is possible to flip between them by very slight changes in the conditions.

J. Paul

Well that's right, but I would make this point that if this mechanism exists in the cell, an absolute switch, it is an unusual mechanism simply because most genes in most cells are transcribed. That is what the complexity experiments tell us.

U. Schibler

I just don't believe that.

A. Radbruch

But it has nothing to do with differentiation.

J. Paul

It forces us to consider the mechanism. You can call them housekeeping genes or whatever you like, but you have 10 000 of them expressed as messengers in nearly every cell type you care to look at.

A. Radbruch

I thought we are now looking for mechanisms that can put special genes into function or not. For example if you look at the immunoglobulin genes, they have been looked at to some extent and there you find that there is gene rearrangement. The daughter cells are different from the parental cells, but on the other hand the gene rearrangement very often leads to a situation where you get transcripts and even translation products, that are very labile if they don't have the proper configuration.

U. Schibler

Sometimes nuclear transcripts don't even get out of the nucleus, like in the case of transcripts from non-rearranged K-constant region genes.

J. Paul

Even in that very special case, I am not sure. How special is a special gene and which genes are special? You would say amylase, of course.

U. Schibler

No, I would argue that most genes so far looked at are transcribed very specifically. Globin sequences are certainly not produced in impressive amounts by muscle cells or fibroblasts. Neither is myosin mRNA readily detected in erythroblasts. Strong tissue specificity is even observed for relatively rare mRNAs like α-amylase mRNA in liver (0.02% of polyadenylated RNA). We were unable to detect any liver type α-amylase mRNA in tissues such as brain, kidney, lung, heart-muscle and spleen.

J. Paul

So a special gene is defined as the kind of gene that a biochemist looks at. That reduces it to about half a dozen out of ten thousand.

R. Clayton

That is limited by the number of biochemists in the world.

G. Birnie

I think that, if we look at the question in this way, we are returning to the old ideas of the gene-masking hypothesis. This postulates that there is a spectrum of genes, the 'housekeeping genes', which are transcribed in every type of cell, and superimposed on this expression is that of the specialized genes, such as the globin genes, whose expression determines the phenotype of the cell. However, when the mRNAs from these so-called housekeeping genes in one type of cell are compared with those in another type, it is found that there are large differences in the relative abundances of sequences within this common group. If these genes are being expressed simply because a cell is a cell, why should there be differences in their relative abundances? Is it not likely that many of these genes also play a role in determining cellular phenotype?

U. Schibler

There is a very important technical point. It seems clear that in a highly specialized cell like the exocrine pancreas cell, in which a few mRNA genes contribute 90% of the mNRA mass, the relative proportion of mNRA encoding housekeeping functions is lower than in a relatively unspecialized cell. Not necessarily because the cellular concentrations of 'housekeeping mRNAs' are lower in specialized cells, but because the tissue-specific genes contribute most of the mRNA mass. It would be more meaningful to compare the absolute cellular concentrations of 'housekeeping mRNAs'.

G. Birnie

There is certainly a paucity of data at the moment. However, what is available does indicate that the concentrations of these common mRNAs in different cells do differ relative to each other. The whole group of mRNAs common to two types of cell do not differ in concert: some are increased in concentration, some are decreased (Birnie et al., 1981).

A. Radbruch

There has been an experiment by Mark M. Davis (personal communication) from the N.I.H. who has absorbed the messenger RNA from T cells on B cells and vice versa and has estimated on that basis, very roughly, the number of different RNAs to about 200, between these two very narrow cell types.

J. Paul

That is about 200 out of about 10 000.

A. Radbruch

But they are very narrow cell types.

U. Schibler

The question really is: does the complexity of the housekeeping function make up 99% and the complexity of the cell- or tissue-specific functions 1% or is it the other way round?

J. Paul

That is the question, or is the difference a quantitative difference. They are the extremes.

M. Buckingham

Could we consider a question that we in the muscle system are particularly aware of, that there may be isogene families or iso-forms of things which all cross-hybridize. That is certainly true of the muscle contractile proteins and there are also indications that there may be different non-muscle contractile proteins. I think Ruth raised the possibility for crystallins too, that what one picks up as crystallin at low levels in other cell types may, in fact, be an iso-form or a product of a different gene anyway.

R. Clayton

In tissues which don't show cross-reaction we cannot disprove that the sensitivity might be at issue. In ones which do cross-react and transdifferentiate we are defining whether it is true crystallin by the product, when we have got enough to analyse, but you couldn't I suppose be certain that if you back-tracked to the earliest stages there wasn't in fact a switch from one iso-form to another. It seems an unnecessary postulate but it can't at present be excluded.

J. Paul

So can we come back to the question are housekeeping genes special genes or are special genes just extreme forms of housekeeping genes?

M. Buckingham

What is a housekeeping globin?

J. Paul

Well, I think if instead of being multicellular we were a large lump of jelly we might still have use for haemoglobin.

R. Clayton

You'd probably use cytochrome....

U. Schibler

I think David Yaffe and other people have shown that during muscle cell differentiation there is a switch from α to β actin. There is no obvious reason why β actin could not do the job all the time.

R. Clayton

There is some evidence that if you haven't got the right product something not too dissimilar will do: For instance there are some fish which have lost one of the lactate dehydrogenase genes, but they do still manage to have the function though they don't have the flexibility.

J. Paul

The peptide hormones, of course, are a good example of what we are talking about because presumably when a particular endocrine function becomes concentrated in certain cells there is an increase in efficiency. Yet that presumably is a function of all cells, and in fact in very many tissues if you look carefullly enough you will find that there is a significant production of most peptide hormones and this brings us back to the question about the lung tumours and whether these arise by transdifferentiation. Now I wish we had a really good embryologist here who knew about human embryology, but, it seems to me from what little I recollect, that there is a good deal of migration if not in the mature animal, certainly during embryological development, and I think that it is not at all impossible that some of the cells that come to lie in the bronchial epithelium at one time might have originated from much the same area as anterior pituitary cells. This may be particularly true of K cells of neuroendocrine type in the lung.

R. Clayton

I can't pretend to remember in sufficient detail the experiments using quail cells to see where neural crest migration does end up but I rather think that some of the sites where you have ectopic hormone tumours are not entered by neural crest cells. John Ratcliffe described in detail a particular set of examples from lung but as he explained, tumours producing inappropriate but genetically normal products are by no means uncommon and such products might include the so-called carcino-embryonic antigens. However they arise, what happens is that the tumour cells have a superabundant product which they shouldn't have and which does characterise other times or other places. I think we could easily find amongst this group of conditions some tissues which might be transdifferentiation-like, because neural crest migration doesn't account for it. Is that not right, Tokindo, that Le Douarin's experiments show that you cannot explain everything by saying "I like to think it must be a neural crest stem cell because otherwise I am uncomfortable about it." Did she ever find that neural crest cell ever got into, for example, trachaea?

T. Okada

No.

J. Paul

But the fact of the matter is that if you remove every endocrine gland you can think of you do not completely abolish the production of peptide hormones. They are nearly all coming from some other place, in small if inadequate amounts.

R. Clayton

That doesn't in the least disagree with the thesis I am putting forward, just for the hell of it, which is that there are cells which are ready to swing into production, their ability related proportionately to what they are already doing. Now whether they do this by producing some new transcripts or whether they do this by some other mechanism, is a separate question. But I don't think that you can explain away everything by hoping that a migrating neural crest cell will save the old theories.

D. McDevitt

Let me answer your question about the anterior pituitary. It arises from the roof of the stomodaeum, very far, in other words from where you might form the branching of the primordial lung. They are both endodermal of course, and also the part of the pituitary that will elaborate those kind of hormones is in the endodermal part.

P. Harrison

You have to remember that these are malignant cells and they could have done some gene rearrangement that allows other gene sets to be amplified or switched on - whatever mechanism you want. The question is, these cells that produce the endocrine hormones, what other differentiation markers do they show? Because if they have switched gene set because of the malignant transformation they might show other endocrine functions, even if they are derived originally from a lung cell, but by a malignant transformation by chance they have switched back to something.

R. Clayton

We have got a genuine pathologist here who perhaps can tell us. Can you, Andrew?

A. Wylie

I am not certain that I quite follow the question.

P. Harrison

We went through this before, but do the cells in the lung that produce these hormones, do they have differentiation markers of lung cells or of endocrine cells or neither or what? What is the situation, because that helps to interpret what is going on.

A. Wylie

I think the immediate situation is that I cannot answer the question, but I don't know of any work which looked specifically for other differentiation markers. What was very surprising from the early mixed cell population studies was that the normal cells of the lung, as I remember from the bronchio carcinoma tissues, you can trace abnormal product in them as well, but in the normal lung tissue of normal people. But that is a different question from the one we are asking.

R. Clayton

Really it suggests that there is a sort of quantitative range of expression that some sort of process, whatever it is, has provoked a large area of lung tissue and that a certain sub-group of these cells has been further provoked into a totally different pathway, in fact that fits in with the analogue theory rather.

J. Paul

Let me put on another hat. I intend to say a little bit about what people are thinking about carcinogenesis, because this is relevant to the whole discussion. In viral carcinogenesis it is now known that the oncorno-viruses have a genome which when it is inserted into the host genome has at its ends what are known as Long Terminal Repeats, which repeat the 3' end of the gene sequence in front of the gene and the 5' end at the end. These 5' ends contain promoter sites, so they obviously promote the transcription of endogenous viral genes, but when they are inserted into the genome of the host they also promote the transcription of adjacent host sequences. This is one way in which genes are actually switched on - by inserting sequences of this kind - and as I mentioned at the beginning, they can work in different configurations. Now does that only apply in the abnormal situation where you have retroviruses? Well, we don't know how abnormal retroviruses are. The fact is that sequences similar to viral oncogenes turn up in species in which retroviruses haven't been described. Moreover, there have been some very interesting recent experiments which show that you can take DNA from tumours and can transform cells. And in human tumours it has been shown by restriction analysis that the bits of DNA that are involved are always the same. So here we have a suggestion that in tumour cells you actually do have a genomic rearrangement a little bit like the rearrangement you get in the production of immunoglobulins. As a final question perhaps we should ask how widespread this kind of mechanism might be, because if it is normally used it is very easy to see how genes can be switched on that are not normally expressed in a cell and this of course would then fit into the category of the digital models.

R. Clayton

There is one experiment that we have done recently which really needs repeating. It is only very preliminary, but we found that MNNG exposure apparently can shift cell differentiation in lens and neural retina and we used it because it produced not tumours but a shift in differentiation when Dr. Eguchi tried it in vivo (Eguchi and Watanabe, 1973). We had a look at what happened to lens cells treated with MNNG after zero hours, 2 hours, 4 hours, 8 hours and so on. The first thing that seemed to happen was that all the polysomes came apart but translatable message was still in the cells: but after a period of time the polysomes had reassembled and what they were carrying was still the messages that were there before, but now in totally different proportions, and the cells were no longer making δ as their main product (Clayton et al., 1980). This resembles findings of Chakrabarty and Schneider, 1978, after thioacetamide treatment of liver cells. Eventually there were other changes including crystallin loss, but the initial change and the initial event may by no means be the same as what you ultimately see as having happened. It may be that if you bash a cell about you make it more susceptible to a whole range of modifications but when you finally analyse a tumour you are looking at the end result of a number of different events. Finally it is possible that we really must not look for one mechanism, and we must not look for one mode of explanation for something as tremendously heterogeneous a phenomenon as the changes in differentiation, transdifferentiation and tumour formation.

REFERENCES

Arking, R., 1978, Tissue-, age- and stage-specific patterns of protein synthesis during the development of Drosophila melanogaster, Develop. Biol., 63:118.
Beckmann, S. L., Chikaraishi, D. M., Deeb, S. S., and Sueoka, N., 1981, Sequence complexity of nuclear and cytoplasmic ribonucleic acids from clonal neurotumor cell lines and brain sections of the rat, Biochemistry, N.Y., 20:2684.
Birnie, G. D., Balmain, A., Jacobs, H., Shott, R., Wilkes, P. R., and Paul, J., 1981, Post-transcription control of messenger abundance, Molec. Biol. Rep., 7:159.

Chakrabarty, P. R., and Schneider, W. C., 1978, Increased activity of rat liver messenger RNA and of albumin messenger RNA modulated by thioacetamide, Cancer Res., 38:2043.

Chikarashi, D. M., Deeb, S. S., and Sueoka, N., 1978, Sequence complexity of nuclear RNAs in adult rat tissues, Cell, 13:11.

Clayton, R. M., Bower, D. J., Clayton, P. R., Patek, C. E., Randall, F. E., Sime, C., Wainwright, N. R., and Zehir, A., 1980, Cell culture in the investigation of normal and abnormal differentiation of eye tissues, in:"Tissue Culture in Medical Research (II)," R. J. Richards and K. T. Rajan, eds., Pergamon Press, Oxford and New York, p.185.

Darmon, M., Rizzino, A, and Sato, G., 1981, Isolation of myoblastic, fibroadipogenic and fibroblastic clonal cell line from a common precursor and the study of their requirements for growth and differentiation, Expl. Cell Res., 133:313.

Eguchi, G., and Okada, T. S., 1973, Differentiation of lens tissue from the progeny of chick retinal pigment cells cultured in vitro: a demonstration of a switch of cell types in clonal cell culture, Proc. Natl. Acad. Sci. U.S.A., 70:1495.

Eguchi, G., and Watanabe, K., 1973, Elicitation of lens formation from the 'ventral iris' epithelium of the newt by a carcinogen, N-methyl-N-nitro-N-nitrosoguanidine, J. Embryol. exp. Morph., 30:63.

Iwig, M., Glaesser, D., and Hieke, H., 1978, Re-activation of nuclei containing lens fibre cells to mitotic growth. Biochemical and immunochemical analysis, Cell Differentiation, 7:159.

Okada, T. S., 1977, A demonstration of lens-forming cells in neural retina in clonal cell culture, Develop. Growth and Differ., 19:47.

Okada, T. S., Yasuda, K., Araki, M., and Eguchi, G., 1979, Possible demonstration of multipotential nature of embryonic neural retina by clonal cell culture, Develop. Biol., 68:600.

Rodgers, M. E., and Shearn, A., 1977, Patterns of protein synthesis in imaginal discs of Drosophila melanogaster, Cell, 12:915.

Schmid, V., 1975, Cell transformation in isolated striated muscle of hydromedusae independent of DNA synthesis, Expl. Cell Res., 94:401.

Schmid, V., 1980, The in vitro regeneration of a functional medusa organ from two differentiated cell types, in:"Invertebrate Systems In Vitro," Kurstak et al., eds., Elsevier/North Holland Biomedical Press, Amsterdam, p.92.

Seybold, W. D., and Sullivan, D. T., 1978, Protein synthetic patterns during differentiation of imaginal discs in vitro, Develop. Biol., 65:69.

Thomson, I., de Pomerai, D. I., Jackson, J. F., and Clayton, R. M., 1979, Lens-specific mRNA in transdifferentiating cultures of embryonic chick neural retina and pigmented epithelium, Expl. Cell Res., 122:73.

Thomson, I., Yasuda, K., de Pomerai, D. I., Clayton, R. M., and Okada, T. S., 1981, The accumulation of lens-specific protein and mRNA in cultures of neural retina from 3½ day chick embryos, Expl. Cell Res., 135:445.

de Vitry, F., Picart, R., Jacque, C., Legault. L., Dupouey, P., and Tixier-Vidal, A., 1980, Presumptive common precursor for neuronal and glial cell lineages in mouse hypothalamus, Proc. Natl. Acad. Sci. U.S.A., 77:4165.

Yasuda, K., Okada, T. S., and Eguchi. G., 1980, Age-dependent changes in the capacity of transdifferentiation of retinal pigment cells as revealed in clonal cell culture, Cell Differentiation, 10:3.

DEMONSTRATION:

HIGH LEVELS OF δ-CRYSTALLIN PRECURSOR RNA IN EARLY DEVELOPMENT

OF CHICK LENS CELLS

D. J. Bower, B. J. Pollock and R. M. Clayton

Department of Genetics
University of Edinburgh
Edinburgh   EH9 3JN

δ-crystallin is the first lens-specific protein detected in bird lens development and the most abundant in the adult bird lens (reviewed Clayton, 1974).

RNA extracted from 6-day embryo chicks has a far higher proportion of nuclear RNA complementary to δ-crystallin coding sequence than RNA extracted from day-old post hatch chicks. Nuclear RNA from 6 day embryos, carrying δ-crystallin coding sequences, consists predominantly of high molecular weight species, although there is also a fair amount of the fully-processed mRNA sequence of 2 000 nucleotides which is the major species found in the cytoplasm (Bower et al., 1981). These results are shown in Fig. 1. Earlier stages of lens development have been studied by in situ hybridisation to sections of three and a half day chick embryo eyes (Fig. 2). At this stage, labelling by the ^{3}H-cDNA probe is in both nuclear and cortical regions of the lens, indicating high levels of δ-crystallin coding sequence. Control sections hybridised to ^{3}H-pBR322, the plasmid vector for the cDNA probe, showed no labelling (not shown). Fig. 3 shows an area of the lens cortex at higher magnification, in which it appears, as in other heavily-labelled regions, that nuclei are more heavily labelled than cytoplasm.

We have not yet obtained satisfactory sections of the earliest stages of development of the eye, with intact RNA, but we have studied the earliest stages of lentoid differentiation in transdifferentiating neural retina from 8-day chick embryos (reviewed Clayton, 1978; Eguchi, 1979; Okada, 1980) cultured on glass slides (see Fig. 4).

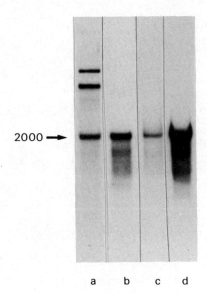

Figure 1. Polysomal and nuclear RNA was extracted from 6 day embryo and day old post hatch chicks, 15 μg per lane was electrophoresed through a 1.2% agarose-formaldehyde gel, and blotted onto nitrocellulose filters. The blots were hybridised to ^{32}P-labelled pM56, a chick δ-crystallin cDNA clone (Bower et al., 1981). Methods as described in Bower et al. (1981). Lane a) 6-day embryo nuclear RNA, b) 6-day embryo polysomal RNA, c) day-old chick nuclear RNA, d) day-old chick polysomal RNA. The band of fully-processed 2 000 nucleotide δ-crystallin mRNA is indicated.

These results suggest that there is an accumulation of δ-crystallin high molecular weight mRNA precursors in the nucleus at an early stage of lentoid differentiation whether in the lens or in transdifferentiating cultures. It is not clear if there is also an increased rate of transcription of the δ-crystallin genes, or whether the high concentration of δ-crystallin sequence is due entirely to the build-up of species which would otherwise be rapidly processed or degraded. The RNA transfer data (Fig. 1) indicate that in the day old lens, large quantities of δ-crystallin mRNA are associated with polysomes and presumably being translated, but the proportion of δ-crystallin precursors in the nuclear RNA is very low. This could be due to either rapid processing of transcripts, or a fall off in transcription of the δ-crystallin genes.

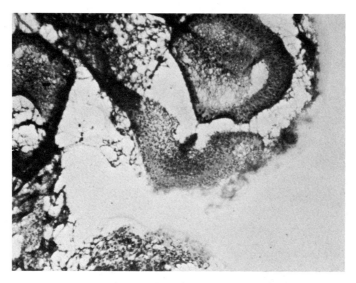

Figure 2. Section of a three and a half day chick embryo eye (x100), fixed and hybridised to a ^{3}H-labelled, cloned, δ-crystallin cDNA, pM56, in 38% formamide, 2 x SSC* for 16 hours at 45°C. After extensive washing, slides were dried and dipped in Ilford nuclear emulsion, then laid down in the dark for eight weeks.
*(1 x SSC = 0.15 M NaCl, 0.015 M Sodium citrate.)

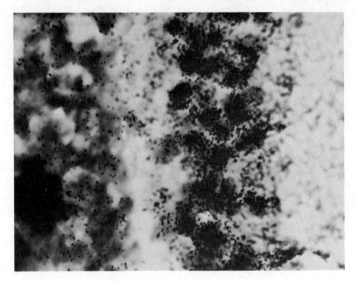

Figure 3. Detail (x 400) from Fig. 3, showing an area of the lens cortex. Black dots are silver grains, indicating a high, local concentration of δ-crystallin coding sequence complementary to the probe pM56. Nuclei are more heavily labelled than cytoplasm.

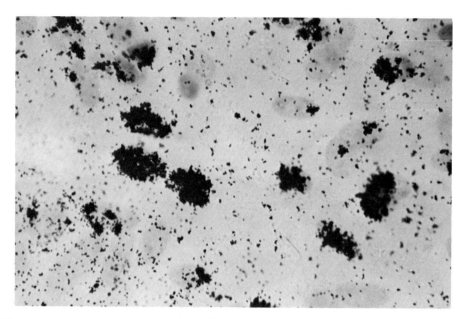

Figure 4. Neural retina (x 90) from 8 day chick embryos, dissociated and cultured in vitro for 25 days, then fixed and hybridised to ^{3}H-pM56, a δ-crystallin cDNA probe, as described for Fig. 3. Regions which are beginning to transdifferentiate into lentoids show very heavy labelling of some nuclei, and very light labelling of cytoplasm. Control slides hybridised to ^{3}H-pBR322 (the control plasmid vector) showed no labelling (not shown).

REFERENCES

Bower, D. J., Errington, L. H., Wainwright, N. R., Sime, C., Morris, S., and Clayton, R. M., 1981, Cytoplasmic RNA sequences complementary to cloned chick δ-crystallin cDNA show size heterogeneity, Biochem. J., 201:339.

Clayton, R. M., 1974, Comparative aspects of lens proteins, in: "The Eye," H. Davson, ed., Academic Press, London, p.339.

Clayton, R. M., 1978, Divergence and convergence in lens cell differentiation: regulation of the formation and specific content of lens fibre cells, in:"Stem Cells and Tissue Homeostasis," B. I. Lord, C. S. Potten and R. J. Cole, eds., Cambridge University Press, Cambridge, London, p.115.

Eguchi, G., 1979, "Transdifferentiation" in pigmented epithelial cells of vertebrate eyes in vitro, in:"Mechanisms of Cell Change," J. D. Ebert and T. S. Okada, eds., John Wiley and Sons, New York, p.273.

Okada, T. S., 1980, Cellular metaplasia or transdifferentiation as a model for retinal differentiation, Current Topics in Develop. Biol., 16:349.

DEMONSTRATION:

HYBRIDIZATION SELECTION AND CELL FREE TRANSLATION OF mRNA FROM A δ-CRYSTALLIN CLONE

L. H. Errington, C. Sime, and R. M. Clayton

Department of Genetics
University of Edinburgh
Edinburgh EH9 3JN

We have translated mRNA which specifically hybridizes to a cDNA δ-crystallin clone, pV89 (Bower et al., 1981). The RNA used was the polyA$^+$ fraction from day old chick lens. Fig. 1 shows that thermal elution of the RNA hybrids at increments of 10°C releases all the translatable RNA at one temperature (70°C). The major and largest of the translation products co-migrates with the δ-crystallin polypeptides (M.wt. 48-50 000 daltons) produced by in vitro translation of total RNA extracted from day old chick lens (Fig. 1) and with δ-crystallin polypeptides extracted from day old chick lens (not shown). There are also four smaller detectable translation products from the hybrid selected RNA with molecular weights of approximately 47 Kd, 44 Kd, 31 Kd and 25 Kd. The largest of these runs slightly faster than the 48 Kd and 50 Kd δ-crystallins, and, along with the 44 Kd and 31 Kd translation products, does not co-migrate with accumulated (cold) proteins from day old chick lens. 47 Kd, 44 Kd and 31 Kd polypeptides therefore appear to be δ-crystallin related products, which if they are synthesised in vivo are rapidly turned over. Only the 25 Kd protein co-migrates with accumulated proteins from day old chick lens.

The same translation products have been obtained with hybrid selected RNA from another δ-crystallin clone pM56 (result not shown) which has been shown to carry a cDNA insert which is only partly homologous to pV89 (Bower et al., 1981).

δ-crystallin cDNA appears therefore to hybridize to RNA which codes for the 48-50 Kd δ-crystallin polypeptides plus several smaller proteins. It has already been shown that although there are only

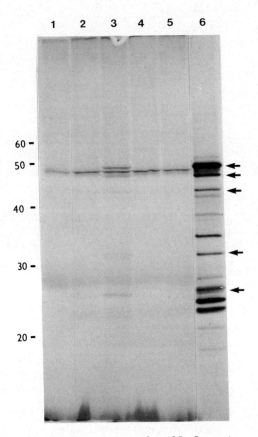

Fig. 1. Hybridization selection and cell free translation of RNA complementary to cloned cDNA sequence in pV89. 10 µg of linearized pV89 plasmid cDNA was bound to nitrocellulose using the 'dot-blot' method of Kafatos et al. (1979). Filters were hybridized to 4 µg of PolyA+ RNA from day old chick lens and washed extensively using a method similar to Riccardi et al. (1979). Thermal elution of hybridized RNA with 2 mM EDTA were performed in increments of 10°C from 60°C to 90°C, eluates at each step being precipitated with 1/15th volume of 3 m Sodium Acetate pH 6.5, containing 75 µg/ml yeast tRNA and 2 volumes of ethanol. RNA from thermal eluates were separately translated in the reticulocyte system (Amersham) according to manufacturers instructions. Polypeptides were fractionated in SDS 12% polyacrylamide gels (Laemmli, 1970) and fluorography was carried out by the method of Bonner and Lasky (1974). Lane 1. Endogenous products of translation system. Lane 2. 60° RNA eluate. Lane 3. 70° RNA eluate. Lane 4. 80° RNA eluate. Lane 5. 90° eluate. Lane 6. Polysomal RNA from day old chick lens. Molecular weights indicated in daltons ($\times 10^{-3}$) on left of figure. Position of hybrid selected products indicated on right.

two δ-crystallin genes (Bhat et al., 1980), there are more than two δ-crystallin mRNAs (Bower et al., 1981). In addition to the major 2 000 nucleotide δ-crystallin mRNA there are several smaller complementary RNA species. We have discussed (Bower et al., 1981) the possibilities that these shorter sequences are translated into functional polypeptides or alternatively have no functional role e.g. they are untranslatable products of errors in transcription or processing, or are translated into partial non-functional polypeptides. The hybrid selection experiments, in detecting δ-crystallin related proteins smaller than the 48-50 Kd δ-crystallins, raises the possibility that these minor RNA species are responsible for their synthesis. The suggestion that a single gene can synthesize more than one RNA product and one protein is not unique. The μ gene of IGM in B-lymphocytes has been shown to encode two separate mRNAs of different sizes, each of which synthesises a protein with slightly different amino acid sequences (Rogers et al., 1982).

ACKNOWLEDGEMENTS

This work was supported by the MRC. One of us (L.E.) was supported by a grant from the Melville Trust before being supported by a grant from the MRC.

REFERENCES

Bhat, S. P., Jones, R. E., Sullivan, M., and Piatigorsky, J., 1980, Chicken lens crystallin DNA sequences show at least two δ-crystallin genes, Nature, Lond., 284:234.

Bonner, W. M., and Laskey, R. A., 1974, A film detection method for tritium labelled proteins and nucleic acids in polyacrylamide gels, Eur. J. Biochem., 46:83.

Bower, D. J., Errington, L. H., Wainwright, N. R., Sime, C., Morris, S., and Clayton, R. M., 1981, Cytoplasmic RNA sequences complementary to cloned chick δ-crystallins show size heterogeneity, Biochem. J., 201:339.

Kafatos, F. C., Jones, C. W., and Efstratiadis, A., 1979, Determination of nucleic acid sequence homologies and relative concentrations by a dot hybridization procedure, Nucleic Acids Res., 7:1541.

Laemmli, U. K., 1970, Cleavage of structural proteins during the assembly of the head of bacteriophage T4, Nature New Biol., 227:680.

Riccardi, R. P., Miller, J. S., and Roberts, B. F., 1979, Purification and mapping of specific mRNAs by hybridization-selection and cell-free translation, Proc. Natl. Acad. Sci. U.S.A., 76:4927.

Rogers, J., Early, P., Carter, C., Calane, K., Bond, M., Hood, L., and Wall, R., 1980, Two mRNAs with different 3' ends encode membrane-bound and secreted froms of immunoglobulin μ chains, Cell, 20:303.

DEMONSTRATION:

THE DEDIFFERENTIATION OF CHICK LENS EPITHELIUM IN CELL CULTURE AND

THE EFFECT ON THIS PROCESS OF EXPOSURE TO A CARCINOGEN IN VITRO

C. E. Patek and R. M. Clayton

Department of Genetics
University of Edinburgh
Edinburgh EH9 3JN

The appearance of fibroblast-like morphology in long term cell cultures of lens epithelium (LE) and the loss of crystallin expression in these cultures has been reported for rat and bovine LE cells (Tassin et al., 1976; Courtois et al., 1978; van Venrooij et al., 1974; Hamada et al., 1979; Rink and Vornhagen, 1979), but long term avian LE cultures have not been studied. Day old chick LE cell cultures can differentiate to give lentoids which are terminally differentiated lens fibre cell masses (Okada et al., 1971), and lentoids have been obtained even in tertiary cultures, the growth rate of which was often lower than that of primary cultures (Eguchi et al., 1975; Clayton et al., 1976). Lentoid formation is accompanied by an increase in α, β and δ crystallins, but the possibility that longer term cultures might lead to crystallin loss and fibroblast-like morphology similar to those found for mammalian LE cells requires investigation. The exposure of LE cells in culture to a carcinogen, N-methyl-N'-nitro-N-nitrosoguanidine (MNNG) on either day 0 or day 7 of culture was found to lead to delayed effects in late primary and in secondary cultures including the appearance of some fibroblast-like cells. If the LE cells were treated with MNNG on the day of plating (day 0), then other unexpected cell types also appeared (Clayton et al., 1980; Clayton and Patek, these proceedings; Patek and Clayton, submitted for publication). We report here on the changes observed in long term chick LE cultures and compare these with the long term effects of exposure to MNNG on the 7th day of primary culture.

Day old chick LE cells from the Hy-2 strain (reviewed Clayton, 1979) were seeded at 0.8×10^5 cells/ml and grown as previously described (Clayton et al., 1980). Cultures exposed to MNNG, which has

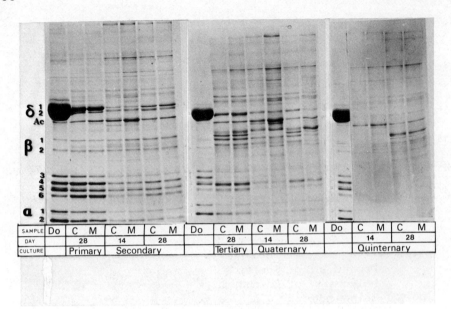

Figure 1. SDS polyacrylamide slab gel electrophoresis (55 µg protein per lane) of day old chicken lens standard (DO) and crystallin lens proteins (α, β and δ) and actin (Ac) accumulated by a series of day old chicken lens epithelium (Hy2 genotype) subcultures and the influence of MNNG (M; 7.5 µg/ml) added once on day 7 of the primary culture.

a half life of 90 minutes in solution (Bertram and Heidelberger, 1974) were treated once on day 7 of primary culture for 60 minutes with medium containing MNNG at 7.5 µg/ml, a level found to be subcytotoxic. The cultures were washed and fresh medium added. The plates were subcultured every 28 days and seeded at the orginal cell density. Sample cultures were harvested at 14 day intervals and the water-soluble proteins extracted and analysed by SDS polyacrylamide slab gel electrophoresis as previously described (Clayton et al., 1980). Crystallins were identified by reference to day-old chick lens crystallin standards as shown in Fig. 1. These are the δ crystallins: $\delta 1$ (50K) and $\delta 2$ (48K); β crystallins, 1-6, $\alpha 1$ (αB) and $\alpha 2$ (αA). Actin (Ac) was identified in protein extracts and its position identified on the gels by using affinity chromatography on DNAase 1 sepharose 4B columns (Lazarides and Lindberg, 1974).

The crystallin proteins accumulated by a series of control and MNNG-treated LE subcultures are shown in Fig. 1. Primary LE control cultures contain only epithelial cells and from 9-11 days onwards differentiated lentoid bodies. By 28 days the cultures had accumulated high levels of α, β and δ crystallins and had a high $\delta 1:\delta 2$ ratio. Secondary cultures were morphologically similar but showed reduced levels of $\alpha 1$ and 2, $\beta 3$ and 6 and $\delta 1$ crystallins and some reduction in the number of lentoid bodies. Tertiary cultures contained a small proportion of fibroblast-like cells and less than half the number of lentoid bodies than primary cultures while the accumulation of $\delta 1$, $\beta 3$ and $\beta 4$ crystallins was either further reduced to very low levels or in some samples was undetectable. Quaternary cultures contained a proportion of fibroblast-like cells and roughly a fifth of the number of lentoids found in primary cultures while crystallin accumulation was further reduced. The levels of α, $\beta 1$ to $\beta 6$ were low and $\delta 1$ was undetectable. Quinternary cultures which showed no lentoid differentiation contained a high proportion of fibroblast-like cells, some of which were vacuolated, and undifferentiated epithelial cells. Only $\delta 2$ and $\beta 1$ and $\beta 2$ crystallins were detectable, but at very low levels, and actin was a major component. The cell number decreased steadily from primary to quinternary cultures, so that an increasing number of dishes had to be harvested for each subculture. By the 5th subculture the growth rate as judged by final cell number per dish was reduced to 10% of that of primary cultures. The small lentoids which appeared in tertiary and quaternary cultures formed 7-12 days later than those in primary and secondary cultures.

MNNG treated primary cultures, examined at 28 days, were morphologically similar to controls, had only half the number of cells and roughly half the number of lentoid bodies. The accumulation of $\alpha 1$, $\beta 3$, $\beta 5$ and $\delta 2$ crystallins was depressed. Secondary cultures appeared to make some recovery, in that they were more similar to controls in cell number, the number of lentoids, and the general pattern of crystallin loss, although quantitative differences in the balance of crystallins was observed. In subsequent subcultures the proportion of fibroblast-like cells was higher and the level of actin was higher in all MNNG-treated cultures than in corresponding controls, but epithelial cells were also present. In both control and MNNG-treated cultures, the actin levels and fibroblast-like cell population was highest at 14 days. When lentoids developed, they contributed the major part of the protein. In contrast to quinternary control cultures, quinternary MNNG-treated cultures contained a very small number of lentoid bodies and had very slightly higher levels of accumulated δ and $\beta 1$ and 2 crystallins.

DISCUSSION

Our data show that day old chick lens cells which have long been known to be able to differentiate to fibre cells in culture (Okada et al., 1971) can also dedifferentiate in culture and that when this happens, crystallins are diminished or lost in a particular sequence, and there is a gradual reduction in the number and size of lentoids.

The regular ontogenic changes in the crystallin composition of both fibres and epithelial cells show that these cells do not have an obligate crystallin composition (Clayton, 1974). Their differential response to various media and growth conditions in primary culture also implies non-coordinate regulation (Clayton et al., 1976), and their differential susceptibility to loss following both ageing in vitro and MNNG treatment, is in agreement with this. Comparison of the diminution or loss of crystallin expression in several genotypes, in ageing in vitro, and following MNNG treatment shows that although the rate of loss varies the sequence does not (Patek and Clayton, in preparation). Some of the changes during long term culture appear to be the reverse of those occuring during differentiation. For example the ratio of 50K to 48K δ crystallin rises during differentiation but falls in long term culture, and $\beta 3$, one of the later ones to appear in differentiation is also amongst the earliest lost. Whether the sequence of changes is related to transcriptional or post transcriptional sequences to cell selection or both remains to be investigated.

Although we have not analysed lentoid composition independently of the remainder of the cells, they contribute the major part of the total terminal culture protein (de Pomerai et al., 1977), so that the small lentoids in quaternary and quinternary cultures would seem to differ from the large ones formed in primary and secondary cultures. The size difference may be related to the crystallin composition, or to the properties or the number of cells able to form lentoids.

REFERENCES

Bertram, J. S., and Heidelberger, C., 1974, Cell cycle dependancy of oncogenic transformation induced by N-methyl-N'-nitro-N-nitrosoguanidine in culture, Cancer Res., 34:526.

Clayton, R. M., 1974, Comparative aspects of lens proteins, in:"The Eye," H. Davson, Ed., p.399, Academic Press, London.

Clayton, R. M., 1979, Genetic regulation in the vertebrate lens cell, in:"Mechanisms of Cell Change," J. D. Ebert and T. S. Okada, eds., p.129, John Wiley, New York.

Clayton, R. M., Eguchi, G., Truman, D. E. S., Perry, M. M., Jacob, J., and Flint, O. P., 1976, Abnormalities in the differentiation and cellular properties of hyperplastic lens epithelium from strains of chickens selected for high growth rate, J. Embryol. exp. Morph., 35:1.

Clayton, R. M., Bower, D. J., Clayton, P. R., Patek, C. E., Randall, F. E., Sime, C., Wainwright, N. R., and Zehir, A., 1980, Cell culture in the investigation of normal and abnormal differentiation of eye tissues, in:"Tissue Culture in Medical Research (II)," R. J. Richards and K. T. Rajan, eds., p.185, Pergamon Press, Oxford and New York.

Clayton, R. M., and Patek, C. E., Apparent redifferentiation of chicken lens epithelium by N-methyl-N'-nitro-N-nitrosoguanidine in vitro, these proceedings.

Courtois, Y., Simmonneau, L., Tassin, J., Laurent, M., and Malaise, E., 1978, Spontaneous transformation of bovine lens epithelial cells, Differentiation, 10:23.

Eguchi, G., Clayton, R. M., and Perry, M. M., 1975, Comparison of the growth and differentiation of epithelial cells in in vitro cell culture, Develop. Growth and Differ., 17:395.

Hamada, Y., Watanabe, K. Y., Aoyama, H., and Okada, T. S., 1979, Differentiation and dedifferentiation of rat lens epithelial cells in short and long-term cultures, Develop. Growth and Differ., 21:205.

Lazarides, U., and Lindberg, U., 1974, Actin is the naturally occurring inhibitor of deoxyribonuclease I., Proc. Natl. Acad. Sci. U.S.A., 71:4742

Okada, T. S., Eguchi, G., and Takeichi, M., 1971, The expression of differentiation by chicken lens epithelium in in vitro cell culture, Develop. Growth and Differ., 13:323.

Patek, C. E., and Clayton, R. M., The dedifferentiation and redifferentiation of chicken lens epithelium by N-methyl-N'nitro-N-nitrosoguanidine in vitro, submitted for publication.

de Pomerai, D. I., Pritchard, D. J., and CLyaton, R. M., 1977, Biochemical and immunological studies of lentoid formation in cultures of embryonic chick neural retina and day-old chick lens epithelium, Develop. Biol., 60:416.

Rink, H., and Vornhagen, R., 1979, Crystallins of lens epithelium cells during ageing and differentiation, Ophthal. Res., 11:355.

Tassin, J., Simmonneau, L., and Courtois, Y., 1976, Epithelial lens cells: a model for studying in vitro ageing and differentiation, in:"Biology of the Epithelial Lens Cell in Relation to Development, Ageing and Cataract," Y. Courtois and F. Regnault, eds., p.145, INSERM, Paris.

DEMONSTRATION:

CELL DELETION IN DIFFERENTIATION

A. H. Wyllie, R. G. Morris and A. J. Strain

Department of Pathology
University of Edinburgh
Edinburgh   EH8 9AG

There are many instances in morphogenesis where cell deletion plays a role complementary to mitosis and differentiation (reviewed by Wyllie, Kerr and Currie, 1980). In some situations there is clear evidence that the death of certain cells at appropriate periods of development or in response to physiological stimuli is genetically determined (Hinchliffe and Thorogood, 1974; Murphy, 1974; Sibley and Tomkins, 1974; Coffino, Bourne and Tomkins, 1975; Robertson and Thomson, 1982). The question thus arises whether the initiation of cell death in the course of development is the result of selective gene expression analogous to other processes in differentiation discussed in this volume.

Studies have been hindered by the fact that the dying cells are usually surrounded and outnumbered by their viable neighbours. It has been possible, however to demonstrate that there is a characteristic mophology associated with cell death in development of vertebrates and some, but not all, invertebrates; this differs markedly from the morphology usually observed in death due to non-physiological stimuli (Wyllie, 1981). This distinctive mode of death, called _apoptosis_, is characterised by nucleolar disruption, generalised condensation of nuclear chromatin, and cytoplasmic shrinkage (Wyllie et al., 1980). Several studies have also shown that this type of cell death may be averted by inhibitors of RNA or protein synthesis (Weber, 1965; Tata, 1966; Lockshin and Beaulaton, 1974; Pratt and Greene, 1976).

Rat cortical thymocytes provide a useful model system for the study of developmentally regulated cell death. The lifespan of these cells, normally short (Joel et al., 1977), is further truncated by glucocorticoid hormones in low pharmacological concentrations (Lundin

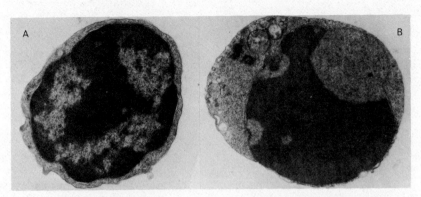

Figure 1. Comparison of the ultrastructure of a normal thymocyte (A) and a thymocyte rendered apoptotic by treatment with 10^{-5} M methylprednisolone in vitro (B). Whereas the heterochromatin of the normal cell is arranged in granular peripheral aggregates and a central body around the nucleolus, that of the apoptotic cell extends widely across much of the nucleus. In the apoptotic cell endoplasmic reticulum is dilated but otherwise cytoplasmic architecture is relatively well preserved; the centrioles are evident in this section. (Uranyl acetate and lead citrate, x11 000.)

and Schelin, 1965; van Haelst, 1967). The morphology of steroid-induced thymocyte death is characteristic of apoptosis and this can be faithfully reproduced by as little as 1-2 h exposure to the hormone in vitro (Fig. 1 and Wyllie, 1980). A small number of new protein species are induced in these cells coincident with the appearance of chromatin condensation (Voris and Young, 1981). Although there is at present no proof that these new proteins cause any of the changes of apoptosis, it is suggestive that concentrations of cycloheximide and actinomycin D which inhibit protein synthesis also completely abrogate the chromatin changes (Wyllie, 1982). Thymocyte apoptosis also involves a substantial rise in cellular buoyant density (Fig. 2a). This allows homogeneous populations of apoptotic cells

Figure 2. (a) Separation of apoptotic cells by isopyknic centrifugation. The cell numbers recovered from consecutive fractions of 2 linear Percoll gradients are shown. The gradients were loaded with equal numbers of thymocytes treated with 10^{-5} M methylprednisolone for 2 h (●) or their untreated controls (o) and centrifuged for 10 min at 2 000 g. The refractive index (proportional to the Percoll density) of the gradients is also shown (stippled line). The new peak in the treated population, with modal density of 1.094 g/ml, consisted of apoptotic cells.

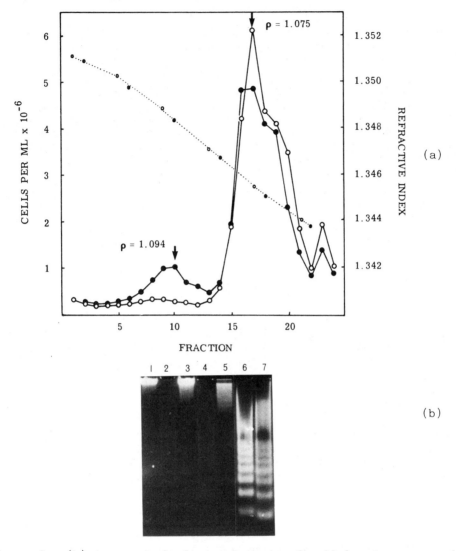

Figure 2. (b) Agarose gel electrophoresis of cell lysates prepared from control, untreated thymocytes (1, 2), treated thymocytes of normal density and morphology (3, 4), and treated, apoptotic thymocytes (5, 6). Lysates from equal numbers of cells were centrifuged at 27 000 g for 10 minutes to separate sedimentable (1, 3, 5) from non-sedimentable (2, 4, 6) chromatin. DNA from these chromatin fractions were compared to that obtained from a micrococcal nuclease digest of freshly prepared normal nuclei (7).

to be isolated and studied biochemically.

Using such populations we have shown that most of the chromatin in apoptotic thymocytes is fragmented to oligonucleosome chains (Fig. 2b). In nucleosome periodicity, sedimentation velocity and buoyant density, this material is identical to authentic, normal chromatin digested by micrococcal nuclease. Studies on the ultrastructure of nuclei deproteinised in 0.2 M HCl confirm that in apoptosis there is collapse of the normal intranuclear architecture of chromatin fibres, but at least the peripheral pore lamina of the nuclear matrix appears intact. It thus seems probable that one major biochemical feature of apoptosis is activation of an endogenous intranuclear nuclease.

The enzyme responsible has not yet been identified. There is however a substantial body of evidence from several systems that lysosomal activation is not involved in apoptosis (Kerr, 1965; Searle et al., 1973; Pannese, 1976; Pipana and Sterle, 1979; Wyllie, 1980, 1982). The nuclei of thymocytes and several other cell types are known to contain nucleases capable of effecting the required pattern of double-stranded DNA cleavage (Wang and Furth, 1977; McKenna et al., 1981; Nakamura et al., 1981). Moreover, similar chromatin cleavage has been observed in cells undergoing chromatin condensation in the course of terminal differentiation, namely normoblasts (Williamson, 1971) and lens cells (Appleby and Modak, 1977). Nuclease activity correlating with the presence of chromatin condensation has been isolated in crude cell fractions from developing lens tissue (B. Herve, personal communication).

In conclusion, rodent thymocytes demonstrate endogenous endonuclease activation as they undergo apoptosis in response to glucocorticoid hormones. This is associated with evidence for new gene expression, although a causal relationship is not yet proven. Similar nuclease activation, with chromatin condensation, occurs in other instances of cell death in development and morphogensis.

ACKNOWLEDGEMENTS

This work was supported by a grant from the Cancer Research Campaign.

REFERENCES

Appleby, D. W., and Modak, S. P., 1977, DNA degradation in terminally differentiating lens fibre cells from chick embryos, Proc. Natl. Acad. Sci. U.S.A., 74:5579.
Coffino, P., Bourne, H. R., and Tomkins, G. M., 1975, Mechanisms of lymphoma cell death induced by cyclic AMP, Amer. J. Path.,

81:199.

Hinchliffe, J. R., and Thorogood, P. V., 1974, Genetic inhibition of mesenchymal cell death and the development of form and skeletal pattern in the limbs of talpid[3] mutant chick embryos, J. Embryol. Exp. Morphol., 31:747.

Joel, D. D., Chanana, A. D., Cottier, H., Cronkite, E. P., and Laissue, J. A., 1977, Fate of thymocytes: studies with ^{125}I-iododeoxyuridine and ^{3}H-thymidine in mice, Cell Tissue Kinet., 10:57.

Kerr, J. F. R., 1965, A histochemical study of hypertrophy and ischaemic injury of rat liver with special reference to changes in lysosomes, J. Path. Bact., 90:419.

Lockshin, R. A., and Beaulaton, J., 1974, Programmed cell death. Cytochemical appearance of lysosomes when the death of the intersegmental muscles is prevented, J. Ultrastruct. Res., 46:63.

Lundin, P. M., and Schelin, U., 1966, The effect of steroids on the histology and ultrastructure of lymphoid tissue. I. Acute thymic involution, Path. Europ., 1:15.

McKenna, W. G., Maio, J. J., and Brown, F. L., 1981, Purification and properties of a mammalian endonuclease showing site-specific cleavage of DNA, J. Biol. Chem., 256:6435.

Murphy, C., 1974, Cell death and autonomous gene action in lethals affecting imaginal discs in Drosophila melanogaster, Develop. Biol., 39:23.

Nakamura, M., Sakaki, Y., Watanabe, N., and Takagi, Y., 1981, Purification and characterization of the Ca^{2+} plus Mg^{2+}-dependent endo-deoxyribonuclease from calf thymus chromatin, J. Biochem., 89:143.

Pannese, E., 1976, An electron microscopic study of cell degeneration in chick embryo spinal ganglia, Neuropathol. Appl. Neurobiol., 2:247.

Pipan, N., and Sterle, M., 1979, Cytochemical analysis of organelle degradation in phagosomes and apoptotic cells of the mucoid epithelium of mice, Histochemistry, 59:225.

Pratt, R. M., and Greene, R. M., 1976, Inhibition of palata epithelial cell death by altered protein synthesis, Develop. Biol., 54:135.

Robertson, A. M. G., and Thomson, J. N., 1982, The morphology of programmed cell death in the ventral nerve cord of Caenorhabditis elegans larvae, J. Embryol. Exp. Morph., in press.

Searle, J., Collines, D. J., Harmon, B., and Kerr, J. F. R., 1973, The spontaneous occurrence of apoptosis in squamous carcinomas of the uterine cervix, Pathology, 5:163.

Sibley, C. H., and Tomkins, G. M., 1974, Mechanisms of steroid resistance Cell, 2:221.

Tata, J. R., 1966, Requirement for RNA and protein synthesis for induced regression of tadpole tail in organ culture, Develop. Biol., 13:77.

Van Haelst, U., 1967, Light and electron microscopic study of the normal and pathological thymus of the rat. II. The acute thymic involution, Z. Zellforsch. Mikrosk. Anat., 80:153.

Voris, B. P., and Young, D. A., 1981, Glucocorticoid-induced proteins in rat thymus cells, J. Biol. Chem., 11319.

Wang, E.-C., and Furth, J. J., 1977, Mammalian endonuclease, DNAse V. Purification and properties of enzyme of calf thymus, J. biol. Chem., 252:116.

Weber, R., 1965, Inhibitory effect of actinomycin D on tail atrophy in Xenopus larvae at metamorphosis, Experientia, 21:665.

Williamson, R., 1970, Properties of rapidly labelled deoxyribonucleic acid fragments from the cytoplasm of primary cultures of embryonic mouse liver, J. Mol. Biol., 51:157.

Wyllie, A. H., 1980, Glucocorticoid-induced thymocyte apoptosis is associated with endogenous endonuclease activation, Nature, 284:555.

Wyllie, A. H., 1981, Cell death: a new classification separating apoptosis from necrosis, in:"Cell Death in Biology and Pathology," I. D. Bowen and R. A. Lockshin, eds., Chapman and Hall, London, New York, p.9.

Wyllie, A. H., 1982, An in vitro model for programmed cell death: steroid-induced apoptosis of thymocytes and T-cell lines, Differentiation, in press.

Wyllie, A. H., Kerr, J. F. R., and Currie, A. R., 1980, Cell death: the significance of apoptosis, Int. Rev. Cytol., 68:251.

CLONING AND STRUCTURE OF δ-CRYSTALLIN GENES

Kunio Yasuda, Noboru Nakajima, Kiyokazu Agata,
Hisato Kondoh, T. S. Okada and Yoshir Shimura

Department of Biophysics
Kyoto University
Kyoto 606
Japan

Total chick DNA was limit-digested with EcoRI and subjected to agarose gel electrophoresis. DNA fragments showed a broad smeary banding pattern in the gel after staining with ethidium bromide. The DNA fragments were transferred to a nitrocellulose filter by the method of Southern (1975). The filter was treated according to the procedure of Denhardt (1966) and those DNA fragments containing the δ-crystallin genes (Clayton, 1974; Piatigorsky, 1982) were identified by hybridization with ^{32}P-labeled δ-crystallin cDNA followed by autoradiography. The autoradiograms show that the δ-crystallin gene sequences are contained in at least four major EcoRI fragments of 11 kb, 7.6 kb, 4.0 and 3.0 kb (Fig. 1). The fragments hybridizable to the cDNA probe were extracted from the agarose gel slices, ligated to the 'arms' of λgt WES λB, and subsequently packaged into phages. The recombinant phages containing the δ-crystallin gene sequences were screened by hybridization to the cDNA probe. These phages were cultured and their DNAs were prepared. EcoRI digestion of DNA of the clones, λgδ28, λgδ1, λgδ43, and λgδ44 produced fragments of 11 kb, 7.6 kb, 4.0 kb, and 3.0 kb, respectively. Those fragments were hybridized to ^{32}P-labeled cDNA (Fig. 1).

To identify those DNA fragments containing the right and left halves of the genomic δ-crystallin gene sequences, the EcoRI digests of the DNA of the clones were electrophoresed, blotted to the filter and hybridized to the $pB\delta7_R$ and $pB\delta7_L$ probes. pBδ7 is a recombinant plasmid clone containing δ-crystallin cDNA of approximately 1.0 kb. Digestion of insert of pBδ7 DNA with EcoRI resulted in two fragments containing the right ($pB\delta7_R$) and left ($pB\delta7_L$) halves of the cloned cDNA. The 11 kb and 4 kb DNA fragments were hybridizable to the pB probes, thereby indicating that these EcoRI fragments represent the 3' terminal regions of the δ-crystallin gene sequences. An analysis

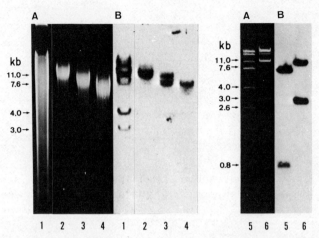

Figure 1. Agarose gel electrophoresis of various DNA fragments and identification of δ-crystallin DNA by Southern hybridization. A shows the ethidium bromide staining of the gel and B the corresponding autoradiograms after the DNA was blotted to a nitrocellulose filter and hybridized to ^{32}P-cDNA (1, 2, 3 and 4) and ^{32}P-pBδ7$_L$ DNA (5 and 6).
1; 20 μg of EcoRI-digests of total chick DNA.
2-4: 2 μg each of EcoRI-digested chick DNA extracted from various gel slices of a preparative gel.
5; 0.5 μg of the EcoRI-digested λCδ106 DNA.
6; 0.5 μg of the EcoRI-digested λCδ109 DNA.

of heteroduplex formed between λgδ28 and λgδ43 revealed the existence of about 2 kb homologous region in the 11 kb and 4 kb fragments under the stringent hybridization conditions. This indicates that each of these genomic clones contains the 3'-proximal portion of the different δ-crystallin genes and suggests that there are at least two different δ-crystallin genes in the chicken genome.

To isolate the complete structural δ-crystallin genes, we have recently obtained a chick genomic DNA library using the Charon 4A vector and chick DNA which was partially digested with EcoRI. About 2×10^5 phage plaques from this chicken gene library were screened for the δ-crystallin genes using ^{32}P-labeled nick-translated pBδ2 DNA, another cDNA clone of about 0.4 kb insert, as a probe and several positive plaques were obtained. One of these clones, designated λCδ106, contained an insert of 15 kb, which yields 7.6 kb, 4.0 kb, 2.6 kb and 0.8 kb fragments by limit EcoRI digestion. When the EcoRI digests of λCδ106 DNA were electrophoresed on a 0.8% agarose gel, transferred to a nitrocellulose filter and hybridized to ^{32}P-labeled δ-crystallin cDNA of nearly full length, the 7.6 kb, 4.0 kb and 0.8 kb DNA fragments were hybridizable to the cDNA probe but the 2.6 kb frag-

CLONING AND STRUCTURE OF δ-CRYSTALLIN GENES

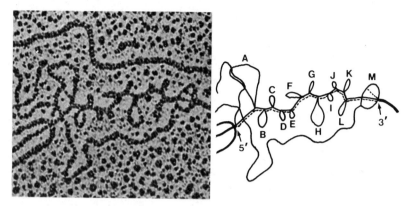

Figure 2. An electron micrograph and its line drawing of a hybrid molecule formed between δ-crystallin mRNA and the δ-crystallin gene (λCδ106). In the line drawing, the dashed line represents the mRNA, the solid line the DNA. The letters refer to the intervening sequences. The 5' and 3' ends of the δ-crystallin mRNA are indicated with arrows.

ment was not. This indicates that the 7.6 kb, 4.0 kb and 0.8 kb DNA fragments contain the δ-crystallin gene. The alignment of these EcoRI fragments was determined by partial EcoRI digestion of λCδ106 DNA. An 8.4 kb fragment consisting of the 7.6 kb and 0.8 kb fragments and another fragment consisting of 4.0 kb and 0.8 kb fragments were detected in the partial digests. A Charon 4A clone carrying the EcoRI fragments of the 8.0 kb, 2.6 kb and 7.6 kb fragments has been isolated. Therefore, the alignment of the four EcoRI fragments in λCδ106 DNA is concluded to be 2.6 kb-7.6 kb-0.8 kb-4.0 kb. The 4.0 kb fragment is identical with that cloned in λgδ43, judging from restriction endonuclease mapping and heteroduplex analysis. Since the 4.0 kb fragment represents the 3' terminal region of the δ-crystallin mRNA sequences, the alignment of these four EcoRI digests in λCδ106 is 2.6 kb-7.6 kb- 0.8 kb-4.0 kb in the direction of transcription.

In an attempt to characterize the organization of the δ-crystallin gene more thoroughly, hybrid molecules formed between δ-crystallin mRNA and λCδ106 DNA were examined by electron microscopy. DNA was thermally denatured and incubated with δ-crystallin mRNA in 70% formamide, 10 mM tricine (pH 8.8), 0.5 M NaCl and 10 mM EDTA at 53° for 2 hr. Hybrid molecules were spread by the method of Schafer et al. (1980) using formamide and urea. Fig. 2 shows such a hybrid molecule formed between δ-crystallin mRNA and λCδ106 DNA. Thirteen intervening DNA loops of heterogeneous sizes are apparent in the figure. This indicates discontinuity of the gene sequences complementary to δ-crystallin mRNA. The R-loop analysis of hybrid molecules formed between δ-crystallin mRNA and λgδ1 DNA, which contains the

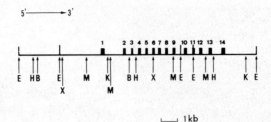

Figure 3. Organization of the complete δ-crystallin gene (λCδ106). δ-crystallin mRNA gene sequences are represented by ■, flanking DNA sequences and intervening sequences by ———. The arrows indicate various restriction endonuclease sites on the δ-crystallin gene. B, Bam HI; E, EcoRI; H, Hind III; K, Kpn I; M, Msp I; X, XBA I.

7.6 kb EcoRI fragment of λCδ106 DNA as an insert, shows that the largest loop in the figure (loop A) represents the 5' terminal intron of the δ-crystallin gene. The RNA tail in the intervening sequence M in the figure represents the 3' terminal poly(A) sequence of the mRNA. As the majority of δ-crystallin mRNA molecules prepared in the present study have the 3' terminal poly(A) tail and are capped at their 5' termini (data not shown), they are largely intact. Thus, the R-loop analysis of the δ-crystallin mRNA-λCδ106 hybrids led us to conclude that the λCδ106 DNA contains the entire mRNA gene sequence and their flanking regions and that the δ-crystallin gene contains at least thirteen intervening sequences (Yasuda et al., submitted; Fig 3).

We have isolated another Charon 4A clone, λCδ109, which is hybridizable to pBδ2 probe. This clone has an insert of 14 kb DNA fragment, which is cleaved into two fragments of 11 kb and 3 kb by EcoRI digestion. The heteroduplex analysis and restriction endonuclease mapping show that this 11 kb DNA fragment is identical with that of λgδ28 DNA. Thus, this λCδ109 DNA has the 3' terminal sequence of the other δ-crystallin gene. We compared the gene sequence of λCδ106 with that of λCδ109 by the heteroduplex analysis. Although the gene sequences in the two genomic clones had a considerable homology spanning about 5 kb, they were different in that a stretch of about 2 kb long from the 5' terminus of the gene in λCδ106 did not hybridize to λCδ109 DNA and four small heterogeneous loops were found in the heteroduplexes. It is apparent that λCδ109 does not contain the entire δ-crystallin gene sequence. This shows that each clone represents different genes for the δ-crystallin. It is worth noting that these EcoRI fragments have also been detected in DNA isolated from an inbred chicken. Therefore, we conclude that there are at least two non-allelic δ-crystallin genes in the chicken (Jones et al., 1980).

To determine whether the mRNA present in pBδ7 is transcribed from the gene of λCδ106 or from that of λCδ109, we performed the following experiments. The EcoRI digests of λCδ106 and λCδ109 DNAs were hybridized with pBδ7$_L$ and pBδ7$_R$ separately. The pBδ7$_L$ insert hybridized to the 7.6 kb and 0.8 kb DNA fragments of λCδ106 and the 3.0 kb and 11 kb DNA fragments of λCδ109. On the other hand, pBδ7$_R$ hybridized to the 4.0 kb DNA fragment of λCδ106 and the 11 kb DNA fragment of λCδ109. These results suggest that the EcoRI site in the insert of pBδ7 corresponds to that between the 0.8 kb and 4.0 kb fragments of λCδ106 DNA but not to that between the 3.0 kb and 11 kb fragments of λCδ109 DNA. Therefore, the mRNA sequence cloned in pBδ7 appears to be transcribed from the δ-crystallin gene in λCδ106 (Yasuda et al., submitted).

REFERENCES

Clayton, R. M., 1974, Comparative aspects of lens proteins, in: "The Eye," H. Davson and L. T. Graham, eds., Academic Press, London, Vol. 5, p.399.

Denhardt, D. T., 1966, A membrane-filter technique for the detection of complementary DNA, Biochem. Biophys. Res. Commun., 23:641.

Jones, R. E., Bhat, S. P., Sullivan, M. A., and Piatigorsky, J., 1980, Comparison of two δ-crystallin genes in the chicken, Proc. Natl. Acad. Sci. U.S.A., 77:5879.

Piatigorsky, J., 1982, Lens differentiation in vertebrates: A review of cellular and molecular features, Differentiation (in press).

Schafer, M. P., Boyd, C. D., Tolstoshev, P., and Crystal, R. G., 1980, Structural organization of a 17 kb segment of the α2 collagen gene: evaluation by R loop mapping, Nucleic Acids Res., 8:2241.

Southern, E. M., 1975, Detection of specific sequences among DNA fragments separated by gel electrophoresis, J. molec. Biol., 98:503.

Yasuda, K., Kondoh, H., Okada, T. S., Nakajima, N., and Shimura, Y., (submitted) Organization of δ-crystallin genes in the chicken.

CONTRIBUTORS AND PARTICIPANTS

*Indicates attendance at the meeting in Edinburgh, September 1981.

Affara, N., Beatson Institute for Cancer Research, Glasgow
 G61 1BD, U.K.
Agata, K., Department of Biophysics, Kyoto University, Kyoto 606,
 Japan.
Alonso, S., Department of Molecular Biology, Pasteur Institute,
 25 rue du Dr. Roux, Paris, France.
Arruti, C., Laboratorio de Cultivo de Tejidos, Dpto de Histologia
 Y Embryologia, Faculdad de Medicina, Montevideo, Uraguay.
*Balazs, R., MRC Developmental Neurobiology Unit, 33 John's Mews,
 London WC1N 2NS.
Banks, J., Memorial Sloan-Kettering Cancer Centre, New York,
 NY 10021, U.S.A.
Barritault, D., INSERM U118 - CNRS ERA 842, 29 rue Wilhelm, 75016
 Paris, France.
Bernardi, G., Laboratoire de Genetique Moléculaire, Institut de
 Recherche en Biologie Moléculaire, F-75221 Paris, Cedex 05,
 France.
*Bird, A. P., MRC Mammalian Genome Unit, Edinburgh EH9 3JT, U.K.
*Birnie, G. D., Beatson Institute for Cancer Research, Glasgow
 G61 1BD, U.K.
Black, E., Beatson Institute for Cancer Research, Glasgow G61 1BD,
 U.K.
*Bloemendal, H., Department of Biochemistry, University of Nijmegen,
 6525 EZ Nijmegen, The Netherlands.
*Bower, D. J., Department of Genetics, University of Edinburgh,
 Edinburgh EH9 3JN, U.K.
Bruggemann, M., Institute for Genetics, University of Cologne,
 D-5000 Cologne 41, Federal Republic of Germany.
*Buckingham, M. E., Department of Molecular Biology, Pasteur
 Institute, 25 rue du Dr. Roux, Paris, France.
Busslinger, M., National Institute for Medical Research, London
 NW7 1AA, U.K.
Calothy, G., Institut Curie Biologie, 91405 Orsay Cedex, France.
*Campbell, J. C., Department of Genetics, University of Edinburgh,
 Edinburgh EH9 3JN, U.K.
Caravatti, M., Friedrich Miescher Institute, Postfach 273, Basel
 CH-4002, Switzerland.

CONTRIBUTORS AND PARTICIPANTS

Carmon, Y., Department of Cell Biology, The Weizmann Institute of
  Science, Rehovot 76100, Israel.
*Chader, G. J., National Eye Institute, National Institutes of Health,
  Bethesda, MD 20205, U.S.A.
Chen, Z-X, Memorial Sloan-Kettering Cancer Centre, New York,
  NY 10021, U.S.A.
Chow, T., Department of Molecular, Cellular and Developmental
  Biology, University of Colorado, Boulder, CO 80309, U.S.A.
*Clayton, R. M., Department of Genetics, University of Edinburgh,
  Edinburgh EH9 3JN, U.K.
Cohen, A., Department of Molecular Biology, Pasteur Institute,
  25 rue du Dr. Roux, Paris, France.
*Courtois, Y., INSERM U118 - CNRS ERA 842, 27 rue Wilhelm, 75016
  Paris, France.
Crisanti-Combes, P., CNRS ER 231 and INSERM U178, Hôpital Broussais,
  75674 Paris, Cedex 14, France.
Czosnek, H., Department of Cell Biology, The Weizmann Institute of
  Science, Rehovot 76100, Israel.
Daubas, P., Department of Molecular Biology, Pasteur Institute,
  25 rue du Dr. Roux, Paris, France.
Degenstein, L., Laboratory for Developmental Biology, Cummings Life
  Science Center, University of Chicago, Chicago, Illinois 60637,
  U.S.A.
Dodemont, H., Department of Biochemistry, University of Nijmegen,
  6525 EZ Nijmegen, The Netherlands.
Droms, K., Department of Molecular, Cellular and Developmental
  Biology, University of Colorado, Boulder, CO 80309, U.S.A.
*Eguchi, G., Institute of Molecular Biology, Nagoya University,
  Nagoya 464, Japan.
Epner, E., Memorial Sloan-Kettering Cancer Centre, New York,
  NY 10021, U.S.A.
*Errington, L., Department of Genetics, University of Edinburgh,
  Edinburgh, EH9 3JN, U.K.
Flavell, R. A., National Institute for Medical Research, London
  NW7 1AA, U.K.
Fleming, J., Beatson Institute for Cancer Research, Glasgow G61 1BD,
  U.K.
Gali, M. A. H., Department of Zoology, University of Nottingham,
  Nottingham NG7 2RD, U.K.
Girard, A., CNRS ER 231 and INSERM U178, Hopital Broussais, 75674
  Paris, Cedex 14, France.
Goldfarb, P. S., Beatson Institute for Cancer Research, Glasgow
  G61 1BD, U.K.
Goldman, A., Depts. of Ophthalmology and Anatomy, Medical College
  of Wisconsin, Milwaukee, U.S.A.
Groffen, J., National Institute for Medical Research, London NW7 1AA,
  U.K.
*Grosveld, F., National Institute for Medical Research, London
  NW7 1AA, U.K.

CONTRIBUTORS AND PARTICIPANTS

Hagenbuchle, O., Swiss Institute for Experimental Cancer Research, CH-1066 Epalinges, Switzerland.
*Harrison, P. R., Beatson Institute for Cancer Research, Glasgow G61 1BD, U.K.
Imada, M., Department of Pathology, Denver Medical Center, University of Colorado, Denver, Co 80262, U.S.A.
*Iscove, N. N., Basel Institute for Immunology, 4005 Basel, Switzerland.
Itoh, Y., Biological Laboratory, Aichi Medical University, Nagakute, Aichi 480-11, Japan.
Jacobs, H., Beatson Institute for Cancer Research, Glasgow G61 1BD, U.K.
*John, H. A., Department of Genetics, University of Edinburgh, Edinburgh EH9 3JN.
Karasawa, Y., Institute of Molecular Biology, Nagoya University, Nagoya 464, Japan.
Kasturi, K., Beatson Institute for Cancer Research, Glasgow G61 1BD, U.K.
de Kleine, A., National Institute for Medical Research, London NW7 1AA, U.K.
Kodama, R., Institute of Molecular Biology, Nagoya University, Nagoya 464, Japan.
Kondoh, H., Department of Biophysics, Kyoto University, Kyoto 606, Japan.
Laurent, M., INSERM U235, 27 rue du Faubourg Saint Jacques, 75014 Paris, France.
Leighton, T., Department of Molecular, Cellular and Developmental Biology, University of Colarado, Boulder, CO 80309, U.S.A.
Lenstra, J. A., Department of Biochemistry, University of Nijmegen, 6525 EZ Nijmegen, The Netherlands.
Liesegang, B., Institute for Genetics, University of Cologne, D-5000 Cologne 41, Federal Republic of Germany.
*Lord, B. I., Paterson Laboratories, Christie Hospital and Holt Radium Institute, Manchester M20 9BX, U.K.
Lorinet, A.-M., CNRS ER 231 and INSERM U178, Hopital Broussais, 75674 Paris Cedex 14, France.
Lyons, A., Beatson Institute for Cancer Research, Glasgow G61 1BD, U.K.
*McDevitt, D. S., University of Pennsylvania, School of Veterinary Medicine, U.S.A.
Macleod, D., MRC Mammalian Genome Unit, Edinburgh EH9 3JT, U.K.
Marks, P. A., Memorial Sloan-Kettering Cancer Centre, New York, NY 10021, U.S.A.
Masterson, E., National Eye Institute, National Institutes of Health, Bethesda, MD 20205, U.S.A.
Masuda, A., Institute of Molecular Biology, Nagoya University, Nagoya 464, Japan.
Minty, A., Department of Molecular Biology, Pasteur Institute, 25 rue du Dr. Roux, Paris, France.
*Mirsky, R., MRC Neuroimmunology Group, University College London, London WC1E 6BT, U.K.

Moormann, R., Department of Molecular Biology, University of Nijmegen, 6525 EZ Nijmegen, The Netherlands.
*Moscona, A. A., Laboratory for Developmental Biology, Cummings Life Science Center, University of Chicago, Chicago, Illinois 60637, U.S.A.
Moskal, J. R., National Heart, Lung and Blood Institute, National Institutes of Health, Bethesda, Md 20205, U.S.A.
Nakajima, N., Department of Biophysics, Kyoto University, Kyoto 606, Japan.
Nichols, R., Beatson Institute for Cancer Research, Glasgow G61 1BD, U.K.
Nirenberg, M., National Heart, Lung and Blood Institute, National Institutes of Health, Bethesda, Md 20205, U.S.A.
Nomura, K., Institute of Biophysics, Faculty of Science, University of Kyoto, Kyoto 606, Japan.
Nudel, U., Department of Cell Biology, The Weizmann Institute of Science, Rehovot 76100, Israel.
*Okada, T. S., Institute of Biophysics, Faculty of Science, University of Kyoto, Kyoto 606, Japan.
Olivie, M., INSERM U118 - CNRS ERA 842, 29 rue Wilhelm, 75016 Paris, France.
O'Prey, J., Beatson Institute for Cancer Research, Glasgow G61 1BD, U.K.
*Owen, J. J. T., Department of Anatomy, Medical School, Birmingham B15 2TJ, U.K.
*Patek, C., Department of Genetics, University of Edinburgh, Edinburgh EH9 3JN, U.K.
*Paul, J., Beatson Institute for Cancer Research, Glasgow G61 1BD, U.K.
Perry, M., ARC Poultry Research Centre, Roslin, Midlothian EH25 9PS, U.K.
*Pessac, B., CNRS ER 231 and INSERM U178, Hopital Broussais, 75674 Paris, Cedex 14, France.
Plouet, J., INSERM U118 - CNRS ERA 842, 29 rue Wilhelm, 75016 Paris, France.
*Pollock, B. J., Department of Genetics, University of Edinburgh, Edinburgh EH9 3JN, U.K.
*de Pomerai, D. I., Department of Zoology, University of Nottingham, Nottingham NG7 2RD, U.K.
Quax, W. J., Department of Biochemistry, University of Nijmegen, 6525 EZ Nijmegen, The Netherlands.
*Radbruch, A., Institute for Genetics, University of Cologne, D-5000 Cologne 41, Federal Republic of Germany.
Rajewsky, K., Institute for Genetics, University of Cologne, D-5000 Cologne 41, Federal Republic of Germany.
Ramaekers, F. C. S., Department of Biochemistry, University of Nijmegen, 6525 EZ Nijmegen, The Netherlands.
*Ratcliffe, J. G., Department of Biochemistry, Royal Infirmary, Glasgow G4 0SF

CONTRIBUTORS AND PARTICIPANTS

Rattray, S., MRC Neuroimmunology Group, University College London, London WC1E 6BT, U.K.
*Rifkind, R. A., Memorial Sloan-Kettering Cancer Centre, New York, NY 10021, U.S.A.
Robert, B., Department of Molecular Biology, Pasteur Institute, 25 rue du Dr. Roux, Paris, France.
*Schibler, U., Swiss Institute for Experimental Cancer Research, CH-1066 Epalinges, Switzerland.
Schneider, M. D., National Heart, Lung and Blood Institute, National Institutes of Health, Bethesda, Md 20205, U.S.A.
Schoenmakers, J. G. G., Department of Molecular Biology, University of Nijmegen, 6525 EZ Nijmegen, The Netherlands.
Shani, M., Department of Cell Biology, The Weizmann Institute of Science, Rehovot 76100, Israel.
Shimura, Y., Department of Biophysics, Kyoto University, Kyoto 606, Japan.
Shott, R., Beatson Institute for Cancer Research, Glasgow G61 1BD, U.K.
Soriano, P., Laboratoire de Genetique Moléculaire, Institut de Recherche en Biologie Moléculaire, F-75221 Paris, Cedex 05, France.
Spandidos, D. A., Beatson Institute for Cancer Research, Glasgow G61 1BD, U.K.
*Sueoka, N., Department of Molecular, Cellular and Developmental Biology, University of Colorado, Boulder, Co 80309, U.S.A.
Syme, C. M., Department of Genetics, University of Edinburgh, Edinburgh EH9 3JN, U.K.
Taggart, M. H., MRC Mammalian Genome Unit, Edinburgh EH9 3JT, U.K.
Takagi, S., Institute of Biophysics, Faculty of Science, University of Kyoto, Kyoto 606, Japan.
Tassin, J., INSERM U118 - CNRS ERA 842, 29 rue Wilhelm, 75016 Paris, France.
Tomozawa, Y., Department of Molecular, Cellular and Developmental Biology, University of Colorado, Boulder, Co 80309, U.S.A.
Tosi, M., Swiss Institute for Experimental Cancer Research, CH-1066 Epalinges, Switzerland.
*Trisler, G. D., National Heart, Lung, and Blood Institute, National Institutes of Health, Bethesda, Md 20205, U.S.A.
*Truman, D. E. S., Department of Genetics, University of Edinburgh, Edinburgh EH9 3JN, U.K.
Vulliamy, T., MRC Neuroimmunology Group, University College London, London WC1E 6BT, U.K.
Wasseff, M., INSERM U106, Centre Medico-chirurgical Foch, 92150 Suresnes, France.
Wellauer, P.K. Swiss Institute for Experimental Cancer Research, CH-1066, Epalinges, Switzerland.
Weydert, A., Department of Molecular Biology, Pasteur Institute, 25 rue du Dr. Roux, Paris, France.
Wilkes, P. R., Beatson Institute for Cancer Research, Glasgow G61 1BD, U.K.

*Woodruff, M. F. A., MRC Clinical and Population Cytogenetics Unit, Western General Hospital, Edinburgh EH4 2XU, U.K.

Van Workum, M. E. S., Department of Biochemistry, University of Nijmegen, 6525 EZ Nijmegen, The Netherlands.

*Wyllie, A. H., Department of Pathology, University of Edinburgh, Edinburgh EH8 9AG, U.K.

*Yaffe, D., Department of Cell Biology, The Weizmann Institute of Science, Rehovot 76100, Israel.

Yang, Q.-S., Beatson Institute for Cancer Research, Glasgow G61 1BD, U.K.

Yasuda, K., Institute of Biophysics, Faculty of Science, University of Kyoto, Kyoto 606, Japan.

Young, R. A., Department of Biochemistry, Stanford University School of Medicine, Stanford, Ca 94305, U.S.A.

Zakut, R., Department of Cell Biology, The Weizmann Institute of Science, Rehovot 76100, Israel.

INDEX

Actinomycin D, 264
Acetylcholine receptors, 268 331
Acetylcholinesterase, 268
Actin, 43, 44, 56, 59, 127-136 232, 235, 297, 331-344, 349-355
Actinin, 297
Adenohypophysis, 29-31
Adipose cell, 344, 406
Adrenal medulla, 108
Adrenocorticotropic hormone, ACTH, 26, 157, 158
Albumin, 433
Alu sequences, 146
α-Aminoadipic acid, 194
α-Amylase, 246, 381-384, 431, 435
Antibodies, 422
  monoclonal, 86, 107-112
Antidiuretic hormone, ADH, 157
Antigen, cell surface, 107, 123, 174
Anterior pituitary, 440
Apoptosis, 461
APUD (amine precursor uptake and decarboxylation) cells, 144, 150, 157, 162
Ascorbic acid, 215
Astrocyte, 59, 111, 140, 324, 423
5-Azacytidine, 397, 406

Basement membrane, 403
B-lymphocyte, 85, 93, 148, 421, 437

Bronchial carcinoma, 144 (see also Lung)
Brain, 57
Branching mechanism, 165

Calcitonin, 157
Calcium, regulatory effects, 262
Canalisation, 400
Carcino-embryonic antigens, 8, 240, 439
Carcinogenesis
  and mutation, 144
  viral, 441
Carcinogens, 8, 10, 33, 55, 144, 423, 442, 455
Cardiac isoforms of actin, 339
Cardiac muscle, see Muscle
Cartilage, 408
Cell
  adhesion, 218
  aggregation, 188, 307
  -cell interactions, 29, 403, 414, 418
    of myoblasts, 133, 142, 259-270
  contact, 248
  cycle, 414
    and differentiation, 253-255, 405, 406
    and transdifferentiation, 412
  death, 422, 461
  deletion, 461
  dissociation, 188
  division, 253-255, 275-281, 415-417
  fusion, of myoblasts, 133, 142, 259-270

Cell (continued)
  hybrids, 433-434
  membrane, 86-87, 321
    proteins, 86
  proliferation, 253-255, 275-281, 415-417
    regulators of, 285, 289-303
  surface, 87, 140, 248, 298, 321
    antigens, 107, 123, 174, 321
    role in transdifferentiation, 195, 218
  type conversion, 166, 419
Cerebellum, 123
Cerebrum, 123
CFU-S, see Spleen colony forming unit
Chalone, 248
Choline acetyltransferase, 115-121, 199-206, 224-227
Choroid, 291, 307-318
Chromaffin cell, 108
Chromatin
  digestion by nuclease, 133
  structure, 2, 17, 429-433
    and gene activity, 133-136, 246-247
Chromosomal abnormalities, 148, 161, 169
Classification of cell types, 45-52
Class-switching in immunoglobulins, 93-102, 140
Clonal cell culture, 223, 242
Collagen, 248, 298
  influence on differentiation, 267
Commitment, 253-255
Competence, 23-33, 61-63
Conditioned medium, 265
Conversion of cell type, 166, 419
Conversion coupling, 170
Cornea, 49, 303, 423
Corneal endothelium, 299
Corticosteroids, 157
Creatine phosphokinase (CPK), 261, 262, 343

Crystallin, 26, 28, 31, 32, 42, 43, 59, 177-184, 199-206, 210, 215, 224, 232, 244, 307, 317-318, 359-365, 404, 409, 445, 451, 467
Cyclic AMP, 249, 261, 411
Cytoskeleton, 40, 42, 58, 249, 297, 349-355

Dedifferentiation, 455
Demethylation of DNA, 378
Determination, 2, 3, 23-33, 425
Dexamethasone, 324, 368
Diabetes, 323
DNAaseI, 17, 56, 63, 127, 133, 247, 377, 410, 429, 431
DNA
  control sequences, 18
  flanking sequences, 68, 128
  intergenic, 66
  methylation, 17, 63, 375-379
  recombinant, 68-77, 82-87, 89-91, 127-136, 333-344, 349-355, 359-365
  replication, 426
  synthesis, 253
Dysdifferentiation, 162

Ectopic hormones, 8, 148, 155-163, 235, 239, 414
Embryo extract, 200, 268
Embryonic induction, 245, 247
Epigenetic formation of tumours, 8
Epithelial growth factor (EGF), 294, 303, 323
Epithelial-mesenchymal interactions, 245, 247
Erythroblast, 83, 85, 86, 89, 133
Erythrocyte, 81-87, 276-281, 367-371, 402, 413
Erythroid precursor cell (BFU-E), 81
Erythroleukaemia cell, 89-92, 253-255, 367-371
Erythropoiesis, 81-87, 323, 410
Erythropoietin, 81, 158, 323
Evolution
  of differentiation, 51
  of introns, 131
Eye, 29, 46, 49 (see also Choroid, Cornea, Iris, Lens, Retina)

Eye cup, 58, 123
Eye derived growth factor
    (EDGF), 289-303, 323,
    423

α-Fetoprotein, 151
Fibroblast, 266, 323, 406
Fibroblast growth factor
    (FGF), 294
Fibroblast-like cell, 232,
    290, 317, 344, 455
Fibronectin, 248, 298
Friend cell, 81, 85, 89-91,
    139, 235, 404, 405,
    433

Galactocerebroside, 111-112
Gene,
    actin, 40, 127, 334, 349
    amplification, 62
    α-amylase, 381-384
    conalbumin, 40
    crystallin, 453, 467
    derepression of, 161
    families, 40, 334
    globin, 15, 40, 56, 62, 65,
        82, 89, 133
    histone, 18, 40
    immunoglobulin, 40, 422
    insulin, 40
    interferon, 40
    qaq, 146
    regulatory, 50
    ribosomal RNA, 375-379
    src, 115, 145
    structure in active state,
        17, 133-136
    switching, 13
    vitellogenin, 40
    vimentin, 40, 349
Genomic constancy, 62
Glial cell, 142, 165, 168, 172,
    187-195, 234, 407, 419,
    424
Glial fibrillary acidic protein (GFAP), 111, 116,
    424
Globin, 89, 91, 139, 253-255,
    368
    genes, 15, 40, 56, 62, 65,

Globin (continued)
    genes (continued), 82, 89, 133
    synthesis,
        induction of, 139, 253-255
Glutamic acid decarboxylase, 115
Glutamine synthetase, 116, 199-206
Glycophorin, 86
Glycoprotein, 141
Gonadotrphin, chorionic, 27, 156,
    157
Gradients, glycoprotein, 123-125,
    141
Granulocyte, 277, 402
Growth factors, 81, 245, 249, 289-
    303, 323, 418, 420-421

Haemin, 368, 404
Haemoglobin, 13, 65, 253-255, 367-
    371, 404, 405, 411, 438
    (see also Globin)
Haemopoiesis, 46, 81, 85, 275-281,
    285, 323, 367-371, 399-400,
    410
Heat shock in Drosophila, 14
Hepatoma, 387-392
Hexamethylene bisacetamide (HMBA),
    367
Histone gene, 18, 40
Histamine, 157
Hormone, 245, 439
    ectopic, 8, 148, 155-163, 235,
        239, 414
    glucocorticoid, 461
    influence on tumours, 11
    peptide, 148, 156-163, 439
    receptors, 246-247
Hyaluronidase, 215
5-Hydroxytryptamine, 157
Hypophysis, 29-31

Idiotope, 100
Idiotype, 100
Imaginal discs, 421
Immune system, 93-102, 421
Immunoglobulin, 93-102, 140
    gene, constant region, 422
    heavy chain, 95
    variable region, 99, 422
Induction,
    embryonic, 23, 245, 247

Initiation
   factors in protein synthesis, 265
   of transcription, 18, 384
Insulin, 26, 40
Intercellular matrix, 248, 299, 421
Intermediate filaments, 41-43, 349
Intron, 70, 73, 127, 131
Iris, 32, 49, 149, 177-184, 209, 234, 291, 412

Kidney, 48, 57
Karyotype, 169
   abnormalities of, 148, 161

Lactate dehydrogenase, 438
L-cell, 89
Lens, 40-44, 149, 177-185, 187-195, 199-206, 209-219, 223-227, 229, 242, 290-303, 318, 349, 359-365, 399, 403, 406, 408, 410, 413, 416-418, 442, 445, 446, 464
Lentectomy, 177-185, 244
Lentoid, 29, 187-195, 211-213, 225, 230, 234, 244, 307, 318
Lentropin, 243
Leukaemia, 81-87, 89-92, 240, 253-255, 367
Liver, 48, 387-392, 406
Long terminal repeat sequences, 19, 441
Lung tumours, 144, 157, 239, 441
Lymphocyte, 57, 85, 93, 148, 421, 437
   differentiation, 287
Lymphoblastoma, 235
Lymphoma, 82, 85, 235

Macrophage, 324, 402
Malignancy, reversal of, 55
Malignant transformation, 7, 55
Mammary gland, 57, 246
Megacaryocyte, 277

Melanin, 309, 317
Melanogenesis, 212
Melanosome, 211
Membrane, cell, 86-87, 321
Merocyanine-540, 224
Messenger RNA, (see also RNA)
   actin, 331-344
   abundance, 24, 25, 57, 387-392
   crystallin, 29, 359-365, 409, 451
   in differentiation, 14
   globin, 91, 253-255, 371, 427
   histone, 328
   infrequent, 57
   liver-specific, 387
   muscle-specific, 134
   myosin, 331-344
   sequestration, 264
   stability, 15, 392
Metaplasia, 47, 51, 209
Metastasis, 143, 239
Methylcholanthrene, 234
5-methyl cytosine, 375-379
Methylation of DNA, 17, 63, 375-379
N-Methyl-N-nitro-N-nitrosoguanidine, (MNNG), 33, 229-235, 243, 406, 442, 455
Microtubule, 349
Müller cell, 115, 116, 142, 188-195
Murine erythroleukaemia cell (MELC), 367-371
Muscle, 48, 399, 406, 408, 437 (see also Myoblast, Myofibril, Myogenesis, Myosin, Myotube)
Mutagens, 8, 229, 406
Myelination, 112
Myeloma, 9, 82, 93
Myoblast, 331-344
   fusion, 133, 142, 259, 331-344
   growth factor (MGF), 303
Myofibril, 404
Myogenic cell line, 127-136
Myogenesis, 127-136, 259-270, 331-344
Myosin, 56, 127, 260-270
   heavy chain, 334
   light chain, 335, 343
   mRNP, 264
Myotube, 260-270, 427

Neoblast, 400

INDEX

Neoplasia, see Tumour
Nervous system, 399, 423
Neural crest, 59, 111, 240, 414, 439
Neural retina, 32, 115, 223, 229, 411 (see also Retina)
Neural influences on myogenesis, 268
Neuroblast, 226
Neuroblastoma, 235
Neurone, 107-112, 142, 165-175, 188, 321, 419, 421, 424, 427
Neurone-like cells, 232
Neurotrophic factors, 268
Neurotumour, 165, 419, 423
Non-histone proteins, 17
Normoblast, 464
Nuclear fusion, 260
Nuclear transplantation, 2, 62, 408
Nucleosome, 426

Oestrogen, 63, 246
Oligodendrocyte, 111, 423
Oncofoetal antigens, 8, 240, 439
Oncogenic viruses, 144, 441
Oocyte, 89, 408
Optic cup, 58, 123
Optic tectum, 123, 141
Osteolysin, 158
Ovalbumin, 26, 63, 246, 429
Oviduct, 26, 63

Pancreas, 26, 48, 157, 248, 436
Peptide hormones, 148, 156-163, 439
Phaeochromocytoma, 108, 157
Phenylthiourea, 212-213, 216
Phorbol esters, 324
Phosphorylation of proteins, 261
Pigment, 307
Pigment epithelium, 32, 209-219, 232, 411, 416-417
Pineal, 29-31

Pituitary, 27, 58
Platelet derived growth factor (PDGF), 303
Placenta, 26, 156
Positional information, 123
Post-transcriptional control, 15, 63, 387-392
Post-translational control, 39-40, 262-264
Prekeratin, 42
Prolactin, 158, 246
Promoter sequences, 18, 73, 74, 384
Prostaglandin, 261
Protein
  degredation, 14
  in differentiation, 13
  isoforms, 331-344, 437
  kinase, 120, 261
  phosphorylation, 261
  synthesis, 14
    specific initiation of, 265
  S-100, 168, 169, 172, 424
Puffs, chromosomal, 16

qaq-gene, 146
Quantal mitosis, 420
Quantitative regulation, 13, 407

Receptor
  acetyl choline, 268, 321, 331
  β-adrenergic, 324
  hormone, 246-247
Recombinant DNA, see DNA
Red blood cell, see Erythrocyte
Regeneration, 177-184, 209-210, 399, 410, 413
Regulatory genes, 50
Reticulocyte, 83
Retina, 30-32, 49, 115, 123-125, 141, 149, 187-195, 199-206, 209-219, 229-235, 242, 289, 407-411, 416-418
  neural, 32, 115-121, 223, 229, 411
  pigmented, 32, 209-219, 223, 232, 411, 416-417
Retinitis pigmentosa, 141
Ribosomal RNA, 375-379
RNA
  abundance, 387-392
  complexity, 57, 427

RNA (continued)
  messenger, see Messenger RNA
  nuclear, 390, 445
  polyadenylation, 390
  processing, 15, 390, 446
  ribosomal, 375-379
  splicing, 73, 381
Rous sarcoma virus, 115-121

S_1 nuclease, 70, 381
Sarcomere, 333
Satellite cells, 111
Schwann cells, 111-112
Serum, influence on different-
    iation, 119-206, 267,
    420
Sickle-cell anaemia, 65
Sodium influx, 168, 169, 172
Somatostatin, 157
Spectrin, 86, 405
Spleen colony forming unit
    (CFU-S), 275-281, 401
Splicing of RNA, 73, 381
src gene, 115, 145
Stem cells, 239-240, 400-407,
    416-417, 424
  haemopoietic, 275-281, 285
  inhibitors of, 277
  stimulators of, 277
  and tumour formation, 239
Stilboestrol, 234
Synapses, 115-121

Taxonomy of differentiation,
    45-52
Teratocarcinoma, 8, 55, 420
Teratoma, 147
Terminal differentiation,
    403
Tetanus toxin binding, 116, 140
Thalamus, 123
Thalassaemia, 65-66, 89
Thymus, 287, 461
Thymocyte, 461
Tissue specificity, 24, 381
T-lymphocytes, 81, 85, 93, 148,
    421, 437

Transdetermination, 241, 405
Transdifferentiation, 23-33, 47,
    49-51, 61, 143, 149,
Transdifferentiation (continued),
    150, 165, 187-195, 199-206,
    209-219, 223, 229, 240, 318,
    400, 405, 409, 412, 416
  factors affecting, 199-206, 209-
    219
Transcription, 57, 388, 427, 446
  control, 16, 19, 39, 62, 262,
    375, 401
  initiation, 73, 74
  rate, 16, 18
Transformation, 7-9, 90, 146
  malignant, reversibility, 8
Translational control, 29, 39, 262,
    324
Transposible elements, 143, 146
Tropomyosin, 56, 297
Tubulin, 59, 349
Tumour, 7-11, 52, 143-151, 155-163
  adrenal medullary, 108
  bronchial, 144
  cells, unity and diversity, 7, 9
  cervical, 148, 234
  hormone production by, 8, 148,
    155-163, 235, 239, 414
  lung, 144, 157, 239, 441
  ovary, 144
  progression, 145
  promoters, 368
  regression, 11
  regulability, 7, 8, 11
  spontaneous, 8
  vaginal, 234
Tumourigenesis, 144, 169, 423, 441

Vasopressin, 157
Vimentin, 41-44, 59, 349-355
Vitreous body, 291

Wolffian regneration, 211
Wound healing, 303

Xeroderma pigmentosum, 144
Xenopus hybrids, 379